String Theory,

An Introduction to the Bosonic String

The two volumes that comprise *String Theory* provide an up-to-date, comprehensive, and pedagogic introduction to string theory.

Volume I, *An Introduction to the Bosonic String*, provides a thorough introduction to the bosonic string, based on the Polyakov path integral and conformal field theory. The first four chapters introduce the central ideas of string theory, the tools of conformal field theory and of the Polyakov path integral, and the covariant quantization of the string. The next three chapters treat string interactions: the general formalism, and detailed treatments of the tree-level and one loop amplitudes. Chapter eight covers toroidal compactification and many important aspects of string physics, such as T-duality and D-branes. Chapter nine treats higher-order amplitudes, including an analysis of the finiteness and unitarity, and various nonperturbative ideas. An appendix giving a short course on path integral methods is also included.

Volume II, *Superstring Theory and Beyond*, begins with an introduction to supersymmetric string theories and goes on to a broad presentation of the important advances of recent years. The first three chapters introduce the type I, type II, and heterotic superstring theories and their interactions. The next two chapters present important recent discoveries about strongly coupled strings, beginning with a detailed treatment of D-branes and their dynamics, and covering string duality, M-theory, and black hole entropy. A following chapter collects many classic results in conformal field theory. The final four chapters are concerned with four-dimensional string theories, and have two goals: to show how some of the simplest string models connect with previous ideas for unifying the Standard Model; and to collect many important and beautiful general results on world-sheet and spacetime symmetries. An appendix summarizes the necessary background on fermions and supersymmetry.

Both volumes contain an annotated reference section, emphasizing references that will be useful to the student, as well as a detailed glossary of important terms and concepts. Many exercises are included which are intended to reinforce the main points of the text and to bring in additional ideas.

An essential text and reference for graduate students and researchers in theoretical physics, particle physics, and relativity with an interest in modern superstring theory.

Joseph Polchinski received his Ph.D. from the University of California at Berkeley in 1980. After postdoctoral fellowships at the Stanford Linear Accelerator Center and Harvard, he joined the faculty at the University of Texas at Austin in 1984, moving to his present position of Professor of Physics at the University of California at Santa Barbara, and Permanent Member of the Institute for Theoretical Physics, in 1992.

Professor Polchinski is not only a clear and pedagogical expositor, but is also a leading string theorist. His discovery of the importance of D-branes in 1995 is one of the most important recent contributions in this field, and he has also made significant contributions to many areas of quantum field theory and to supersymmetric models of particle physics.

From reviews of the hardback editions:

Volume 1

'... This is an impressive book. It is notable for its consistent line of development and the clarity and insight with which topics are treated ... It is hard to think of a better text in an advanced graduate area, and it is rare to have one written by a master of the subject. It is worth pointing out that the book also contains a collection of useful problems, a glossary, and an unusually complete index.'

Physics Today

'... the most comprehensive text addressing the discoveries of the superstring revolutions of the early to mid 1990s, which mark the beginnings of "modern" string theory.'

Donald Marolf, University of California, Santa Barbara, *American Journal of Physics*

'Physicists believe that the best hope for a fundamental theory of nature – including unification of quantum mechanics with general relativity and elementary particle theory – lies in string theory. This elegant mathematical physics subject is expounded by Joseph Polchinski in two volumes from Cambridge University Press ... Written for advanced students and researchers, this set provides thorough and up-to-date knowledge.'

American Scientist

'We would like to stress the pedagogical value of the present book. The approach taken is modern and pleasantly systematic, and it covers a broad class of results in a unified language. A set of exercises at the end of each chapter complements the discussion in the main text. On the other hand, the introduction of techniques and concepts essential in the context of superstrings makes it a useful reference for researchers in the field.'

Mathematical Reviews

'It amply fulfils the need to inspire future string theorists on their long slog and is destined to become a classic. It is a truly exciting enterprise and one hugely served by this magnificent book.'

David Bailin, *The Times Higher Education Supplement*

Volume 2

'In summary, these volumes will provide ... the standard text and reference for students and researchers in particle physics and relativity interested in the possible ramifications of modern superstring theory.'

Allen C. Hirshfeld, *General Relativity and Gravitation*

'Polchinski is a major contributor to the exciting developments that have revolutionised our understanding of string theory during the past four years; he is also an exemplary teacher, as Steven Weinberg attests in his foreword. He has produced an outstanding two-volume text, with numerous exercises accompanying each chapter. It is destined to become a classic ... magnificent.'

David Bailin, *The Times Higher Education Supplement*

'The present volume succeeds in giving a detailed yet comprehensive account of our current knowledge of superstring dynamics. The topics covered range from the basic construction of the theories to the most recent discoveries on their non-perturbative behaviour. The discussion is remarkably self-contained (the volume even contains a useful appendix on spinors and supersymmetry in several dimensions), and thus may serve as an introduction to the subject, and as an excellent reference for researchers in the field.'

Mathematical Reviews

CAMBRIDGE MONOGRAPHS ON MATHEMATICAL PHYSICS

General editors: P. V. Landshoff, D. R. Nelson, S. Weinberg

[†] Issued as a paperback

STRING THEORY
VOLUME I

An Introduction to the Bosonic String

JOSEPH POLCHINSKI

Institute for Theoretical Physics
University of California at Santa Barbara

CAMBRIDGE
UNIVERSITY PRESS

CAMBRIDGE UNIVERSITY PRESS
Cambridge, New York, Melbourne, Madrid, Cape Town, Singapore, São Paulo

Cambridge University Press
The Edinburgh Building, Cambridge CB2 8RU, UK

Published in the United States of America by Cambridge University Press, New York

www.cambridge.org
Information on this title: www.cambridge.org/9780521672276

First published 1998
Reprinted 1999, 2000 (with corrections), 2002 (twice), 2003
Paperback edition first published 2005

A catalogue record for this publication is available from the British Library

ISBN 978-0-521-67227-6 paperback

Transferred to digital printing 2007

TO MY PARENTS

Contents

Contents xi

Outline of volume II

Foreword

From the beginning it was clear that, despite its successes, the Standard Model of elementary particles would have to be embedded in a broader theory that would incorporate gravitation as well as the strong and electroweak interactions. There is at present only one plausible candidate for such a theory: it is the theory of strings, which started in the 1960s as a not-very-successful model of hadrons, and only later emerged as a possible theory of all forces.

There is no one better equipped to introduce the reader to string theory than Joseph Polchinski. This is in part because he has played a significant role in the development of this theory. To mention just one recent example: he discovered the possibility of a new sort of extended object, the 'Dirichlet brane,' which has been an essential ingredient in the exciting progress of the last few years in uncovering the relation between what had been thought to be different string theories.

Of equal importance, Polchinski has a rare talent for seeing what is of physical significance in a complicated mathematical formalism, and explaining it to others. In looking over the proofs of this book, I was reminded of the many times while Polchinski was a member of the Theory Group of the University of Texas at Austin, when I had the benefit of his patient, clear explanations of points that had puzzled me in string theory. I recommend this book to any physicist who wants to master this exciting subject.

Steven Weinberg
Series Editor
Cambridge Monographs on Mathematical Physics
1998

Preface

When I first decided to write a book on string theory, more than ten years ago, my memories of my student years were much more vivid than they are today. Still, I remember that one of the greatest pleasures was finding a text that made a difficult subject accessible, and I hoped to provide the same for string theory.

Thus, my first purpose was to give a coherent introduction to string theory, based on the Polyakov path integral and conformal field theory. No previous knowledge of string theory is assumed. I do assume that the reader is familiar with the central ideas of general relativity, such as metrics and curvature, and with the ideas of quantum field theory through non-Abelian gauge symmetry. Originally a full course of quantum field theory was assumed as a prerequisite, but it became clear that many students were eager to learn string theory as soon as possible, and that others had taken courses on quantum field theory that did not emphasize the tools needed for string theory. I have therefore tried to give a self-contained introduction to those tools.

A second purpose was to show how some of the simplest four-dimensional string theories connect with previous ideas for unifying the Standard Model, and to collect general results on the physics of four-dimensional string theories as derived from world-sheet and spacetime symmetries. New developments have led to a third goal, which is to introduce the recent discoveries concerning string duality, M-theory, D-branes, and black hole entropy.

In writing a text such as this, there is a conflict between the need to be complete and the desire to get to the most interesting recent results as quickly as possible. I have tried to serve both ends. On the side of completeness, for example, the various path integrals in chapter 6 are calculated by three different methods, and the critical dimension of the bosonic string is calculated in seven different ways in the text and exercises.

On the side of efficiency, some shorter paths through these two volumes are suggested below.

A particular issue is string perturbation theory. This machinery is necessarily a central subject of volume one, but it is somewhat secondary to the recent nonperturbative developments: the free string spectrum plus the spacetime symmetries are more crucial there. Fortunately, from string perturbation theory there is a natural route to the recent discoveries, by way of T-duality and D-branes.

One possible course consists of chapters 1–3, section 4.1, chapters 5–8 (omitting sections 5.4 and 6.7), chapter 10, sections 11.1, 11.2, 11.6, 12.1, and 12.2, and chapters 13 and 14. This sequence, which I believe can be covered in two quarters, takes one from an introduction to string theory through string duality, M theory, and the simplest black hole entropy calculations. An additional shortcut is suggested at the end of section 5.1.

Readers interested in T-duality and related stringy phenomena can proceed directly from section 4.1 to chapter 8. The introduction to Chan–Paton factors at the beginning of section 6.5 is needed to follow the discussion of the open string, and the one-loop vacuum amplitude, obtained in chapter 7, is needed to follow the calculation of the D-brane tension.

Readers interested in supersymmetric strings can read much of chapters 10 and 11 after section 4.1. Again the introduction to Chan–Paton factors is needed to follow the open string discussion, and the one-loop vacuum amplitude is needed to follow the consistency conditions in sections 10.7, 10.8, and 11.2.

Readers interested in conformal field theory might read chapter 2, sections 6.1, 6.2, 6.7, 7.1, 7.2, 8.2, 8.3 (concentrating on the CFT aspects), 8.5, 10.1–10.4, 11.4, and 11.5, and chapter 15. Readers interested in four-dimensional string theories can follow most of chapters 16–19 after chapters 8, 10, and 11.

In a subject as active as string theory — by one estimate the literature approaches 10 000 papers — there will necessarily be important subjects that are treated only briefly, and others that are not treated at all. Some of these are represented by review articles in the lists of references at the end of each volume. The most important omission is probably a more complete treatment of compactification on curved manifolds. Because the geometric methods of this subject are somewhat orthogonal to the quantum field theory methods that are emphasized here, I have included only a summary of the most important results in chapters 17 and 19. Volume two of Green, Schwarz, and Witten (1987) includes a more extensive introduction, but this is a subject that has continued to grow in importance and clearly deserves an introductory book of its own.

This work grew out of a course taught at the University of Texas at Austin in 1987–8. The original plan was to spend a year turning the lecture notes into a book, but a desire to make the presentation clearer and more complete, and the distraction of research, got in the way. An early prospectus projected the completion date as June 1989 $\pm$ one month, off by 100 standard deviations. For eight years the expected date of completion remained approximately one year in the future, while one volume grew into two. Happily, finally, one of those deadlines didn't slip.

I have also used portions of this work in a course at the University of California at Santa Barbara, and at the 1994 Les Houches, 1995 Trieste, and 1996 TASI schools. Portions have been used for courses by Nathan Seiberg and Michael Douglas (Rutgers), Steven Weinberg (Texas), Andrew Strominger and Juan Maldacena (Harvard), Nathan Berkovits (São Paulo) and Martin Einhorn (Michigan). I would like to thank those colleagues and their students for very useful feedback. I would also like to thank Steven Weinberg for his advice and encouragement at the beginning of this project, Shyamoli Chaudhuri for a thorough reading of the entire manuscript, and to acknowledge the support of the Departments of Physics at UT Austin and UC Santa Barbara, the Institute for Theoretical Physics at UC Santa Barbara, and the National Science Foundation.

During the extended writing of this book, dozens of colleagues have helped to clarify my understanding of the subjects covered, and dozens of students have suggested corrections and other improvements. I began to try to list the members of each group and found that it was impossible. Rather than present a lengthy but incomplete list here, I will keep an updated list at the erratum website

http://www.itp.ucsb.edu/~joep/bigbook.html.

In addition, I would like to thank collectively all who have contributed to the development of string theory; volume two in particular seems to me to be largely a collection of beautiful results derived by many physicists. String theory (and the entire base of physics upon which it has been built) is one of mankind's great achievements, and it has been my privilege to try to capture its current state.

Finally, to complete a project of this magnitude has meant many sacrifices, and these have been shared by my family. I would like to thank Dorothy, Steven, and Daniel for their understanding, patience, and support.

Joseph Polchinski
Santa Barbara, California
1998

Notation

This book uses the $+++$ conventions of Misner, Thorne, and Wheeler (1973). In particular, the signature of the metric is $(-++\ldots+)$. The constants $\hbar$ and c are set to 1, but the Regge slope α' is kept explicit.

A bar $^-$ is used to denote the conjugates of world-sheet coordinates and moduli (such as z, τ, and q), but a star * is used for longer expressions. A bar on a spacetime fermion field is the Dirac adjoint (this appears only in volume two), and a bar on a world-sheet operator is the Euclidean adjoint (defined in section 6.7). For the degrees of freedom on the string, the following terms are treated as synonymous:

holomorphic = left-moving,

antiholomorphic = right-moving,

as explained in section 2.1. Our convention is that the supersymmetric side of the heterotic string is right-moving. Antiholomorphic operators are designated by tildes $^\sim$; as explained in section 2.3, these are not the adjoints of holomorphic operators. Note also the following conventions:

$$d^2z \equiv 2dxdy \,, \quad \delta^2(z,\bar{z}) \equiv \frac{1}{2}\delta(x)\delta(y) \,,$$

where $z = x + iy$ is any complex variable; these differ from most of the literature, where the coefficient is 1 in each definition.

Spacetime actions are written as $\boldsymbol{S}$ and world-sheet actions as S. This presents a problem for D-branes, which are T-dual to the former and S-dual to the latter; S has been used arbitrarily. The spacetime metric is $G_{\mu\nu}$, while the world-sheet metric is γ_{ab} (Minkowskian) or g_{ab} (Euclidean). In volume one, the spacetime Ricci tensor is $\boldsymbol{R}_{\mu\nu}$ and the world-sheet Ricci tensor is R_{ab}. In volume two the former appears often and the latter never, so we have changed to $R_{\mu\nu}$ for the spacetime Ricci tensor.

The following are used:

$\equiv$ defined as

$\cong$ equivalent to

$\approx$ approximately equal to

$\sim$ equal up to nonsingular terms (in OPEs), or rough correspondence.

1
A first look at strings

1.1 Why strings?

One of the main themes in the history of science has been unification. Time and again diverse phenomena have been understood in terms of a small number of underlying principles and building blocks. The principle that underlies our current understanding of nature is quantum field theory, quantum mechanics with the basic observables living at spacetime points. In the late 1940s it was shown that quantum field theory is the correct framework for the unification of quantum mechanics and electromagnetism. By the early 1970s it was understood that the weak and strong nuclear forces are also described by quantum field theory. The full theory, the $SU(3) \times SU(2) \times U(1)$ *Model* or *Standard Model*, has been confirmed repeatedly in the ensuing years. Combined with general relativity, this theory is consistent with virtually all physics down to the scales probed by particle accelerators, roughly 10^{-16} cm. It also passes a variety of indirect tests that probe to shorter distances, including precision tests of quantum electrodynamics, searches for rare meson decays, limits on neutrino masses, limits on axions (light weakly interacting particles) from astrophysics, searches for proton decay, and gravitational limits on the couplings of massless scalars. In each of these indirect tests new physics might well have appeared, but in no case has clear evidence for it yet been seen; at the time of writing, the strongest sign is the solar neutrino problem, suggesting nonzero neutrino masses.

The Standard Model (plus gravity) has a fairly simple structure. There are four interactions based on local invariance principles. One of these, gravitation, is mediated by the spin-2 graviton, while the other three are mediated by the spin-1 $SU(3) \times SU(2) \times U(1)$ gauge bosons. In addition, the theory includes the spin-0 Higgs boson needed for symmetry breaking, and the quarks and leptons, fifteen multiplets of spin-$\frac{1}{2}$ fermions in three generations of five. The dynamics is governed by a Lagrangian that depends upon roughly twenty free parameters such as the gauge and Yukawa couplings.

In spite of its impressive successes, this theory is surely not complete. First, it is too arbitrary: why does this particular pattern of gauge fields and multiplets exist, and what determines the parameters in the Lagrangian? Second, the union of gravity with quantum theory yields a nonrenormalizable quantum field theory, a strong signal that new physics should appear at very high energy. Third, even at the classical level the theory breaks down at the singularities of general relativity. Fourth, the theory is in a certain sense unnatural: some of the parameters in the Lagrangian are much smaller than one would expect them to be. It is these problems, rather than any positive experimental evidence, that presently must guide us in our attempts to find a more complete theory. One seeks a principle that unifies the fields of the Standard Model in a simpler structure, and resolves the divergence and naturalness problems.

Several promising ideas have been put forward. One is grand unification. This combines the three gauge interactions into one and the five multiplets of each generation into two or even one. It also successfully predicts one of the free parameters (the weak mixing angle) and possibly another (the bottom-tau mass ratio). A second idea is that spacetime has more than four dimensions, with the additional ones so highly curved as to be undetectable at current energies. This is certainly a logical possibility, since spacetime geometry is dynamical in general relativity. What makes it attractive is that a single higher-dimensional field can give rise to many four-dimensional fields, differing in their polarization (which can point along the small dimensions or the large) and in their dependence on the small dimensions. This opens the possibility of unifying the gauge interactions and gravity (the Kaluza–Klein mechanism). It also gives a natural mechanism for producing generations, repeated copies of the same fermion multiplets. A third unifying principle is supersymmetry, which relates fields of different spins and statistics, and which helps with the divergence and naturalness problems.

Each of these ideas — grand unification, extra dimensions, and supersymmetry — has attractive features and is consistent with the various tests of the Standard Model. It is plausible that these will be found as elements of a more complete theory of fundamental physics. It is clear, however, that something is still missing. Applying these ideas, either singly or together, has not led to theories that are substantially simpler or less arbitrary than the Standard Model.

Short-distance divergences have been an important issue many times in quantum field theory. For example, they were a key clue leading from the Fermi theory of the weak interaction to the Weinberg–Salam theory. Let

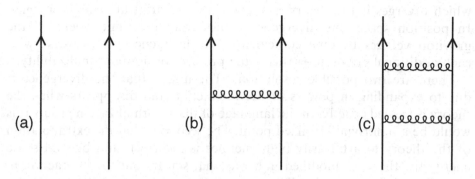

Fig. 1.1. (a) Two particles propagating freely. (b) Correction from one-graviton exchange. (c) Correction from two-graviton exchange.

us look at the short-distance problem of quantum gravity, which can be understood from a little dimensional analysis. Figure 1.1 shows a process, two particles propagating, and corrections due to one-graviton exchange and two-graviton exchange. The one-graviton exchange is proportional to Newton's constant G_N. The ratio of the one-graviton correction to the original amplitude must be governed by the dimensionless combination $G_N E^2 \hbar^{-1} c^{-5}$, where E is the characteristic energy of the process; this is the only dimensionless combination that can be formed from the parameters in the problem. Throughout this book we will use units in which $\hbar = c = 1$, defining the Planck mass

$$M_P = G_N^{-1/2} = 1.22 \times 10^{19} \, \text{GeV} \tag{1.1.1}$$

and the Planck length

$$M_P^{-1} = 1.6 \times 10^{-33} \, \text{cm} . \tag{1.1.2}$$

The ratio of the one-graviton to the zero-graviton amplitude is then of order $(E/M_P)^2$.

From this dimensional analysis one learns immediately that the quantum gravitational correction is an *irrelevant* interaction, meaning that it grows weaker at low energy, and in particular is negligible at particle physics energies of hundreds of GeV. By the same token, the coupling grows stronger at high energy and at $E > M_P$ perturbation theory breaks down. In the two-graviton correction of figure 1.1(c) there is a sum over intermediate states. For intermediate states of high energy E', the ratio of the two-graviton to the zero-graviton amplitude is on dimensional grounds of order

$$G_N^2 E^2 \int dE' \, E' = \frac{E^2}{M_P^4} \int dE' \, E' , \tag{1.1.3}$$

which diverges if the theory is extrapolated to arbitrarily high energies. In position space this divergence comes from the limit where all the graviton vertices become coincident. The divergence grows worse with each additional graviton — this is the problem of nonrenormalizability.

There are two possible resolutions. The first is that the divergence is due to expanding in powers of the interaction and disappears when the theory is treated exactly. In the language of the renormalization group, this would be a nontrivial UV fixed point. The second is that the extrapolation of the theory to arbitrarily high energies is incorrect, and beyond some energy the theory is modified in a way that smears out the interaction in spacetime and softens the divergence. It is not known whether quantum gravity has a nontrivial UV fixed point, but there are a number of reasons for concentrating on the second possibility. One is history — the same kind of divergence problem in the Fermi theory of the weak interaction was a sign of new physics, the contact interaction between the fermions resolving at shorter distance into the exchange of a gauge boson. Another is that we need a more complete theory in any case to account for the patterns in the Standard Model, and it is reasonable to hope that the same new physics will solve the divergence problem of quantum gravity.

In quantum field theory it is not easy to smear out interactions in a way that preserves the consistency of the theory. We know that Lorentz invariance holds to very good approximation, and this means that if we spread the interaction in space we spread it in time as well, with consequent loss of causality or unitarity. Moreover we know that Lorentz invariance is actually embedded in a local symmetry, general coordinate invariance, and this makes it even harder to spread the interaction out without producing inconsistencies.

In fact, there is presently only one way known to spread out the gravitational interaction and cut off the divergence without spoiling the consistency of the theory. This is string theory, illustrated in figure 1.2. In this theory the graviton and all other elementary particles are one-dimensional objects, strings, rather than points as in quantum field theory. Why this should work and not anything else is not at all obvious *a priori*, but as we develop the theory we will see how it comes about.[1] Perhaps we merely suffer from a lack of imagination, and there are many other consistent theories of gravity with a short-distance cutoff. However, experience has shown that divergence problems in quantum field theory

---

[1] There is an intuitive answer to at least one common question: why not membranes, two- or higher-dimensional objects? The answer is that as we spread out particles in more dimensions we reduce the spacetime divergences, but encounter new divergences coming from the increased number of *internal* degrees of freedom. One dimension appears to be the unique case where both the spacetime and internal divergences are under control. However, as we will discuss in chapter 14, the membrane idea has resurfaced in somewhat transmuted form as *matrix theory*.

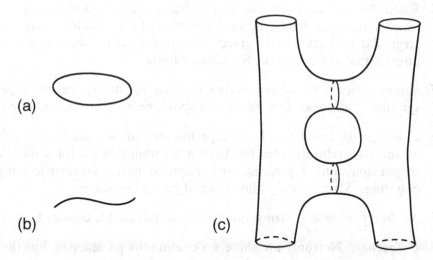

Fig. 1.2. (a) Closed string. (b) Open string. (c) The loop amplitude of fig. 1.1(c) in string theory. Each particle world-line becomes a cylinder, and the interactions no longer occur at points. (The cross-sections on the intermediate lines are included only for perspective.)

are not easily resolved, so if we have even one solution we should take it very seriously. Indeed, we are fortunate that consistency turns out to be such a restrictive principle, since the unification of gravity with the other interactions takes place at such high energy, M_P, that experimental tests will be difficult and indirect.

So what do we find if we pursue this idea? In a word, the result is remarkable. String theory dovetails beautifully with the previous ideas for explaining the patterns in the Standard Model, and does so with a structure more elegant and unified than in quantum field theory. In particular, if one tries to construct a consistent relativistic quantum theory of one-dimensional objects one finds:

1. *Gravity.* Every consistent string theory must contain a massless spin-2 state, whose interactions reduce at low energy to general relativity.

2. *A consistent theory of quantum gravity, at least in perturbation theory.* As we have noted, this is in contrast to all known quantum *field* theories of gravity.

3. *Grand unification.* String theories lead to gauge groups large enough to include the Standard Model. Some of the simplest string theories lead to the same gauge groups and fermion representations that arise in the unification of the Standard Model.

4. *Extra dimensions.* String theory requires a definite number of space-time dimensions, ten.[2] The field equations have solutions with four large flat and six small curved dimensions, with four-dimensional physics that resembles the Standard Model.

5. *Supersymmetry.* Consistent string theories require spacetime super-symmetry, as either a manifest or a spontaneously broken symmetry.

6. *Chiral gauge couplings.* The gauge interactions in nature are parity asymmetric (chiral). This has been a stumbling block for a number of previous unifying ideas: they required parity symmetric gauge couplings. String theory allows chiral gauge couplings.

7. *No free parameters.* String theory has no adjustable constants.

8. *Uniqueness.* Not only are there no continuous parameters, but there is no discrete freedom analogous to the choice of gauge group and representations in field theory: there is a unique string theory.

In addition one finds a number of other features, such as an axion, and hidden gauge groups, that have played a role in ideas for unification.

This is a remarkable list, springing from the simple supposition of one-dimensional objects. The first two points alone would be of great interest. The next four points come strikingly close to the picture one arrives at in trying to unify the Standard Model. And as indicated by the last two points, string theory accomplishes this with a structure that is tighter and less arbitrary than in quantum field theory, supplying the element missing in the previous ideas. The first point is a further example of this tightness: string theory *must* have a graviton, whereas in field theory this and other fields are combined in a mix-and-match fashion.

String theory further has connections to many areas of mathematics, and has led to the discovery of new and unexpected relations among them. It has rich connections to the recent discoveries in supersymmetric quantum field theory. String theory has also begun to address some of the deeper questions of quantum gravity, in particular the quantum mechanics of black holes.

Of course, much remains to be done. String theory may resemble the real world in its broad outlines, but a decisive test still seems to be far away. The main problem is that while there is a unique *theory*, it has an enormous number of classical solutions, even if we restrict attention

---

[2] To be precise, string theory modifies the notions of spacetime topology and geometry, so what we mean by a dimension here is generalized. Also, we will see that ten dimensions is the appropriate number for weakly coupled string theory, but that the picture can change at strong coupling.

to solutions with four large flat dimensions. Upon quantization, each of these is a possible ground state (vacuum) for the theory, and the four-dimensional physics is different in each. It is known that quantum effects greatly reduce the number of stable solutions, but a full understanding of the dynamics is not yet in hand.

It is worth recalling that even in the Standard Model, the dynamics of the vacuum plays an important role in the physics we see. In the electroweak interaction, the fact that the vacuum is less symmetric than the Hamiltonian (spontaneous symmetry breaking) plays a central role. In the strong interaction, large fluctuating gauge fields in the vacuum are responsible for quark confinement. These phenomena in quantum field theory arise from having a quantum system with many degrees of freedom. In string theory there are seemingly many more degrees of freedom, and so we should expect even richer dynamics.

Beyond this, there is the question, 'what is string theory?' Until recently our understanding of string theory was limited to perturbation theory, small numbers of strings interacting weakly. It was not known how even to define the theory at strong coupling. There has been a suspicion that the degrees of freedom that we use at weak coupling, one-dimensional objects, are not ultimately the simplest or most complete way to understand the theory.

In the past few years there has been a great deal of progress on these issues, growing largely out of the systematic application of the constraints imposed by supersymmetry. We certainly do not have a complete understanding of the dynamics of strongly coupled strings, but it has become possible to map out in detail the space of vacua (when there is enough unbroken supersymmetry) and this has led to many surprises. One is the absolute uniqueness of the theory: whereas there are several weakly coupled string theories, all turn out to be limits in the space of vacua of a single theory. Another is a limit in which spacetime becomes eleven-dimensional, an interesting number from the point of view of supergravity but impossible in weakly coupled string theory. It has also been understood that the theory contains new extended objects, D-branes, and this has led to the new understanding of black hole quantum mechanics. All this also gives new and unexpected clues as to the ultimate nature of the theory.

In summary, we are fortunate that so many approaches seem to converge on a single compelling idea. Whether one starts with the divergence problem of quantum gravity, with attempts to account for the patterns in the Standard Model, or with a search for new symmetries or mathematical structures that may be useful in constructing a unified theory, one is led to string theory.

Outline

The goal of these two volumes is to provide a complete introduction to string theory, starting at the beginning and proceeding through the compactification to four dimensions and to the latest developments in strongly coupled strings.

Volume one is an introduction to bosonic string theory. This is not a realistic theory — it does not have fermions, and as far as is known has no stable ground state. The philosophy here is the same as in starting a course on quantum field theory with a thorough study of scalar field theory. That is also not the theory one is ultimately interested in, but it provides a simple example for developing the unique dynamical and technical features of quantum field theory before introducing the complications of spin and gauge invariance. Similarly, a thorough study of bosonic string theory will give us a framework to which we can in short order add the additional complications of fermions and supersymmetry.

The rest of chapter 1 is introductory. We present first the action principle for the dynamics of string. We then carry out a quick and heuristic quantization using light-cone gauge, to show the reader some of the important aspects of the string spectrum. Chapters 2–7 are the basic introduction to bosonic string theory. Chapter 2 introduces the needed technical tools in the world-sheet quantum field theory, such as conformal invariance, the operator product expansion, and vertex operators. Chapters 3 and 4 carry out the covariant quantization of the string, starting from the Polyakov path integral. Chapters 5–7 treat interactions, presenting the general formalism and applying it to tree-level and one-loop amplitudes. Chapter 8 treats the simplest compactification of string theory, making some of the dimensions periodic. In addition to the phenomena that arise in compactified field theory, such as Kaluza–Klein gauge symmetry, there is also a great deal of 'stringy' physics, including enhanced gauge symmetries, T-duality, and D-branes. Chapter 9 treats higher order amplitudes. The first half outlines the argument that string theory in perturbation theory is finite and unitary as advertised; the second half presents brief treatments of a number of advanced topics, such as string field theory. Appendix A is an introduction to path integration, so that our use of quantum field theory is self-contained.

Volume two treats supersymmetric string theories, focusing first on ten-dimensional and other highly symmetric vacua, and then on realistic four-dimensional vacua.

In chapters 10–12 we extend the earlier introduction to the supersymmetric string theories, developing the type I, II, and heterotic superstrings and their interactions. We then introduce the latest results in these subjects. Chapter 13 develops the properties and dynamics of D-branes, still using the tools of string perturbation theory as developed earlier in the book. Chapter 14 then uses arguments based on supersymmetry to understand strongly coupled strings. We find that the strongly coupled limit of any string theory is described by a dual weakly coupled string theory, or by a new eleven-dimensional theory known provisionally as M-theory. We discuss the status of the search for a complete formulation of string theory and present one promising idea, m(atrix) theory. We briefly discuss the quantum mechanics of black holes, carrying out the simplest entropy calculation. Chapter 15 collects a number of advanced applications of the various world-sheet symmetry algebras.

Chapters 16 and 17 present four-dimensional string theories based on orbifold and Calabi–Yau compactifications. The goal is not an exhaustive treatment but rather to make contact between the simplest examples and the unification of the Standard Model. Chapter 18 collects results that hold in wide classes of string theories, using general arguments based on world-sheet and spacetime gauge symmetries. Chapter 19 consists of advanced topics, including (2,2) world-sheet supersymmetry, mirror symmetry, the conifold transition, and the strong-coupling behavior of some compactified theories.

Annotated reference lists appear at the end of each volume. I have tried to assemble a selection of articles, particularly reviews, that may be useful to the student. A glossary also appears at the end of each volume.

1.2 Action principles

We want to study the classical and quantum mechanics of a one-dimensional object, a string. The string moves in D flat spacetime dimensions, with metric $\eta_{\mu\nu} = \text{diag}(-, +, +, \cdots, +)$.

It is useful to review first the classical mechanics of a zero-dimensional object, a relativistic point particle. We can describe the motion of a particle by giving its position in terms of $D-1$ functions of time, $\mathbf{X}(X^0)$. This hides the covariance of the theory though, so it is better to introduce a parameter τ along the particle's world-line and describe the motion in spacetime by D functions $X^\mu(\tau)$. The parameterization is arbitrary: a different parameterization of the same path is physically equivalent, and all physical quantities must be independent of this choice. That is, for any monotonic function $\tau'(\tau)$, the two paths X'^μ and X^μ are the same, where

$$X'^\mu(\tau'(\tau)) = X^\mu(\tau) \,. \tag{1.2.1}$$

We are trading a less symmetric description for a more symmetric but redundant one, which is often a useful step. Figure 1.3(a) shows a parameterized world-line.

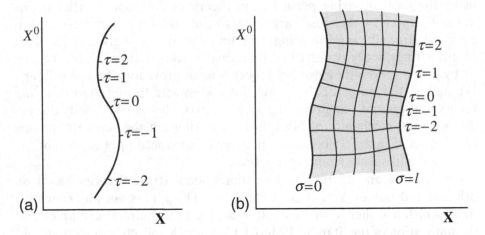

Fig. 1.3. (a) Parameterized world-line of a point particle. (b) Parameterized world-sheet of an open string.

The simplest Poincaré-invariant action that does not depend on the parameterization would be proportional to the proper time along the world-line,

$$S_{\text{pp}} = -m \int d\tau \, (-\dot{X}^\mu \dot{X}_\mu)^{1/2} , \qquad (1.2.2)$$

where a dot denotes a τ-derivative. The variation of the action, after an integration by parts, is

$$\delta S_{\text{pp}} = -m \int d\tau \, \dot{u}_\mu \delta X^\mu , \qquad (1.2.3)$$

where

$$u^\mu = \dot{X}^\mu (-\dot{X}^\nu \dot{X}_\nu)^{-1/2} \qquad (1.2.4)$$

is the normalized D-velocity. The equation of motion $\dot{u}^\mu = 0$ thus describes free motion. The normalization constant m is the particle's mass, as can be checked by looking at the nonrelativistic limit (exercise 1.1).

The action can be put in another useful form by introducing an additional field on the world-line, an independent world-line metric $\gamma_{\tau\tau}(\tau)$. It will be convenient to work with the tetrad $\eta(\tau) = (-\gamma_{\tau\tau}(\tau))^{1/2}$, which is defined to be positive. We use the general relativity term *tetrad* in any number of dimensions, even though its root means 'four.' Then

$$S'_{\text{pp}} = \frac{1}{2} \int d\tau \, \left(\eta^{-1} \dot{X}^\mu \dot{X}_\mu - \eta m^2 \right) . \qquad (1.2.5)$$

This action has the same symmetries as the earlier action S_{pp}, namely Poincaré invariance and world-line reparameterization invariance. Under the latter, $\eta(\tau)$ transforms as

$$\eta'(\tau')d\tau' = \eta(\tau)d\tau .\qquad(1.2.6)$$

The equation of motion from varying the tetrad is

$$\eta^2 = -\dot{X}^\mu \dot{X}_\mu / m^2 .\qquad(1.2.7)$$

Using this to eliminate $\eta(\tau)$, the action S'_{pp} becomes the earlier S_{pp}. Incidentally, S'_{pp} makes perfect sense for massless particles, while S_{pp} does not work in that case.

Although the actions S_{pp} and S'_{pp} are equivalent classically, it is hard to make sense of S_{pp} in a path integral because of its complicated form, with derivatives inside the square root. On the other hand S'_{pp} is quadratic in derivatives and its path integral is fairly easily evaluated, as we will see in later chapters. Presumably, any attempt to define the quantum theory for S_{pp} will lead to a result equivalent to the S'_{pp} path integral, and we will take the latter as our starting point in defining the quantum theory.

A one-dimensional object will sweep out a two-dimensional world-sheet, which can be described in terms of two parameters, $X^\mu(\tau,\sigma)$ as in figure 1.3(b). As in the case of the particle, we insist that physical quantities such as the action depend only on the embedding in spacetime and not on the parameterization. Not only is it attractive that the theory should have this property, but we will see that it is necessary in a consistent relativistic quantum theory. The simplest invariant action, the Nambu–Goto action, is proportional to the area of the world-sheet. In order to express this action in terms of $X^\mu(\tau,\sigma)$, define first the induced metric h_{ab} where indices $a, b, \ldots$ run over values (τ,σ):

$$h_{ab} = \partial_a X^\mu \partial_b X_\mu .\qquad(1.2.8)$$

Then the Nambu–Goto action is

$$S_{NG} = \int_M d\tau d\sigma \, \mathscr{L}_{NG} ,\qquad(1.2.9a)$$

$$\mathscr{L}_{NG} = -\frac{1}{2\pi\alpha'}(-\det h_{ab})^{1/2} ,\qquad(1.2.9b)$$

where M denotes the world-sheet. The constant α', which has units of spacetime-length-squared, is the Regge slope, whose significance will be seen later. The tension T of the string is related to the Regge slope by

$$T = \frac{1}{2\pi\alpha'}\qquad(1.2.10)$$

(exercise 1.1).

Let us now consider the symmetries of the action, transformations of $X^\mu(\tau,\sigma)$ such that $S_{NG}[X'] = S_{NG}[X]$. These are:

1. The isometry group of flat spacetime, the D-dimensional Poincaré group,

$$X'^{\mu}(\tau,\sigma) = \Lambda^{\mu}_{\nu} X^{\nu}(\tau,\sigma) + a^{\mu} \qquad (1.2.11)$$

with Λ^{μ}_{ν} a Lorentz transformation and a^{μ} a translation.

2. Two-dimensional coordinate invariance, often called diffeomorphism *(diff)* invariance. For new coordinates $(\tau'(\tau,\sigma),\sigma'(\tau,\sigma))$, the transformation is

$$X'^{\mu}(\tau',\sigma') = X^{\mu}(\tau,\sigma) . \qquad (1.2.12)$$

The Nambu–Goto action is analogous to the point particle action S_{pp}, with derivatives in the square root. Again we can simplify it by introducing an independent world-sheet metric $\gamma_{ab}(\tau,\sigma)$. Henceforth 'metric' will always mean γ_{ab} (unless we specify 'induced'), and indices will always be raised and lowered with this metric. We will take γ_{ab} to have Lorentzian signature $(-,+)$. The action is

$$S_{\text{P}}[X,\gamma] = -\frac{1}{4\pi\alpha'} \int_M d\tau\, d\sigma\, (-\gamma)^{1/2} \gamma^{ab} \partial_a X^{\mu} \partial_b X_{\mu} , \qquad (1.2.13)$$

where $\gamma = \det\gamma_{ab}$. This is the Brink–Di Vecchia–Howe–Deser–Zumino action, or Polyakov action for short. It was found by Brink, Di Vecchia, and Howe and by Deser and Zumino, in the course of deriving a generalization with local world-sheet supersymmetry. Its virtues, especially for path integral quantization, were emphasized by Polyakov.

To see the equivalence to S_{NG}, use the equation of motion obtained by varying the metric,

$$\delta_{\gamma} S_{\text{P}}[X,\gamma] = -\frac{1}{4\pi\alpha'} \int_M d\tau\, d\sigma\, (-\gamma)^{1/2}\, \delta\gamma^{ab} \left(h_{ab} - \tfrac{1}{2}\gamma_{ab}\gamma^{cd} h_{cd} \right) , \qquad (1.2.14)$$

where h_{ab} is again the induced metric (1.2.8). We have used the general relation for the variation of a determinant,

$$\delta\gamma = \gamma\gamma^{ab}\delta\gamma_{ab} = -\gamma\gamma_{ab}\delta\gamma^{ab} . \qquad (1.2.15)$$

Then $\delta_{\gamma} S_{\text{P}} = 0$ implies

$$h_{ab} = \tfrac{1}{2}\gamma_{ab}\gamma^{cd} h_{cd} . \qquad (1.2.16)$$

Dividing this equation by the square root of minus its determinant gives

$$h_{ab}(-h)^{-1/2} = \gamma_{ab}(-\gamma)^{-1/2} , \qquad (1.2.17)$$

so that γ_{ab} is proportional to the induced metric. This in turn can be used to eliminate γ_{ab} from the action,

$$S_{\text{P}}[X,\gamma] \to -\frac{1}{2\pi\alpha'} \int d\tau\, d\sigma\, (-h)^{1/2} = S_{\text{NG}}[X] . \qquad (1.2.18)$$

The action S_P has the following symmetries:

1. D-dimensional Poincaré invariance:

$$X'^{\mu}(\tau, \sigma) = \Lambda^{\mu}{}_{\nu} X^{\nu}(\tau, \sigma) + a^{\mu} ,$$
$$\gamma'_{ab}(\tau, \sigma) = \gamma_{ab}(\tau, \sigma) . \qquad (1.2.19)$$

2. Diff invariance:

$$X'^{\mu}(\tau', \sigma') = X^{\mu}(\tau, \sigma) ,$$
$$\frac{\partial \sigma'^c}{\partial \sigma^a} \frac{\partial \sigma'^d}{\partial \sigma^b} \gamma'_{cd}(\tau', \sigma') = \gamma_{ab}(\tau, \sigma) , \qquad (1.2.20)$$

for new coordinates $\sigma'^a(\tau, \sigma)$.

3. Two-dimensional Weyl invariance:

$$X'^{\mu}(\tau, \sigma) = X^{\mu}(\tau, \sigma) ,$$
$$\gamma'_{ab}(\tau, \sigma) = \exp(2\omega(\tau, \sigma)) \gamma_{ab}(\tau, \sigma) , \qquad (1.2.21)$$

for arbitrary $\omega(\tau, \sigma)$.

The Weyl invariance, a local rescaling of the world-sheet metric, has no analog in the Nambu–Goto form. We can understand its appearance by looking back at the equation of motion (1.2.17) used to relate the Polyakov and Nambu–Goto actions. This does not determine γ_{ab} completely but only up to a local rescaling, so Weyl-equivalent metrics correspond to the *same* embedding in spacetime. This is an extra redundancy of the Polyakov formulation, in addition to the obvious diff invariance.

The variation of the action with respect to the metric defines the energy-momentum tensor,

$$T^{ab}(\tau, \sigma) = -4\pi(-\gamma)^{-1/2} \frac{\delta}{\delta \gamma_{ab}} S_P$$

$$= -\frac{1}{\alpha'}(\partial^a X^{\mu} \partial^b X_{\mu} - \tfrac{1}{2}\gamma^{ab} \partial_c X^{\mu} \partial^c X_{\mu}) . \qquad (1.2.22)$$

This has an extra factor of -2π relative to the usual definition in field theory; this is a standard and convenient convention in string theory. It is conserved, $\nabla_a T^{ab} = 0$, as a consequence of diff invariance. The invariance of S_P under arbitrary Weyl transformations further implies that

$$\gamma_{ab} \frac{\delta}{\delta \gamma_{ab}} S_P = 0 \implies T^a_a = 0 . \qquad (1.2.23)$$

The actions S_{NG} and S_P define two-dimensional field theories on the string world-sheet. In string theory, we will see that amplitudes for *space-time* processes are given by matrix elements in the two-dimensional quantum field theory on the *world-sheet*. While our interest is four-dimensional spacetime, most of the machinery we use in string perturbation theory is two-dimensional. From the point of view of the world-sheet, the coordinate transformation law (1.2.12) defines $X^\mu(\tau, \sigma)$ as a scalar field, with μ going along for the ride as an internal index. From the two-dimensional point of view the Polyakov action describes massless Klein–Gordon scalars X^μ covariantly coupled to the metric γ_{ab}. Also from this point of view, the Poincaré invariance is an internal symmetry, meaning that it acts on fields at fixed τ, σ.

Varying γ_{ab} in the action gives the equation of motion

$$T_{ab} = 0 \, . \tag{1.2.24}$$

Varying X^μ gives the equation of motion

$$\partial_a[(-\gamma)^{1/2}\gamma^{ab}\partial_b X^\mu] = (-\gamma)^{1/2}\nabla^2 X^\mu = 0 \, . \tag{1.2.25}$$

For world-sheets with boundary there is also a surface term in the variation of the action. To be specific, take the coordinate region to be

$$-\infty < \tau < \infty \, , \quad 0 \leq \sigma \leq \ell \, . \tag{1.2.26}$$

We will think of τ as a time variable and σ as spatial, so this is a single string propagating without sources. Then

$$\delta S_P = \frac{1}{2\pi\alpha'} \int_{-\infty}^{\infty} d\tau \int_0^\ell d\sigma \, (-\gamma)^{1/2}\delta X^\mu \nabla^2 X_\mu$$
$$- \frac{1}{2\pi\alpha'} \int_{-\infty}^{\infty} d\tau \, (-\gamma)^{1/2}\delta X^\mu \partial^\sigma X_\mu \Big|_{\sigma=0}^{\sigma=\ell} \, . \tag{1.2.27}$$

In the interior, the equation of motion is as above. The boundary term vanishes if

$$\partial^\sigma X^\mu(\tau, 0) = \partial^\sigma X^\mu(\tau, \ell) = 0 \, . \tag{1.2.28}$$

These are *Neumann* boundary conditions on X^μ. Stated more covariantly,

$$n^a \partial_a X_\mu = 0 \quad \text{on } \partial M \, , \tag{1.2.29}$$

where n^a is normal to the boundary ∂M. The ends of the open string move freely in spacetime.

The surface term in the equation of motion will also vanish if we impose

$$X^\mu(\tau, \ell) = X^\mu(\tau, 0) \, , \quad \partial^\sigma X^\mu(\tau, \ell) = \partial^\sigma X^\mu(\tau, 0) \, , \tag{1.2.30a}$$
$$\gamma_{ab}(\tau, \ell) = \gamma_{ab}(\tau, 0) \, . \tag{1.2.30b}$$

That is, the fields are *periodic*. There is no boundary; the endpoints are joined to form a closed loop.

The open string boundary condition (1.2.28) and closed string boundary condition (1.2.30) are the only possibilities consistent with D-dimensional

Poincaré invariance and the equations of motion. If we relax the condition of Poincaré invariance then there are other possibilities which will be important later. Some of these are explored in the exercises.

The Nambu–Goto and Polyakov actions may be the simplest with the given symmetries, but simplicity of the action is not the right criterion in quantum field theory. Symmetry, not simplicity, is the key idea: for both physical and technical reasons we must usually consider the most general local action consistent with all of the symmetries of the theory. Let us try to generalize the Polyakov action, requiring that the symmetries be maintained and that the action be polynomial in derivatives (else it is hard to make sense of the quantum theory).[3] Global Weyl invariance, $\omega(\tau, \sigma) = $ constant, requires that the action have one more factor of γ^{ab} than γ_{ab} to cancel the variation of $(-\gamma)^{1/2}$. The extra upper indices can only contract with derivatives, so each term has precisely two derivatives. The coordinate invariance and Poincaré invariance then allow one additional term beyond the original S_P:

$$\chi = \frac{1}{4\pi} \int_M d\tau \, d\sigma \, (-\gamma)^{1/2} R \, , \qquad (1.2.31)$$

where R is the Ricci scalar constructed from γ_{ab}. Under a local Weyl rescaling,

$$(-\gamma')^{1/2} R' = (-\gamma)^{1/2} (R - 2\nabla^2 \omega) \, . \qquad (1.2.32)$$

The variation is a total derivative, because $(-\gamma)^{1/2} \nabla_a v^a = \partial_a ((-\gamma)^{1/2} v^a)$ for any v^a. The integral (1.2.31) is therefore invariant for a world-sheet without boundary. With boundaries an additional surface term is needed (exercise 1.3).

Since χ is allowed by the symmetries we will include it in the action:

$$S'_P = S_P - \lambda \chi \, .$$

$$= -\int_M d\tau \, d\sigma \, (-\gamma)^{1/2} \left(\frac{1}{4\pi\alpha'} \gamma^{ab} \partial_a X^\mu \partial_b X_\mu + \frac{\lambda}{4\pi} R \right) \, . \qquad (1.2.33)$$

This is the most general (diff×Weyl)-invariant and Poincaré-invariant action with these fields and symmetries. For the moment the discussion of the symmetries is classical — we are ignoring possible quantum anomalies, which will be considered in chapter 3.

The action S'_P looks like the Hilbert action for the metric, $\int (-\gamma)^{1/2} R$, minimally coupled to D massless scalar fields X^μ. However, in two dimen-

---

[3] If we relax this requirement then there are various additional possibilities, such as invariants constructed from the curvature of the induced metric h_{ab}. These would be expected to appear for such one-dimensional objects as electric and magnetic flux tubes, and in that case the path integral makes sense because the thickness of the object provides a cutoff. But for infinitely thin strings, which is what we have in fundamental string theory, this is not possible.

sions, the Hilbert action depends only on the topology of the world-sheet and does not give dynamics to the metric. To see this, we use the result from general relativity that its variation is proportional to $R_{ab} - \frac{1}{2}\gamma_{ab}R$. In two dimensions, the symmetries of the curvature tensor imply that $R_{ab} = \frac{1}{2}\gamma_{ab}R$ and so this vanishes: the Hilbert action is invariant under any continuous change in the metric. It does have some global significance, as we will see in the chapter 3.

There are various further ways one might try to generalize the theory. One, suggested by our eventual interest in $D = 4$, is to allow more than four X^μ fields but require only four-dimensional Poincaré invariance. A second, motivated by the idea that symmetry is paramount, is to keep the same local symmetries plus four-dimensional Poincaré invariance, but to allow a more general field content — that is, two-dimensional fields in different representations of the world-sheet and spacetime symmetry groups. These ideas are both important and we will return to them at various times. A further idea is to enlarge the local symmetry of the theory; we will pursue this idea in volume two.

1.3   The open string spectrum

We discuss the spectrum of the open string in this section and that of the closed in the next, using light-cone gauge to eliminate the diff×Weyl redundancy. This gauge hides the covariance of the theory, but it is the quickest route to the spectrum and reveals important features like the critical dimension and the existence of massless gauge particles. The emphasis here is on the results, not the method, so the reader should not dwell on fine points. A systematic treatment using the covariant conformal gauge will begin in the next chapter.

Define light-cone coordinates in spacetime:

$$x^\pm = 2^{-1/2}(x^0 \pm x^1), \qquad x^i, \ i = 2, \ldots, D - 1. \tag{1.3.1}$$

We are using lower case for the spacetime coordinates x^μ and upper case for the associated world-sheet fields $X^\mu(\sigma, \tau)$. In these coordinates the metric is

$$a^\mu b_\mu = -a^+ b^- - a^- b^+ + a^i b^i, \tag{1.3.2a}$$

$$a_- = -a^+, \quad a_+ = -a^-, \quad a_i = a^i. \tag{1.3.2b}$$

We will set the world-sheet parameter τ at each point of the world-sheet to be equal to the spacetime coordinate x^+. One might have thought to try $\tau = x^0$ but this does not lead to the same simplifications. So x^+ will play the role of time and p^- that of energy. The longitudinal variables x^- and

p^+ are then like spatial coordinates and momenta, as are the transverse x^i and p^i.

We start by illustrating the procedure in the case of the point particle, using the action S'_{pp}. Fix the parameterization of the world-line by

$$X^+(\tau) = \tau . \tag{1.3.3}$$

The action then becomes

$$S'_{pp} = \frac{1}{2} \int d\tau \left(-2\eta^{-1}\dot{X}^- + \eta^{-1}\dot{X}^i\dot{X}^i - \eta m^2 \right) . \tag{1.3.4}$$

The canonical momenta $p_\mu = \partial L / \partial \dot{X}^\mu$ are

$$p_- = -\eta^{-1} , \quad p_i = \eta^{-1}\dot{X}^i . \tag{1.3.5}$$

Also recall the metric $p^i = p_i$ and $p^+ = -p_-$. The Hamiltonian is

$$H = p_-\dot{X}^- + p_i\dot{X}^i - L$$
$$= \frac{p^i p^i + m^2}{2p^+} . \tag{1.3.6}$$

Note that there is no term $p_+\dot{X}^+$ because X^+ is not a dynamical variable in the gauge-fixed theory. Also, $\dot{\eta}$ does not appear in the action and so the momentum p_η vanishes. Since $\eta = -1/p_-$ it is clear that we should not treat η as an independent canonical coordinate; in a more complete treatment we would justify this by appealing to the theory of systems with constraints.

To quantize, impose canonical commutators on the dynamical fields,

$$[p_i, X^j] = -i\delta_i{}^j , \quad [p_-, X^-] = -i . \tag{1.3.7}$$

The momentum eigenstates $|k_-, k^i\rangle$ form a complete set. The remaining momentum component p_+ is determined in terms of the others as follows. The gauge choice relates τ and X^+ translations, so $H = -p_+ = p^-$. The relative sign between H and p_+ arises because the former is active and the latter passive. Then eq. (1.3.6) becomes the relativistic mass-shell condition, and we have obtained the spectrum of a relativistic scalar.

Now turn to the open string, taking again the coordinate region $-\infty < \tau < \infty$ and $0 \leq \sigma \leq \ell$. Again we must make a gauge choice to fix the redundancies in the Polyakov action, and by a good choice we can also make the equations of motion simple. Set

$$X^+ = \tau , \tag{1.3.8a}$$
$$\partial_\sigma \gamma_{\sigma\sigma} = 0 , \tag{1.3.8b}$$
$$\det \gamma_{ah} = -1 . \tag{1.3.8c}$$

That is, we choose light-cone gauge for the world-sheet time coordinate and impose two conditions on the metric. This is three conditions for three local symmetries (the choice of two world-sheet coordinates plus the Weyl scaling).

Let us see how to reach the gauge (1.3.8). First choose τ according to (1.3.8a). Now notice that $f = \gamma_{\sigma\sigma}(-\det\gamma_{ab})^{-1/2}$ transforms as

$$f'd\sigma' = f\,d\sigma \qquad (1.3.9)$$

under reparameterizations of σ where τ is left fixed. Thus we define an invariant length $f\,d\sigma = dl$. Define the σ coordinate of a point to be proportional to its invariant distance $\int dl$ from the $\sigma = 0$ endpoint; the constant of proportionality is determined by requiring that the coordinate of the right-hand endpoint remain at $\sigma = \ell$. In this coordinate system $f = dl/d\sigma$ is independent of σ. It still depends on τ; we would like the parameter region $0 \le \sigma \le \ell$ to remain unchanged, so there is not enough freedom to remove the τ-dependence. Finally, make a Weyl transformation to satisfy condition (1.3.8c). Since f is Weyl-invariant, $\partial_\sigma f$ still vanishes. With condition (1.3.8c) this implies that $\partial_\sigma\gamma_{\sigma\sigma} = 0$, and so condition (1.3.8b) is satisfied as well.[4]

We can solve the gauge condition (1.3.8c) for $\gamma_{\tau\tau}(\tau,\sigma)$, and $\gamma_{\sigma\sigma}$ is independent of σ, so the independent degrees of freedom in the metric are now $\gamma_{\sigma\sigma}(\tau)$ and $\gamma_{\sigma\tau}(\tau,\sigma)$. In terms of these, the inverse metric is

$$\begin{bmatrix} \gamma^{\tau\tau} & \gamma^{\tau\sigma} \\ \gamma^{\sigma\tau} & \gamma^{\sigma\sigma} \end{bmatrix} = \begin{bmatrix} -\gamma_{\sigma\sigma}(\tau) & \gamma_{\tau\sigma}(\tau,\sigma) \\ \gamma_{\tau\sigma}(\tau,\sigma) & \gamma_{\sigma\sigma}^{-1}(\tau)(1 - \gamma_{\tau\sigma}^2(\tau,\sigma)) \end{bmatrix}. \qquad (1.3.10)$$

The Polyakov Lagrangian then becomes

$$L = -\frac{1}{4\pi\alpha'} \int_0^\ell d\sigma \left[\gamma_{\sigma\sigma}(2\partial_\tau x^- - \partial_\tau X^i \partial_\tau X^i) \right.$$
$$\left. - 2\gamma_{\sigma\tau}(\partial_\sigma Y^- - \partial_\tau X^i \partial_\sigma X^i) + \gamma_{\sigma\sigma}^{-1}(1 - \gamma_{\tau\sigma}^2)\partial_\sigma X^i \partial_\sigma X^i \right]. \qquad (1.3.11)$$

Here we have separated $X^-(\tau,\sigma)$ into two pieces $x^-(\tau)$ and $Y^-(\tau,\sigma)$, the first being the mean value of X^- at given τ and the second having a mean value of zero. That is,

$$x^-(\tau) = \frac{1}{\ell} \int_0^\ell d\sigma X^-(\tau,\sigma), \qquad (1.3.12a)$$

$$Y^-(\tau,\sigma) = X^-(\tau,\sigma) - x^-(\tau). \qquad (1.3.12b)$$

---

[4] We have shown how to assign a definite (τ,σ) value to each point on the world-sheet, but we have not shown that this is a good gauge — that different points have different coordinates. Already for the point particle this is an issue: τ is a good coordinate only if the world-line does not double back through an x^+-hyperplane. For the string we need this and also that f is positive everywhere. It is not hard to show that for allowed classical motions, in which all points travel at less than the speed of light, these conditions are met. In the quantum theory, one can show for the point particle that the doubled-back world-lines give no net contribution. Presumably this can be generalized to the string, but this is one of those fine points not to dwell on.

The field Y^- does not appear in any terms with time derivatives and so is nondynamical. It acts as a Lagrange multiplier, constraining $\partial_\sigma^2 \gamma_{\tau\sigma}$ to vanish (extra ∂_σ because Y^- has zero mean). Note further that in the gauge (1.3.8) the open string boundary condition (1.2.28) becomes

$$\gamma_{\tau\sigma}\partial_\tau X^\mu - \gamma_{\tau\tau}\partial_\sigma X^\mu = 0 \quad \text{at } \sigma = 0, \ell . \tag{1.3.13}$$

For $\mu = +$ this gives

$$\gamma_{\tau\sigma} = 0 \quad \text{at } \sigma = 0, \ell \tag{1.3.14}$$

and since $\partial_\sigma^2 \gamma_{\tau\sigma} = 0$, then $\gamma_{\tau\sigma}$ vanishes everywhere. For $\mu = i$, the boundary condition is

$$\partial_\sigma X^i = 0 \quad \text{at } \sigma = 0, \ell . \tag{1.3.15}$$

After imposing the gauge conditions and taking into account the Lagrange multiplier Y^-, we can proceed by ordinary canonical methods. The system is reduced to the variables $x^-(\tau)$, $\gamma_{\sigma\sigma}(\tau)$ and the fields $X^i(\tau, \sigma)$. The Lagrangian is

$$L = -\frac{\ell}{2\pi\alpha'}\gamma_{\sigma\sigma}\partial_\tau x^- + \frac{1}{4\pi\alpha'}\int_0^\ell d\sigma \left(\gamma_{\sigma\sigma}\partial_\tau X^i\partial_\tau X^i - \gamma_{\sigma\sigma}^{-1}\partial_\sigma X^i\partial_\sigma X^i\right) . \tag{1.3.16}$$

The momentum conjugate to x^- is

$$p_- = -p^+ = \frac{\partial L}{\partial(\partial_\tau x^-)} = -\frac{\ell}{2\pi\alpha'}\gamma_{\sigma\sigma} . \tag{1.3.17}$$

As with η in the particle example, $\gamma_{\sigma\sigma}$ is a momentum and not a coordinate. The momentum density conjugate to $X^i(\tau, \sigma)$ is

$$\Pi^i = \frac{\delta L}{\delta(\partial_\tau X^i)} = \frac{1}{2\pi\alpha'}\gamma_{\sigma\sigma}\partial_\tau X^i = \frac{p^+}{\ell}\partial_\tau X^i . \tag{1.3.18}$$

The Hamiltonian is then

$$H = p_-\partial_\tau x^- - L + \int_0^\ell d\sigma\, \Pi_i\partial_\tau X^i$$

$$= \frac{\ell}{4\pi\alpha'p^+}\int_0^\ell d\sigma \left(2\pi\alpha'\Pi^i\Pi^i + \frac{1}{2\pi\alpha'}\partial_\sigma X^i\partial_\sigma X^i\right) . \tag{1.3.19}$$

This is just the Hamiltonian for $D - 2$ free fields X^i, with p^+ a conserved quantity. The equations of motion are

$$\partial_\tau x^- = \frac{\partial H}{\partial p_-} = \frac{H}{p^+} , \quad \partial_\tau p^+ = \frac{\partial H}{\partial x^-} = 0 , \tag{1.3.20a}$$

$$\partial_\tau X^i = \frac{\delta H}{\delta\Pi^i} = 2\pi\alpha'c\Pi^i , \quad \partial_\tau\Pi^i = -\frac{\delta H}{\delta X^i} = \frac{c}{2\pi\alpha'}\partial_\sigma^2 X^i , \tag{1.3.20b}$$

implying the wave equation

$$\partial_\tau^2 X^i = c^2\partial_\sigma^2 X^i \tag{1.3.21}$$

with velocity $c = \ell/2\pi\alpha' p^+$. We will not develop the interactions in the light-cone formalism, but for these it is useful to choose the coordinate length ℓ for each string to be proportional to p^+, so that $c = 1$. Since p^+ is positive and conserved, the total string length is then also conserved.

The transverse coordinates satisfy a free wave equation so it is useful to expand in normal modes. In fact $X^\pm$ also satisfy the free wave equation. For X^+ this is trivial, while for X^- it requires a short calculation. The general solution to the wave equation with boundary condition (1.3.15) is

$$X^i(\tau, \sigma) = x^i + \frac{p^i}{p^+}\tau + i(2\alpha')^{1/2} \sum_{\substack{n=-\infty \\ n \neq 0}}^{\infty} \frac{1}{n}\alpha_n^i \exp\left(-\frac{\pi i n c \tau}{\ell}\right) \cos\frac{\pi n \sigma}{\ell} . \quad (1.3.22)$$

Reality of X^i requires $\alpha_{-n}^i = (\alpha_n^i)^\dagger$. We have defined the center-of-mass variables

$$x^i(\tau) = \frac{1}{\ell} \int_0^\ell d\sigma\, X^i(\tau, \sigma) , \quad (1.3.23a)$$

$$p^i(\tau) = \int_0^\ell d\sigma\, \Pi^i(\tau, \sigma) = \frac{p^+}{\ell} \int_0^\ell d\sigma\, \partial_\tau X^i(\tau, \sigma) , \quad (1.3.23b)$$

which are the average position and the total momentum. These are the Heisenberg operators; the Schrödinger operators $x^i \equiv x^i(0)$ and $p^i \equiv p^i(0)$ appear in the expansion (1.3.22).

To quantize, impose the equal time canonical commutation relations

$$[x^-, p^+] = i\eta^{-+} = -i , \quad (1.3.24a)$$

$$[X^i(\sigma), \Pi^j(\sigma')] = i\delta^{ij}\delta(\sigma - \sigma') , \quad (1.3.24b)$$

with all other commutators between the independent variables vanishing. In terms of the Fourier components,

$$[x^i, p^j] = i\delta^{ij} , \quad (1.3.25a)$$

$$[\alpha_m^i, \alpha_n^j] = m\delta^{ij}\delta_{m,-n} . \quad (1.3.25b)$$

For each m and i the modes satisfy a harmonic oscillator algebra with nonstandard normalization

$$\alpha_m^i \sim m^{1/2}a , \quad \alpha_{-m}^i \sim m^{1/2}a^\dagger , \quad m > 0 , \quad (1.3.26)$$

where $[a, a^\dagger] = 1$. The oscillators are labeled by the direction of oscillation i and the harmonic m.

The state $|0;k\rangle$, where $k = (k^+, k^i)$, is defined to be annihilated by the lowering operators and to be an eigenstate of the center-of-mass momenta,

$$p^+|0;k\rangle = k^+|0;k\rangle , \quad p^i|0;k\rangle = k^i|0;k\rangle , \quad (1.3.27a)$$

$$\alpha_m^i|0;k\rangle = 0 , \quad m > 0 . \quad (1.3.27b)$$

A general state can be built by acting on $|0;k\rangle$ with the raising operators,

$$|N;k\rangle = \left[\prod_{i=2}^{D-1} \prod_{n=1}^{\infty} \frac{(\alpha^i_{-n})^{N_{in}}}{(n^{N_{in}} N_{in}!)^{1/2}} \right] |0;k\rangle . \tag{1.3.28}$$

That is, the independent states can be labeled by the center-of-mass momenta k^+ and k^i, and by the occupation numbers N_{in} for each mode (i,n), where $i = 2, \ldots, D-1$ and $n = 1, \ldots, \infty$. The center-of-mass momenta are just the degrees of freedom of a point particle, while the oscillators represent an infinite number of internal degrees of freedom. Every choice of these occupation numbers corresponds, from the spacetime point of view, to a different particle or spin state.

Note that the states (1.3.28) form the Hilbert space $\mathcal{H}_1$ of a single open string, and in particular the state $|0;0\rangle$ is the ground state of a single string with zero momentum, not the zero-string vacuum state. We will call the latter $|\text{vacuum}\rangle$ to make this clear. The various operators appearing above do not create or destroy strings but act within the space of states of a single string. The n-string Hilbert space $\mathcal{H}_n$ would be formed as the product of n copies of the space (1.3.28); the wavefunction must be symmetrized because, as we will see, all these states have integer spin. The full Hilbert space of the string theory, at least in the free limit that we are considering here, is then the sum

$$\mathcal{H} = |\text{vacuum}\rangle \oplus \mathcal{H}_1 \oplus \mathcal{H}_2 \oplus \ldots . \tag{1.3.29}$$

Inserting the mode expansion (1.3.22) into the Hamiltonian (1.3.19) gives

$$H = \frac{p^i p^i}{2p^+} + \frac{1}{2p^+ \alpha'} \left(\sum_{n=1}^{\infty} \alpha^i_{-n} \alpha^i_n + A \right) . \tag{1.3.30}$$

In the Hamiltonian H, the order of operators is ambiguous. We have put the lowering operators on the right and the raising operators on the left, and included an unknown constant A from the commutators. In a careful treatment of light-cone quantization, this constant is determined as follows. The choice of light-cone gauge has obscured the Lorentz invariance of the theory. It is necessary to check this by finding the operators $M^{\mu\nu}$ that generate Lorentz transformations, and verifying that they have the correct algebra with p^μ and with each other. One finds that this is the case only for the specific value $A = -1$, and only if the spacetime dimensionality is precisely $D = 26$.

We do not wish to spend this much time on light-cone quantization, so we will try to show that the result is plausible without a systematic treatment. We assure the reader that we will obtain the values of A and D honestly in the conformal gauge approach, which will also give us insight into why Lorentz invariance would be lost in the light-cone approach for the wrong A or D.

First, we assert that the operator ordering constant in the Hamiltonian for a free field comes from summing the zero-point energies of each oscillator mode, $\frac{1}{2}\omega$ for a bosonic field like X^μ. Equivalently, it always works out that the natural operator order is averaged, $\frac{1}{2}\omega(aa^\dagger + a^\dagger a)$, which is the same as $\omega(a^\dagger a + \frac{1}{2})$. In H this would give

$$A = \frac{D-2}{2} \sum_{n=1}^{\infty} n \, , \tag{1.3.31}$$

the factor of $D-2$ coming from the sum over transverse directions. The zero-point sum diverges. It can be evaluated by regulating the theory and then being careful to preserve Lorentz invariance in the renormalization. This leads to the odd result

$$\sum_{n=1}^{\infty} n \rightarrow -\frac{1}{12} \, . \tag{1.3.32}$$

To motivate this, insert a smooth cutoff factor

$$\exp(-\epsilon\gamma_{\sigma\sigma}^{-1/2}|k_\sigma|) \tag{1.3.33}$$

into the sum, where $k_\sigma = n\pi/\ell$ and the factor of $\gamma_{\sigma\sigma}^{-1/2}$ is included to make this invariant under σ reparameterizations. The zero-point constant is then

$$\begin{aligned} A &\rightarrow \frac{D-2}{2} \sum_{n=1}^{\infty} n \, \exp\left[-\epsilon n(\pi/2p^+\alpha'\ell)^{1/2}\right] \\ &= \frac{D-2}{2}\left(\frac{2\ell p^+\alpha'}{\epsilon^2\pi} - \frac{1}{12} + O(\epsilon)\right) \, . \end{aligned} \tag{1.3.34}$$

The cutoff-dependent first term is proportional to the length ℓ of the string and can be canceled by a counterterm in the action proportional to $\int d^2\sigma(-\gamma)^{1/2}$. In fact, Weyl invariance *requires* that it be canceled, leaving only the cutoff-independent second term,

$$A = \frac{2-D}{24} \, . \tag{1.3.35}$$

This finite remainder is an example of a Casimir energy, coming from the fact that the string has a finite length.

As for the point particle, $p^- = H$, so

$$m^2 = 2p^+H - p^ip^i = \frac{1}{\alpha'}\left(N + \frac{2-D}{24}\right) , \tag{1.3.36}$$

where N is the *level*

$$N = \sum_{i=2}^{D-1} \sum_{n=1}^{\infty} n N_{in} .$$ (1.3.37)

The mass of each state is thus determined in terms of the level of excitation. Now let us look at some of the light string states. The lightest is

$$|0;k\rangle , \quad m^2 = \frac{2-D}{24\alpha'} .$$ (1.3.38)

The mass-squared is negative if $D > 2$: the state is a tachyon. In field theory the potential energy for a scalar field is $\frac{1}{2}m^2\phi^2$, so the negative mass-squared means that the no-string 'vacuum' is actually unstable, like the symmetric state in a spontaneously broken theory. It is a complicated question whether the bosonic string has any stable vacuum, and the answer is not known. Starting with our study of the superstring in volume two, we will see that there *are* tachyon-free string theories. For now, the simplest way to proceed is to ignore this instability and use the bosonic string theory as our model for developing string technology. Occasionally we will encounter divergences due to the tachyon, but these will not interfere with our purpose.

The lowest excited states of the string are obtained by exciting one of the $n = 1$ modes once:

$$\alpha_{-1}^i|0;k\rangle , \quad m^2 = \frac{26-D}{24\alpha'} .$$ (1.3.39)

Lorentz invariance now requires a specific value of D as follows. The analysis of spin is different for massive and massless particles. For a massive particle, one goes to the rest frame $p^\mu = (m,0,\ldots,0)$. The internal states then form a representation of the spatial rotation group $SO(D-1)$. For a massless particle there is no rest frame; choose the frame $p^\mu = (E,E,0,\ldots,0)$. The $SO(D-2)$ acting on the transverse directions leaves p^μ invariant, and the internal states form a representation of this smaller group. This is familiar from $D = 4$: massive particles are labeled by spin j, the $SO(3)$ representation, and so have $2j+1$ states. Massless particles are labeled by their helicity λ, which is their eigenvalue under the single generator of $SO(2)$. Lorentz invariance alone thus requires only one state, though CPT symmetry takes λ to $-\lambda$ and so requires two states for $\lambda \neq 0$.

In D dimensions, a massive vector particle thus has $D-1$ spin states while a massless vector need have only $D-2$ states. At the first level we found only the $D-2$ states $\alpha_{-1}^i|0;k\rangle$, so these must be massless and

$$A = -1, \quad D = 26 .$$ (1.3.40)

This is a striking and important result: the spectrum is Lorentz-invariant *only* if the number of spacetime dimensions is $D = 26$. The classical theory is Lorentz-invariant for any D, but there is an anomaly — the symmetry is not preserved by the quantization except when $D = 26$. In later chapters we will obtain a deeper understanding of how this comes about and what it implies.

We conclude this section with a few more comments about spin. The light-cone quantization singles out two directions and leaves manifest only the $SO(D-2)$ that acts in the transverse directions. The spin generators for the transverse directions are

$$S^{ij} = -i \sum_{n=1}^{\infty} \frac{1}{n} (\alpha_{-n}^i \alpha_n^j - \alpha_{-n}^j \alpha_n^i) \; ; \tag{1.3.41}$$

antisymmetry on ij together with Lorentz invariance allows no zero-point constant. For a massless particle, the full $SO(D-2)$ spin symmetry is made manifest by choosing the momentum to lie along the 1-direction singled out in the quantization.

For a massive particle, only an $SO(D-2)$ subgroup of the $SO(D-1)$ spin symmetry will be manifest in light-cone quantization, but this is still quite useful. For example, the $(D-1)$-dimensional vector representation of $SO(D-1)$ breaks up into an invariant and a $(D-2)$-vector under the $SO(D-2)$ acting on the transverse directions,

$$\mathbf{v} = (v^1, 0, \ldots, 0) + (0, v^2, \ldots, v^{D-1}) \; . \tag{1.3.42}$$

Thus, if a massive particle is in the vector representation of $SO(D-1)$, we will see a scalar and a vector when we look at the transformation properties under $SO(D-2)$. This idea extends to any representation: one can always reconstruct the full $SO(D-1)$ spin representation from the behavior under $SO(D-2)$.

The higher excited states of the string, which are massive, do form full representations of $SO(D-1)$. One can check this by hand for a few levels, and in a more complete analysis it follows from the existence of the full set of Lorentz generators. We will show this indirectly in the coming chapters, by finding a consistent covariant quantization.

At level N, the maximum eigenvalue of a given spin component, say S^{23}, is N, obtained by acting N times with $\alpha_{-1}^2 + i\alpha_{-1}^3$. Thus,

$$S^{23} \leq 1 + \alpha' m^2 \; . \tag{1.3.43}$$

The slope in this inequality is known as the *Regge slope*. Meson resonances obey a linear relation of this form, with $\alpha' \sim (1 \text{ GeV})^{-2}$. For this and other reasons, string theory was originally proposed in 1970 as a theory of the strong interaction. Within a few years, however, the $SU(3)$ gauge theory QCD was discovered to be the correct theory, with string-like behavior only in certain limits.

Now that we are considering strings as the unified theory of particle physics, gravity, and quantum mechanics, α' will be of the order of the natural scale determined by the fundamental constants of gravity and quantum mechanics, M_P^{-2}. In particular, masses that are not zero in string theory are of order M_P. This is so large compared to experimentally accessible energy scales that these particles appear only in virtual states. Thus, we will be especially concerned with the massless string spectrum, since this must include all the particles of the Standard Model. Of course, most known particles are massive, but these masses are so small compared to M_P that they are zero to first approximation and become non-zero due to small symmetry-breaking effects.

Some of the interest in string theory has been driven by the possibility that the strong interaction theory QCD is equivalent to a string theory, at least in some approximation. There has been cross-fertilization between this idea and the idea of strings as a fundamental theory. The connection between QCD and string theory is outside the focus of this book, but the recent developments described in chapter 14 have also given new insight in this direction.

1.4 Closed and unoriented strings

The light-cone quantization of the closed string is quite similar to that of the open string. Again impose the gauge conditions (1.3.8). In the open string, these determined the gauge completely. In the closed string there is still some extra coordinate freedom,

$$\sigma' = \sigma + s(\tau) \bmod \ell \,, \tag{1.4.1}$$

because the point $\sigma = 0$ can be chosen anywhere along the string. Most of this remaining freedom can be fixed by the additional gauge condition

$$\gamma_{\tau\sigma}(\tau, 0) = 0 \,. \tag{1.4.2}$$

That is, the line $\sigma = 0$ is orthogonal to the lines of constant τ. This determines the line $\sigma = 0$ except for an overall σ-translation. Conditions (1.3.8) and (1.4.2) thus fix all of the gauge freedom except for τ-independent translations of σ,

$$\sigma' = \sigma + s \bmod \ell \,. \tag{1.4.3}$$

We will deal with this extra gauge freedom later.

The analysis is now parallel to that for the open string. The Lagrangian, canonical momenta, Hamiltonian, and equation of motion are as in the open string, eqs. (1.3.16)–(1.3.21). The general periodic solution to the equation of motion is

$$
X^i(\tau, \sigma) = x^i + \frac{p^i}{p^+}\tau + i\left(\frac{\alpha'}{2}\right)^{1/2}
$$
$$
\times \sum_{\substack{n=-\infty \\ n\neq 0}}^{\infty} \left\{ \frac{\alpha_n^i}{n} \exp\left[-\frac{2\pi i n(\sigma + c\tau)}{\ell}\right] + \frac{\tilde{\alpha}_n^i}{n} \exp\left[\frac{2\pi i n(\sigma - c\tau)}{\ell}\right] \right\} .
$$

(1.4.4)

In the closed string there are two independent sets of oscillators, α_n^i and $\tilde{\alpha}_n^i$, corresponding to left-moving and right-moving waves along the string. In the open string the boundary condition at the endpoints tied these together. The independent degrees of freedom are again the transverse oscillators and the transverse and longitudinal center-of-mass variables,

$$
\alpha_n^i, \ \tilde{\alpha}_n^i, \ x^i, \ p^i, \ x^-, \ p^+,
$$

(1.4.5)

with canonical commutators

$$
[x^-, p^+] = -i , \tag{1.4.6a}
$$
$$
[x^i, p^j] = i\delta^{ij} , \tag{1.4.6b}
$$
$$
[\alpha_m^i, \alpha_n^j] = m\delta^{ij}\delta_{m,-n} , \tag{1.4.6c}
$$
$$
[\tilde{\alpha}_m^i, \tilde{\alpha}_n^j] = m\delta^{ij}\delta_{m,-n} . \tag{1.4.6d}
$$

Starting from the state $|0,0;k\rangle$, which has center-of-mass momentum k^μ and is annihilated by α_m^is and $\tilde{\alpha}_m^i$s for $m > 0$, the general state is

$$
|N, \tilde{N}; k\rangle = \left[\prod_{i=2}^{D-1} \prod_{n=1}^{\infty} \frac{(\alpha_{-n}^i)^{N_{in}}(\tilde{\alpha}_{-n}^i)^{\tilde{N}_{in}}}{(n^{N_{in}} N_{in}! n^{\tilde{N}_{in}} \tilde{N}_{in}!)^{1/2}}\right] |0,0;k\rangle .
$$

(1.4.7)

The mass formula is

$$
\begin{aligned}
m^2 &= 2p^+ H - p^i p^i \\
&= \frac{2}{\alpha'}\left[\sum_{n=1}^{\infty}(\alpha_{-n}^i \alpha_n^i + \tilde{\alpha}_{-n}^i \tilde{\alpha}_n^i) + A + \tilde{A}\right] \\
&= \frac{2}{\alpha'}(N + \tilde{N} + A + \tilde{A}) .
\end{aligned}
$$

(1.4.8)

We have broken up the level and zero-point constant into the part from the right-moving modes and the part from the left. Summing zero-point energies again gives

$$
A = \tilde{A} = \frac{2 - D}{24} .
$$

(1.4.9)

There is one further restriction on the state due to the remaining gauge freedom, the σ-translations (1.4.3). The physical spectrum is obtained by restricting to gauge-invariant states. The operator that generates the σ-translations is

$$
P = -\int_0^\ell d\sigma\, \Pi^i \partial_\sigma X^i
$$

$$
= -\frac{2\pi}{\ell} \left[\sum_{n=1}^\infty (\alpha^i_{-n}\alpha^i_n - \tilde{\alpha}^i_{-n}\tilde{\alpha}^i_n) + A - \tilde{A} \right]
$$

$$
= -\frac{2\pi}{\ell}(N - \tilde{N}) . \tag{1.4.10}
$$

States must therefore satisfy

$$
N = \tilde{N} . \tag{1.4.11}
$$

The lightest closed string state is

$$
|0,0;k\rangle , \quad m^2 = \frac{2-D}{6\alpha'} , \tag{1.4.12}
$$

which is again a tachyon. The first excited states are

$$
\alpha^i_{-1}\tilde{\alpha}^j_{-1}|0,0;k\rangle , \quad m^2 = \frac{26-D}{6\alpha'} . \tag{1.4.13}
$$

As with the open string, these states do not add up to complete representations of $SO(D-1)$ and so this level must be massless. Thus,

$$
A = \tilde{A} = -1 , \quad D = 26 . \tag{1.4.14}
$$

The states (1.4.13) transform as a 2-tensor under $SO(D-2)$. This is a reducible representation: it decomposes into a symmetric traceless tensor, an antisymmetric tensor, and a scalar. That is, any tensor e^{ij} can be decomposed

$$
e^{ij} = \frac{1}{2}\left(e^{ij} + e^{ji} - \frac{2}{D-2}\delta^{ij}e^{kk} \right) + \frac{1}{2}\left(e^{ij} - e^{ji} \right) + \frac{1}{D-2}\delta^{ij}e^{kk} , \tag{1.4.15}
$$

and the three separate terms do not mix under rotations.

At any mass level, N_{in} and $\tilde{N}_{in}$ are independent except for the $N = \tilde{N}$ constraint. Thus, the closed string spectrum at $m^2 = 4(N-1)/\alpha'$ is the product of two copies of the $m^2 = (N-1)/\alpha'$ level of the open string.

Let us now discuss some general issues. The two-dimensional diff invariance removed two families of normal modes from the spectrum. If we tried to make a covariant theory without this invariance, we would have to generalize the transverse commutator (1.3.25) to

$$
[\alpha^\mu_m, \alpha^\nu_n] = m\eta^{\mu\nu}\delta_{m,-n} . \tag{1.4.16}
$$

Lorentz invariance forces the timelike oscillators to have the wrong-sign commutator. A state with an odd number of timelike excitations will then have a negative norm. This is inconsistent with quantum mechanics, since the norm is a probability. In fact a theory based on (1.4.16) came before the string theory we have described, and the latter was discovered by requiring that the negative norm states never be produced in physical processes. The commutator (1.4.16) arises in a covariant quantization of string theory (which we shall use beginning in the next chapter), with the coordinate invariance appearing through constraints on the spectrum to eliminate the unphysical states.

The existence of a massless vector in the open string theory and of the massless symmetric and antisymmetric tensors in the closed string theory is striking. General principles require that a massless vector couple to a conserved current and therefore that the theory have a gauge invariance. This is our first example of the massless gauge particles that are present in all fundamental string theories. Actually, this particular gauge boson, call it the photon, is not so interesting, because the gauge group is only $U(1)$ (there is only one photon) and because it turns out that all particles in the theory are neutral. In chapter 6 we will discuss a simple generalization of open string theory, adding *Chan–Paton* degrees of freedom at the string endpoints, which leads to $U(n)$, $SO(n)$, and $Sp(n)$ gauge groups. Similarly, a massless symmetric tensor particle must couple to a conserved symmetric tensor source. The only such source is the energy-momentum tensor; some additional possibilities arise in special cases, but these are not relevant here. Coupling to this in a consistent way requires the theory to have *spacetime* coordinate invariance. Thus the massless symmetric tensor is the graviton, and general relativity is contained as one small piece of the closed string theory. The massless antisymmetric tensor is known as a 2-form gauge boson. As we will see in section 3.7, there is also a local spacetime symmetry associated with it.

These spacetime gauge and coordinate invariances were not evident in our starting point, the Nambu–Goto or Polyakov theory. We have certainly discovered them in a backwards way: adding up zero-point energies, requiring a Lorentz-invariant spectrum, and citing general results about interactions of massless particles. In gauge theory and in general relativity, one starts with a spacetime symmetry principle. In string theory, too, it would seem that we should first figure out what the full spacetime symmetry is, and use this to define the theory. There are various attempts in this direction, but a complete picture has not yet emerged. For now, let us note one thing: we have observed that the local *world-sheet* symmetry diff removes the unphysical timelike and longitudinal normal modes which would lead to negative norm string states. For example, the open string photon would otherwise have D states rather than $D-2$. From the spacetime point of view, it is the local *spacetime* invariances that remove these same unphysical states. So there is at least some connection between the local symmetries of the world-sheet and those of spacetime.

Anticipating our interest in four dimensions, let us look at the massless string states obtained by exciting only oscillators in the 2- and 3-directions, and classifying them by the helicity $\lambda = S^{23}$. One finds that the photon has $\lambda = \pm 1$ and the graviton $\lambda = \pm 2$, justifying the names. The closed string scalar and antisymmetric tensor each give rise to a $\lambda = 0$ state. These are called the dilaton and the axion, respectively, and we will discuss their physics at various points.

We conclude this section with a discussion of one other generalization, the unoriented string theory. The theories we have been discussing are *oriented* string theories. We have not considered the coordinate transformation

$$\sigma' = \ell - \sigma \,, \quad \tau' = \tau \,, \tag{1.4.17}$$

which changes the orientation (handedness) of the world-sheet. This symmetry is generated by the world-sheet parity operator Ω. Carrying out (1.4.17) twice gives the identity, so $\Omega^2 = 1$ and the eigenvalues of Ω are ± 1. From the mode expansions (1.3.22) and (1.4.4) we see that

$$\Omega \alpha_n^i \Omega^{-1} = (-1)^n \alpha_n^i \tag{1.4.18}$$

in the open string and

$$\Omega \alpha_n^i \Omega^{-1} = \tilde{\alpha}_n^i \,, \tag{1.4.19a}$$
$$\Omega \tilde{\alpha}_n^i \Omega^{-1} = \alpha_n^i \tag{1.4.19b}$$

in the closed string. We define the phase of Ω by fixing $\Omega = +1$ for the ground states $|0;k\rangle$ and $|0,0;k\rangle$; later we will see that this choice is required in order for Ω to be conserved by interactions. Then

$$\Omega|N;k\rangle = (-1)^N |N;k\rangle \,, \tag{1.4.20a}$$
$$\Omega|N,\tilde{N};k\rangle = |\tilde{N},N;k\rangle \,. \tag{1.4.20b}$$

There are consistent interacting string theories, the *unoriented* theories, in which only the $\Omega = +1$ states are kept. Focusing on the massless states, the open string photon has $\Omega = -1$ and is absent in the unoriented theory. Acting on the massless tensor states in the closed string, the parity operator Ω takes e^{ij} to e^{ji}, so the graviton and dilaton are present in the unoriented theory while the antisymmetric tensor is not. Notice that both the open and closed string tachyons survive in the unoriented theory; we will have to work harder to remove these.

Let us mention two other constraints that will emerge from the study of the interactions. First, it is possible to have a consistent theory with only closed strings, or with closed and open strings, but not with open strings alone: closed strings can always be produced in the scattering of open strings. Second, oriented or unoriented open strings can only couple to closed strings of the same type. We list the possible combinations together with their massless spectra, with $G_{\mu\nu}$ representing the graviton, $B_{\mu\nu}$ the antisymmetric tensor, Φ the dilaton, and A_μ the photon:

1. Closed oriented bosonic string: $G_{\mu\nu}$, $B_{\mu\nu}$, Φ .

2. Closed unoriented bosonic string: $G_{\mu\nu}$, Φ .

3. Closed plus open oriented bosonic string: $G_{\mu\nu}$, $B_{\mu\nu}$, Φ, A_μ .

4. Closed plus open unoriented bosonic string: $G_{\mu\nu}$, Φ .

All of these have the graviton and dilaton, as well as the tachyon. In chapter 6 we will discuss more general open string theories, with Chan–Paton degrees of freedom at the endpoints. The massless oriented open strings will then be $U(n)$ gauge bosons, and the massless unoriented open strings will be $SO(n)$ or $Sp(n)$ gauge bosons.

Exercises

1.1 (a) Show that in the nonrelativistic limit the action S_{pp} has the usual nonrelativistic form, kinetic energy minus potential energy, with the potential energy being the rest mass.
(b) Show that for a string moving nonrelativistically, the Nambu–Goto action reduces to a kinetic term minus a potential term proportional to the length of the string. Show that the kinetic energy comes only from the *transverse* velocity of the string. Calculate the mass per unit length, as determined from the potential term and also from the kinetic term.

1.2 Show that classical equations of motion imply that the ends of the open string move at the speed of light.

1.3 For world-sheets with boundary, show that

$$\chi = \frac{1}{4\pi} \int_M d\tau\, d\sigma \, (-\gamma)^{1/2} R + \frac{1}{2\pi} \int_{\partial M} ds\, k$$

is Weyl-invariant. Here ds is the proper time along the boundary in the metric γ_{ab}, and k is the geodesic curvature of the boundary,

$$k = \pm t^a n_b \nabla_a t^b,$$

where t^a is a unit vector tangent to the boundary and n^a is an outward pointing unit vector orthogonal to t^a. The upper sign is for a Lorentzian world-sheet and the lower sign for a Euclidean world-sheet.

1.4 Show that the open string states at levels $m^2 = 1/\alpha'$ and $2/\alpha'$ form complete representations of $SO(D-1)$. You will need to work out the $SO(D-2)$ content of various symmetric and antisymmetric tensor representations of $SO(D-1)$. You can carry this to higher levels if you like, but you will need tensors of mixed symmetry and Young tableaux.

1.5 Extend the sum (1.3.32) to the 'twisted' case

$$\sum_{n=1}^{\infty}(n-\theta)$$

with θ a constant. That is, $k_\sigma = (n-\theta)\pi/\ell$. The answer is given in eq. (2.9.19). You should find that the cutoff-dependent term is independent of θ.

In the following exercises one has the usual open or closed string boundary conditions (Neumann or periodic) on X^μ for $\mu = 0,\ldots,24$ but a different boundary condition on X^{25}. Each of these has an important physical interpretation, and will be developed in detail in chapter 8. Find the mode expansion, the mass spectrum, and (for the closed string) the constraint from σ-translation invariance in terms of the occupation numbers. In some cases you need the result of exercise 1.5.

1.6 Open strings with

$$X^{25}(\tau,0) = 0 \,, \quad X^{25}(\tau,\ell) = y$$

with y a constant. This is an open string with both ends on D-branes.

1.7 Open strings with

$$X^{25}(\tau,0) = 0 \,, \quad \partial^\sigma X^{25}(\tau,\ell) = 0 \,.$$

This is an open string with one end on a D-brane and one end free.

1.8 Closed strings with

$$X^{25}(\tau,\sigma+\ell) = X^{25}(\tau,\sigma) + 2\pi R$$

with R a constant. This is a winding string in toroidal (periodic) compactification. In this case p^{25} must be a multiple of $1/R$.

1.9 Closed strings with

$$X^{25}(\tau,\sigma+\ell) = -X^{25}(\tau,\sigma) \,.$$

This is a twisted string in orbifold compactification.

2
Conformal field theory

In this chapter we develop a number of necessary ideas and techniques from the world-sheet quantum field theory, including the operator product expansion, conformal invariance, the Virasoro algebra, and vertex operators. The focus is on conformally invariant field theory in two flat dimensions; as we will see in the next chapter, this is what we are left with after fixing the local symmetries of the string world-sheet.

2.1 Massless scalars in two dimensions

We will start with the example of D free scalar fields in two dimensions, $X^\mu(\sigma^1, \sigma^2)$. We will refer to these two dimensions as the world-sheet, anticipating the application to string theory. The action is

$$S = \frac{1}{4\pi\alpha'} \int d^2\sigma \left(\partial_1 X^\mu \partial_1 X_\mu + \partial_2 X^\mu \partial_2 X_\mu \right) . \tag{2.1.1}$$

This is the Polyakov action (1.2.13), except that the world-sheet metric γ_{ab} has been replaced with a flat Euclidean metric δ_{ab}, signature $(+, +)$. The overall sign change of the action is a result of the Euclidean convention (A.1.31). As we will see, most string calculations are carried out on a Euclidean world-sheet. At least for flat metrics, the relation between Euclidean and Minkowski amplitudes is given by a standard analytic continuation, explained in the appendix. In fact, the results of the present chapter apply equally to a Minkowski world-sheet, and in the first seven sections all equations make sense if σ^2 is replaced with $i\sigma^0$. For the index μ we still take the flat Minkowski metric.

It is straightforward to quantize the action (2.1.1) canonically, finding the spectrum, vacuum expectation values, and so on. We have done essentially this in chapter 1, after having gone to light-cone gauge. Here we will take a somewhat different route, developing first various local properties such as equations of motion, operator products, Ward identities, and conformal invariance, before working our way around to the spectrum. It will be efficient for us to use the path integral formalism. This is reviewed in the appendix. We will be using the path integral representation primarily to derive operator equations (to be defined below); these can also be derived in a Hilbert space formalism.

It is very useful to adopt complex coordinates

$$z = \sigma^1 + i\sigma^2 , \quad \bar{z} = \sigma^1 - i\sigma^2 . \tag{2.1.2}$$

We will use a bar for the complex conjugates of z and other simple variables, and a star for the complex conjugates of longer expressions. Define also

$$\partial_z = \frac{1}{2}(\partial_1 - i\partial_2) , \quad \partial_{\bar{z}} = \frac{1}{2}(\partial_1 + i\partial_2) \tag{2.1.3}$$

These derivatives have the properties

$$\partial_z z = 1 , \quad \partial_z \bar{z} = 0 , \quad \partial_{\bar{z}} z = 0 , \quad \partial_{\bar{z}} \bar{z} = 1 . \tag{2.1.4}$$

It is conventional to abbreviate ∂_z to ∂ and $\partial_{\bar{z}}$ to $\bar{\partial}$ when this is not ambiguous. For a general vector v^a, define in the same way

$$v^z = v^1 + iv^2 , \quad v^{\bar{z}} = v^1 - iv^2 , \quad v_z = \frac{1}{2}(v^1 - iv^2) , \quad v_{\bar{z}} = \frac{1}{2}(v^1 + iv^2) . \tag{2.1.5}$$

For the indices $1, 2$ the metric is the identity and we do not distinguish between upper and lower, while the complex indices are raised and lowered with

$$g_{z\bar{z}} = g_{\bar{z}z} = \frac{1}{2} , \quad g_{zz} = g_{\bar{z}\bar{z}} = 0 , \quad g^{z\bar{z}} = g^{\bar{z}z} = 2 , \quad g^{zz} = g^{\bar{z}\bar{z}} = 0 . \tag{2.1.6}$$

Note also that

$$d^2z = 2d\sigma^1 d\sigma^2 \tag{2.1.7}$$

with the factor of 2 from the Jacobian,[1] and that $d^2z \, |\det g|^{1/2} = d\sigma^1 d\sigma^2$. We define

$$\int d^2z \, \delta^2(z, \bar{z}) = 1 \tag{2.1.8}$$

so that $\delta^2(z, \bar{z}) = \frac{1}{2}\delta(\sigma^1)\delta(\sigma^2)$. Another useful result is the divergence

---

[1] This differs from much of the literature, where d^2z is defined as $d\sigma^1 d\sigma^2$. Correlated with this, the $\frac{1}{2}$ is omitted from the definition of $\delta^2(z, \bar{z})$.

theorem in complex coordinates,

$$\int_R d^2z \left(\partial_z v^z + \partial_{\bar{z}} v^{\bar{z}} \right) = i \oint_{\partial R} \left(v^z \, d\bar{z} - v^{\bar{z}} \, dz \right) , \qquad (2.1.9)$$

where the contour integral circles the region R counterclockwise.

In this notation the action is

$$S = \frac{1}{2\pi\alpha'} \int d^2z \, \partial X^\mu \bar{\partial} X_\mu \qquad (2.1.10)$$

and the classical equation of motion is

$$\partial \bar{\partial} X^\mu(z, \bar{z}) = 0 . \qquad (2.1.11)$$

The notation $X^\mu(z, \bar{z})$ may seem redundant, since the value of z determines the value of $\bar{z}$, but it is useful to reserve the notation $f(z)$ for fields whose equation of motion makes them *analytic* (equivalently *holomorphic*) functions of z. Writing the equation of motion as

$$\partial(\bar{\partial} X^\mu) = \bar{\partial}(\partial X^\mu) = 0 , \qquad (2.1.12)$$

it follows that ∂X^μ is holomorphic and that $\bar{\partial} X^\mu$ is antiholomorphic (holomorphic in $\bar{z}$), hence the notations $\partial X^\mu(z)$ and $\bar{\partial} X^\mu(\bar{z})$.

Under the Minkowski continuation $\sigma^2 = i\sigma^0$, a holomorphic field becomes a function only of $\sigma^0 - \sigma^1$ and an antiholomorphic field a function only of $\sigma^0 + \sigma^1$. We thus use as synonyms

$$\text{holomorphic} = \text{left-moving} , \qquad (2.1.13a)$$

$$\text{antiholomorphic} = \text{right-moving} . \qquad (2.1.13b)$$

This terminology is chosen to have maximal agreement with the literature, though for it to hold literally we would need to draw σ^1 increasing from right to left. We will tend to use the Euclidean terms early on and shift to the Minkowski terms as we discuss more of the spacetime physics.

Expectation values are defined by the path integral,

$$\langle \mathscr{F}[X] \rangle = \int [dX] \exp(-S) \mathscr{F}[X] , \qquad (2.1.14)$$

where $\mathscr{F}[X]$ is any functional of X, such as a product of local operators. The path integral over X^0 is a wrong-sign Gaussian, so it should be understood to be defined by the analytic continuation $X^0 \to -iX^D$. The reader should not be distracted by this; we will discuss it further in the next chapter. We do not normalize $\langle \mathscr{F}[X] \rangle$ by dividing by $\langle 1 \rangle$.

The path integral of a total derivative is zero. This is true for ordinary bosonic path integrals, which can be regarded as the limit of an infinite number of ordinary integrals, as well as for more formal path integrals as with Grassmann variables. Then

$$
0 = \int [dX] \frac{\delta}{\delta X_\mu(z,\bar{z})} \exp(-S)
$$

$$
= -\int [dX] \exp(-S) \frac{\delta S}{\delta X_\mu(z,\bar{z})}
$$

$$
= -\left\langle \frac{\delta S}{\delta X_\mu(z,\bar{z})} \right\rangle
$$

$$
= \frac{1}{\pi \alpha'} \left\langle \partial \bar{\partial} X^\mu(z,\bar{z}) \right\rangle . \tag{2.1.15}
$$

The same calculation goes through if we have arbitrary additional insertions '...' in the path integral, as long as none of these additional insertions is at z. Thus

$$
\left\langle \partial \bar{\partial} X^\mu(z,\bar{z}) \ \dots \ \right\rangle = 0 . \tag{2.1.16}
$$

We can regard the additional insertions as preparing arbitrary initial and final states (or we could do the same thing with boundary conditions). The path integral statement (2.1.16) is thus the same as the statement in the Hilbert space formalism that

$$
\partial \bar{\partial} \hat{X}^\mu(z,\bar{z}) = 0 \tag{2.1.17}
$$

holds for all matrix elements of the operator $\hat{X}^\mu(z,\bar{z})$. Thus we refer to relations that hold in the sense (2.1.16) as *operator equations*. The statement (2.1.17) is Ehrenfest's theorem that the classical equations of motion translate into operator equations.

The notation '...' in the path integral (2.1.16) implicitly stands for insertions that are located away from z, but it is interesting to consider also the case in which there is an insertion that might be coincident with z:

$$
0 = \int [dX] \frac{\delta}{\delta X_\mu(z,\bar{z})} \Big[\exp(-S) X^\nu(z',\bar{z}') \Big]
$$

$$
= \int [dX] \exp(-S) \Big[\eta^{\mu\nu} \delta^2(z-z',\bar{z}-\bar{z}') + \frac{1}{\pi\alpha'} \partial_z \partial_{\bar{z}} X^\mu(z,\bar{z}) X^\nu(z',\bar{z}') \Big]
$$

$$
= \eta^{\mu\nu} \left\langle \delta^2(z-z',\bar{z}-\bar{z}') \right\rangle + \frac{1}{\pi\alpha'} \partial_z \partial_{\bar{z}} \left\langle X^\mu(z,\bar{z}) X^\nu(z',\bar{z}') \right\rangle . \tag{2.1.18}
$$

That is, the equation of motion holds except at coincident points. Again this goes through with arbitrary additional insertions '...' in the path integral, as long as none of these additional fields is at $(z,\bar{z})$ or $(z',\bar{z}')$:

$$
\frac{1}{\pi\alpha'} \partial_z \partial_{\bar{z}} \left\langle X^\mu(z,\bar{z}) X^\nu(z',\bar{z}') \dots \right\rangle = -\eta^{\mu\nu} \left\langle \delta^2(z-z',\bar{z}-\bar{z}') \dots \right\rangle . \tag{2.1.19}
$$

Thus,

$$\frac{1}{\pi\alpha'}\partial_z\partial_{\bar{z}}X^\mu(z,\bar{z})X^\nu(z',\bar{z}') = -\eta^{\mu\nu}\delta^2(z-z',\bar{z}-\bar{z}') \qquad (2.1.20)$$

holds as an operator equation. In the Hilbert space formalism, the product in the path integral becomes a time-ordered product, and the delta function comes from the derivatives acting on the time-ordering. This connection is developed further in the appendix.

In free field theory, it is useful to introduce the operation of *normal ordering*. Normal ordered operators, denoted $:\mathscr{A}:$, are defined as follows,

$$:X^\mu(z,\bar{z}): = X^\mu(z,\bar{z})\,, \qquad (2.1.21a)$$

$$:X^\mu(z_1,\bar{z}_1)X^\nu(z_2,\bar{z}_2): = X^\mu(z_1,\bar{z}_1)X^\nu(z_2,\bar{z}_2) + \frac{\alpha'}{2}\eta^{\mu\nu}\ln|z_{12}|^2\,, $$
$$(2.1.21b)$$

where

$$z_{ij} = z_i - z_j\,. \qquad (2.1.22)$$

The reader may be familiar with normal ordering defined in terms of raising and lowering operators; these two definitions will be related later. The point of this definition is the property

$$\partial_1\bar{\partial}_1 :X^\mu(z_1,\bar{z}_1)X^\nu(z_2,\bar{z}_2): = 0\,. \qquad (2.1.23)$$

This follows from the operator equation (2.1.20) and the differential equation

$$\partial\bar{\partial}\ln|z|^2 = 2\pi\delta^2(z,\bar{z})\,. \qquad (2.1.24)$$

Eq. (2.1.24) is obvious for $z \neq 0$ because $\ln|z|^2 = \ln z + \ln\bar{z}$, and the normalization of the delta function is easily checked by integrating both sides using eq. (2.1.9).

2.2   The operator product expansion

The basic object of interest in string perturbation theory will be the path integral expectation value of a product of local operators,

$$\langle\, \mathscr{A}_{i_1}(z_1,\bar{z}_1)\mathscr{A}_{i_2}(z_2,\bar{z}_2) \ldots \mathscr{A}_{i_n}(z_n,\bar{z}_n) \,\rangle\,, \qquad (2.2.1)$$

where $\mathscr{A}_i$ is some basis for the set of local operators. It is particularly important to understand the behavior of this expectation value in the limit that two of the operators are taken to approach one another. The tool that gives a systematic description of this limit is the *operator product expansion (OPE)*, illustrated in figure 2.1. This states that a product of two

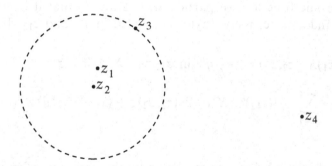

Fig. 2.1. Expectation value of a product of four local operators. The OPE gives the asymptotics as $z_1 \to z_2$ as a series where the pair of operators at z_1 and z_2 is replaced by a single operator at z_2. The radius of convergence is the distance to the nearest other operator, indicated by the dashed circle.

local operators close together can be approximated to arbitrary accuracy by a sum of local operators,

$$\mathscr{A}_i(\sigma_1)\mathscr{A}_j(\sigma_2) = \sum_k c^k{}_{ij}(\sigma_1 - \sigma_2)\mathscr{A}_k(\sigma_2) \,. \qquad (2.2.2)$$

Again this is an operator statement, meaning that it holds inside a general expectation value

$$\langle\, \mathscr{A}_i(\sigma_1)\mathscr{A}_j(\sigma_2) \dots \,\rangle = \sum_k c^k{}_{ij}(\sigma_1 - \sigma_2)\langle\, \mathscr{A}_k(\sigma_2) \dots \,\rangle \qquad (2.2.3)$$

as long as the separation between σ_1 and σ_2 is small compared to the distance to any other operator. The *coefficient functions* $c^k{}_{ij}(\sigma_1 - \sigma_2)$, which govern the dependence on the separation, depend on i, j, and k but not on the other operators in the expectation value; the dependence on the latter comes only through the expectation value on the right-hand side of eq. (2.2.3). The terms are conventionally arranged in order of decreasing size in the limit as $\sigma_1 \to \sigma_2$. This is analogous to an ordinary Taylor series, except that the coefficient functions need not be simple powers and can in fact be singular as $\sigma_1 \to \sigma_2$, as we will see even in the simplest example. Just as the Taylor series plays a central role in calculus, the OPE plays a central role in quantum field theory.

   We will give now a derivation of the OPE for the X^μ theory, using the special properties of free field theory. In section 2.9 we will give a derivation for any conformally invariant field theory.

   We have seen that the normal ordered product satisfies the naive equation of motion. Eq. (2.1.23) states that the operator product is a harmonic function of $(z_1, \bar{z}_1)$. A simple result from the theory of complex variables is that a harmonic function is locally the sum of a holomorphic and an antiholomorphic function. In particular, this means that it is nonsingular as $z_1 \to z_2$ and can be freely Taylor expanded in z_{12} and $\bar{z}_{12}$. Thus,

$$
X^\mu(z_1, \bar{z}_1) X^\nu(z_2, \bar{z}_2) = -\frac{\alpha'}{2} \eta^{\mu\nu} \ln |z_{12}|^2 + \, :X^\nu X^\mu(z_2, \bar{z}_2):
$$

$$
+ \sum_{k=1}^{\infty} \frac{1}{k!} \left[(z_{12})^k \, :X^\nu \partial^k X^\mu(z_2, \bar{z}_2): + (\bar{z}_{12})^k \, :X^\nu \bar{\partial}^k X^\mu(z_2, \bar{z}_2): \right] .
$$

$$(2.2.4)$$

Terms with mixed $\partial\bar{\partial}$ derivatives vanish by the equation of motion. This equation and many others simplify in units in which $\alpha' = 2$, which is the most common convention in the literature. However, several other conventions are also used, so it is useful to keep α' explicit. For example, in *open* string theory equations simplify when $\alpha' = \frac{1}{2}$.

Eq. (2.2.4) has the form of an OPE. Like the equation of motion (2.1.23) from which it was derived, it is an operator statement. For an arbitrary expectation value involving the product $X^\mu(z_1, \bar{z}_1) X^\nu(z_2, \bar{z}_2)$ times fields at other points, it gives the behavior for $z_1 \to z_2$ as an infinite series, each term being a known function of z_{12} and/or $\bar{z}_{12}$ times the expectation value with a local operator replacing the pair.

OPEs are usually used as asymptotic expansions, the first few terms giving the dominant behavior at small separation. Most of our applications will be of this type, and we will often write OPEs as explicit singular terms plus unspecified nonsingular remainders. The use of '$\sim$' in place of '$=$' will mean 'equal up to nonsingular terms.' In fact, OPEs are actually convergent in conformally invariant field theories. This will be very important to us in certain applications: it makes it possible to reconstruct the entire theory from the coefficient functions. As an example, the free-field OPE (2.2.4) has a radius of convergence in any given expectation value which is equal to the distance to the nearest *other* insertion in the path integral. The operator product is harmonic except at the positions of operators, and in particular inside the dashed circle of figure 2.1, and convergence can then be shown by a standard argument from the theory of complex variables.

The various operators on the right-hand side of the OPE (2.2.4) involve products of fields at the same point. Usually in quantum field theory such a product is divergent and must be appropriately cut off and renormalized, but here the normal ordering renders it well-defined. Normal ordering is a convenient way to define composite operators in free field theory. It is of little use in most interacting field theories, because these have additional divergences from interaction vertices approaching the composite operator or one another. However, many of the field theories that we will be interested in are free, and many others can be related to free field theories, so it will be worthwhile to develop normal ordering somewhat further.

The definition of normal ordering for arbitrary numbers of fields can be given recursively as

$$
:X^{\mu_1}(z_1,\bar{z}_1)\ldots X^{\mu_n}(z_n,\bar{z}_n):
$$
$$
= X^{\mu_1}(z_1,\bar{z}_1)\ldots X^{\mu_n}(z_n,\bar{z}_n) + \sum \text{subtractions}\,, \quad (2.2.5)
$$

where the sum runs over all ways of choosing one, two, or more pairs of fields from the product and replacing each pair with $\frac{1}{2}\alpha'\eta^{\mu_i\mu_j}\ln|z_{ij}|^2$. For example,

$$
:X^{\mu_1}(z_1,\bar{z}_1)X^{\mu_2}(z_2,\bar{z}_2)X^{\mu_3}(z_3,\bar{z}_3): = X^{\mu_1}(z_1,\bar{z}_1)X^{\mu_2}(z_2,\bar{z}_2)X^{\mu_3}(z_3,\bar{z}_3)
$$
$$
+ \left(\frac{\alpha'}{2}\eta^{\mu_1\mu_2}\ln|z_{12}|^2 X^{\mu_3}(z_3,\bar{z}_3) + 2 \text{ permutations}\right). \quad (2.2.6)
$$

We leave it to the reader to show that the definition (2.2.5) retains the desired property that the normal ordered product satisfies the naive equation of motion.

The definition can be compactly summarized as

$$
:\mathscr{F}: = \exp\left(\frac{\alpha'}{4}\int d^2z_1 d^2z_2 \ln|z_{12}|^2 \frac{\delta}{\delta X^\mu(z_1,\bar{z}_1)}\frac{\delta}{\delta X_\mu(z_2,\bar{z}_2)}\right)\mathscr{F}\,, \quad (2.2.7)
$$

where $\mathscr{F}$ is any functional of X. This is equivalent to eq. (2.2.5): the double derivative in the exponent contracts each pair of fields, and the exponential sums over any number of pairs with the factorial canceling the number of ways the derivatives can act. As an example of the use of this formal expression, act on both sides with the inverse exponential to obtain

$$
\mathscr{F} = \exp\left(-\frac{\alpha'}{4}\int d^2z_1 d^2z_2 \ln|z_{12}|^2 \frac{\delta}{\delta X^\mu(z_1,\bar{z}_1)}\frac{\delta}{\delta X_\mu(z_2,\bar{z}_2)}\right):\mathscr{F}:
$$
$$
= :\mathscr{F}: + \sum \text{contractions}\,, \quad (2.2.8)
$$

where a contraction is the opposite of a subtraction: sum over all ways of choosing one, two, or more pairs of fields from $:\mathscr{F}:$ and replacing each pair with $-\frac{1}{2}\alpha'\eta^{\mu_i\mu_j}\ln|z_{ij}|^2$.

The OPE for any pair of operators can be generated from

$$
:\mathscr{F}::\mathscr{G}: = :\mathscr{F}\mathscr{G}: + \sum \text{cross-contractions} \quad (2.2.9)
$$

for arbitrary functionals $\mathscr{F}$ and $\mathscr{G}$ of X. The sum now runs over all ways of contracting pairs with one field in $\mathscr{F}$ and one in $\mathscr{G}$. This can also be written

$$
:\mathscr{F}::\mathscr{G}: = \exp\left(-\frac{\alpha'}{2}\int d^2z_1 d^2z_2 \ln|z_{12}|^2 \frac{\delta}{\delta X_F^\mu(z_1,\bar{z}_1)}\frac{\delta}{\delta X_{G\mu}(z_2,\bar{z}_2)}\right):\mathscr{F}\mathscr{G}:,
$$
$$
(2.2.10)
$$

where the functional derivatives act only on the fields in $\mathscr{F}$ or $\mathscr{G}$ respectively. This follows readily from eq. (2.2.7).

As an example,

$$: \partial X^\mu(z)\partial X_\mu(z): \ :\partial' X^\nu(z')\partial' X_\nu(z'):$$

$$= \ :\partial X^\mu(z)\partial X_\mu(z)\partial' X^\nu(z')\partial' X_\nu(z'):$$

$$- 4 \cdot \frac{\alpha'}{2}(\partial\partial' \ln|z-z'|^2) :\partial X^\mu(z)\partial' X_\mu(z'):$$

$$+ 2 \cdot \eta^\mu{}_\mu\left(-\frac{\alpha'}{2}\partial\partial' \ln|z-z'|^2\right)^2$$

$$\sim \frac{D\alpha'^2}{2(z-z')^4} - \frac{2\alpha'}{(z-z')^2} :\partial' X^\mu(z')\partial' X_\mu(z'):$$

$$- \frac{2\alpha'}{z-z'} :\partial'^2 X^\mu(z')\partial' X_\mu(z'): \ . \qquad (2.2.11)$$

The second term in the equality comes from the four ways of forming a single pair and the third from the two ways of forming two pairs. In the final line we have put the OPE in standard form by Taylor expanding inside the normal ordering to express everything in terms of local operators at z' and putting the most singular terms first.

Another important example is

$$\mathscr{F} = e^{ik_1 \cdot X(z,\bar{z})} \ , \quad \mathscr{G} = e^{ik_2 \cdot X(0,0)} \ . \qquad (2.2.12)$$

The variations $\delta/\delta X^\mu_F$ and $\delta/\delta X^\mu_G$ give factors of $ik_{1\mu}$ and $ik_{2\mu}$ respectively, so the general result (2.2.10) becomes

$$: e^{ik_1 \cdot X(z,\bar{z})}: \ : e^{ik_2 \cdot X(0,0)}: \ = \exp\left(\frac{\alpha'}{2}k_1 \cdot k_2 \ln|z|^2\right) :e^{ik_1 \cdot X(z,\bar{z})}e^{ik_2 \cdot X(0,0)}:$$

$$= |z|^{\alpha' k_1 \cdot k_2} :e^{ik_1 \cdot X(z,\bar{z})}e^{ik_2 \cdot X(0,0)}: \ . \qquad (2.2.13)$$

To derive the OPE, Taylor expand inside the normal ordering to give

$$: e^{ik_1 \cdot X(z,\bar{z})}: \ : e^{ik_2 \cdot X(0,0)}: \ = |z|^{\alpha' k_1 \cdot k_2} :e^{i(k_1+k_2)\cdot X(0,0)}[1 + O(z,\bar{z})]: \ . \qquad (2.2.14)$$

The exercises give further practice with normal ordering and the free-field OPE.

Note that the OPEs (2.2.2), (2.2.4), and so on have been written asymmetrically in σ_1 and σ_2, expanding around the latter point. They can also be cast in symmetric form by Taylor expanding the right-hand sides around $(\sigma_1 + \sigma_2)/2$. The coefficient functions for the symmetric form behave simply under interchange of the two operators,

$$c^k_{ij}(\sigma_1 - \sigma_2)_{\text{sym}} = \pm c^k_{ji}(\sigma_2 - \sigma_1)_{\text{sym}} \ , \qquad (2.2.15)$$

where the minus sign appears if $\mathscr{A}_i$ and $\mathscr{A}_j$ are both anticommuting. The asymmetric form is usually more convenient for calculation, so when the symmetry properties of the coefficient functions are needed one can work them out in the symmetric form and then convert to the asymmetric form.

2.3 Ward identities and Noether's theorem

World-sheet symmetries of course play an important role in string theory. In this section we first derive some general consequences of symmetry in field theory.

Consider a general field theory with action $S[\phi]$ in d spacetime dimensions, with $\phi_\alpha(\sigma)$ denoting general fields. Let there be a symmetry

$$\phi'_\alpha(\sigma) = \phi_\alpha(\sigma) + \delta\phi_\alpha(\sigma) \,, \tag{2.3.1}$$

where $\delta\phi_\alpha$ is proportional to an infinitesimal parameter ϵ. The product of the path integral measure and the weight $\exp(-S)$ is invariant,

$$[d\phi'] \exp(-S[\phi']) = [d\phi] \exp(-S[\phi]) \,. \tag{2.3.2}$$

A continuous symmetry in field theory implies the existence of a conserved current (*Noether's theorem*) and also *Ward identities*, which constrain the operator products of the current. To derive these results consider the change of variables,

$$\phi'_\alpha(\sigma) = \phi_\alpha(\sigma) + \rho(\sigma)\delta\phi_\alpha(\sigma) \,. \tag{2.3.3}$$

This is *not* a symmetry, the transformation law being altered by the inclusion of an arbitrary function $\rho(\sigma)$. The path integral measure times $\exp(-S)$ would be invariant if ρ were a constant, so its variation must be proportional to the gradient $\partial_a\rho$,

$$\begin{aligned}&[d\phi'] \exp(-S[\phi']) \\ &= [d\phi] \exp(-S[\phi])\left[1 + \frac{i\epsilon}{2\pi}\int d^d\sigma \, g^{1/2} j^a(\sigma)\partial_a\rho(\sigma) + O(\epsilon^2)\right] . \end{aligned} \tag{2.3.4}$$

The unknown coefficient $j^a(\sigma)$ comes from the variation of the measure and the action, both of which are local, and so it must be a local function of the fields and their derivatives. Take the function ρ to be nonzero only in a small region, and consider a path integral with general insertions '...' *outside* this region; the insertions are therefore invariant under (2.3.3). Invariance of the path integral under change of variables gives

$$\begin{aligned}0 &= \int [d\phi'] \exp(-S[\phi'])\ldots \; - \int [d\phi] \exp(-S[\phi])\ldots \\ &= \frac{\epsilon}{2\pi i}\int d^d\sigma \, g^{1/2}\rho(\sigma)\langle \nabla_a j^a(\sigma) \ldots \rangle \,, \end{aligned} \tag{2.3.5}$$

where the limited support of ρ has allowed us to integrate by parts. Thus we have

$$\nabla_a j^a = 0 \tag{2.3.6}$$

as an operator equation. This is Noether's theorem, which is developed further in exercise 2.5.

To derive the Ward identity, let $\rho(\sigma)$ be 1 in some region R and 0 outside R. Also, include in the path integral some general local operator $\mathcal{A}(\sigma_0)$ at a point σ_0 inside R, and the usual arbitrary insertions '...' outside. Proceeding as above we obtain the operator relation

$$\delta\mathcal{A}(\sigma_0) + \frac{\epsilon}{2\pi i} \int_R d^d\sigma \, g^{1/2} \nabla_a j^a(\sigma)\mathcal{A}(\sigma_0) = 0 \, . \tag{2.3.7}$$

Equivalently,

$$\nabla_a j^a(\sigma)\mathcal{A}(\sigma_0) = g^{-1/2}\delta^d(\sigma - \sigma_0)\frac{2\pi}{i\epsilon}\delta\mathcal{A}(\sigma_0) + \text{total } \sigma\text{-derivative} \, . \tag{2.3.8}$$

The divergence theorem gives

$$\int_{\partial R} dA \, n_a j^a \mathcal{A}(\sigma_0) = \frac{2\pi}{i\epsilon}\delta\mathcal{A}(\sigma_0) \tag{2.3.9}$$

with dA the area element and n^a the outward normal. This relates the integral of the current around the operator to the variation of the operator. Going to two flat dimensions this becomes

$$\oint_{\partial R} (j \, dz - \tilde{j} \, d\bar{z})\mathcal{A}(z_0, \bar{z}_0) = \frac{2\pi}{\epsilon}\delta\mathcal{A}(z_0, \bar{z}_0) \, . \tag{2.3.10}$$

Again we drop indices, $j \equiv j_z$, $\tilde{j} \equiv j_{\bar{z}}$; notice that on a current the omitted indices are implicitly lower. We use a tilde rather than a bar on $\tilde{j}$ because this is *not* the adjoint of j. The Minkowski density j_0 is in general Hermitean, so the Euclidean j_2 with an extra factor of i is anti-Hermitean, and $(j_z)^\dagger = \frac{1}{2}(j_1 - ij_2)^\dagger = j_z$.

It is important that Noether's theorem and the Ward identity are local properties, which do not depend on whatever boundary conditions we might have far away, nor even on whether these are invariant under the symmetry. In particular, since the function $\rho(\sigma)$ is nonzero only inside R, the symmetry transformation need only be defined there.

In conformally invariant theories it is usually the case that j_z is holomorphic and $j_{\bar{z}}$ antiholomorphic (except for singularities at the other fields). In this case the currents $(j_z, 0)$ and $(0, j_{\bar{z}})$ are separately conserved. The integral (2.3.10) then picks out the residues in the OPE,

$$\text{Res}_{z \to z_0} \, j(z)\mathcal{A}(z_0, \bar{z}_0) + \overline{\text{Res}}_{\bar{z} \to \bar{z}_0} \, \tilde{j}(\bar{z})\mathcal{A}(z_0, \bar{z}_0) = \frac{1}{i\epsilon}\delta\mathcal{A}(z_0, \bar{z}_0) \, . \tag{2.3.11}$$

Here 'Res' and '$\overline{\text{Res}}$' are the coefficients of $(z - z_0)^{-1}$ and $(\bar{z} - \bar{z}_0)^{-1}$ respectively. This form of the Ward identity is particularly convenient.

The world-sheet current was defined with an extra factor of $2\pi i$ relative to the usual definition in field theory in order to make this OPE simple.

As an example, return to the free massless scalar and consider the *spacetime* translation $\delta X^\mu = \epsilon a^\mu$. Under $\delta X^\mu(\sigma) = \epsilon \rho(\sigma) a^\mu$,

$$\delta S = \frac{\epsilon a_\mu}{2\pi\alpha'} \int d^2\sigma \, \partial^a X^\mu \partial_a \rho \,. \tag{2.3.12}$$

This is of the claimed form (2.3.5) with Noether current $a_\mu j_a^\mu$, where

$$j_a^\mu = \frac{i}{\alpha'} \partial_a X^\mu \,. \tag{2.3.13}$$

The components are holomorphic and antiholomorphic as expected. For the OPE of this current with the exponential operator one finds

$$j^\mu(z) \, :e^{ik \cdot X(0,0)}: \, \sim \frac{k^\mu}{2z} \, :e^{ik \cdot X(0,0)}: \,, \tag{2.3.14a}$$

$$\tilde{j}^\mu(\bar{z}) \, :e^{ik \cdot X(0,0)}: \, \sim \frac{k^\mu}{2\bar{z}} \, :e^{ik \cdot X(0,0)}: \,, \tag{2.3.14b}$$

from terms with a single contraction. This OPE is in agreement with the general identity (2.3.11).

Another example is the *world-sheet* translation $\delta\sigma^a = \epsilon v^a$, under which $\delta X^\mu = -\epsilon v^a \partial_a X^\mu$. The Noether current is

$$j_a = iv^b T_{ab} \,, \tag{2.3.15a}$$

$$T_{ab} = -\frac{1}{\alpha'} \, :\left(\partial_a X^\mu \partial_b X_\mu - \frac{1}{2}\delta_{ab}\partial_c X^\mu \partial^c X_\mu \right): \,. \tag{2.3.15b}$$

Here T_{ab} is the world-sheet energy-momentum tensor.[2]

2.4 Conformal invariance

The energy-momentum tensor (2.3.15b) is traceless, $T_a^a = 0$. In complex coordinates this is

$$T_{z\bar{z}} = 0 \,. \tag{2.4.1}$$

The conservation $\partial^a T_{ab} = 0$ then implies that in any theory with $T_a^a = 0$,

$$\bar{\partial} T_{zz} = \partial T_{\bar{z}\bar{z}} = 0 \,. \tag{2.4.2}$$

Thus

$$T(z) \equiv T_{zz}(z) \,, \quad \tilde{T}(\bar{z}) \equiv T_{\bar{z}\bar{z}}(\bar{z}) \tag{2.4.3}$$

---

[2] In T_{ab} we have used normal ordering to define the product of operators at a point. The only possible ambiguity introduced by the renormalization is a constant times δ_{ab}, from the subtraction. Adding such a constant gives a different energy-momentum tensor which is also conserved. We choose to focus on the tensor (2.3.15b), for reasons that will be explained more fully in the next chapter.

are respectively holomorphic and antiholomorphic. For the free massless scalar,

$$T(z) = -\frac{1}{\alpha'} :\partial X^\mu \partial X_\mu: \,, \quad \tilde{T}(\bar{z}) = -\frac{1}{\alpha'} :\bar{\partial} X^\mu \bar{\partial} X_\mu: \,, \tag{2.4.4}$$

which are indeed holomorphic and antiholomorphic as a consequence of the equation of motion.

The tracelessness of T_{ab} implies a much larger symmetry. The currents

$$j(z) = iv(z)T(z)\,, \quad \tilde{j}(\bar{z}) = iv(z)^*\tilde{T}(\bar{z}) \tag{2.4.5}$$

are conserved for any holomorphic $v(z)$. For the free scalar theory, one finds the OPE

$$T(z)X^\mu(0) \sim \frac{1}{z}\partial X^\mu(0)\,, \quad \tilde{T}(\bar{z})X^\mu(0) \sim \frac{1}{\bar{z}}\bar{\partial}X^\mu(0)\,. \tag{2.4.6}$$

The Ward identity then gives the transformation

$$\delta X^\mu = -\epsilon v(z)\partial X^\mu - \epsilon v(z)^*\bar{\partial}X^\mu\,. \tag{2.4.7}$$

This is an infinitesimal coordinate transformation $z' = z + \epsilon v(z)$. The finite transformation is

$$X'^\mu(z',\bar{z}') = X^\mu(z,\bar{z})\,, \quad z' = f(z) \tag{2.4.8}$$

for any holomorphic $f(z)$. This is known as a *conformal transformation*.

The conformal symmetry we have found should not be confused with the diff invariance of general relativity. We are in the flat space theory, with no independent metric field to vary, so the transformation $z \to z'$ actually changes the distances between points. We would not ordinarily have such an invariance; it is a nontrivial statement about the dynamics. For the scalar action (2.1.10), the conformal transformation of ∂ and $\bar{\partial}$ just balances that of d^2z. A mass term $m^2 X^\mu X_\mu$ would not be invariant. Obviously there will in the end be a close relation with the diff×Weyl symmetry of the Polyakov string, but we leave that for the next chapter.

Consider the special case

$$z' = \zeta z \tag{2.4.9}$$

for complex ζ. The phase of ζ is a rotation of the system, while its magnitude is a rescaling of the size of the system. Such a scale invariance has occasionally been considered as an approximate symmetry in particle physics, and statistical systems at a critical point are described by scale-invariant field theories.

To get some insight into the general conformal transformation, consider its effect on infinitesimal distances $ds^2 = d\sigma^a d\sigma^a = dz d\bar{z}$. Conformal transformations rescale this by a *position-dependent* factor,

$$ds'^2 = dz' d\bar{z}' = \frac{\partial z'}{\partial z}\frac{\partial \bar{z}'}{\partial \bar{z}}dz d\bar{z}\,. \tag{2.4.10}$$

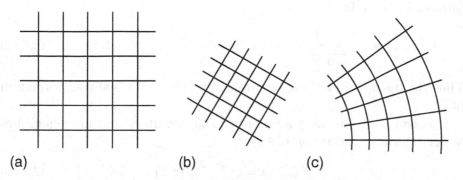

(a) (b) (c)

Fig. 2.2. (a) Two-dimensional region. (b) Effect of the special conformal transformation (2.4.9). (c) Effect of a more general conformal transformation.

Thus, as indicated in figure 2.2, a conformal transformation takes infinitesimal squares into infinitesimal squares, but rescales them by a position-dependent factor. An antiholomorphic function $z' = f(z)^*$ has the same property, but changes the orientation. Most systems that are invariant under the rigid scaling (2.4.9) are actually invariant under the much larger conformal symmetry. A theory with this invariance is called a *conformal field theory (CFT)*. Conformal invariance in more than two dimensions is developed in exercise 2.6.

Conformal invariance and the OPE

Conformal invariance puts strong constraints on the form of the OPE, and in particular on the OPEs of the energy-momentum tensor. Consider the OPE of T with the general operator $\mathscr{A}$. Because $T(z)$ and $\tilde{T}(\bar{z})$ are (anti)holomorphic except at insertions, the corresponding coefficient functions must also have this property. The OPE of T with a general $\mathscr{A}$ is therefore a *Laurent expansion,* in integer but possibly negative powers of z. Further, all the singular terms are determined by the conformal transformation of $\mathscr{A}$. To see this, let us write a general expansion of the singular terms,

$$T(z)\mathscr{A}(0,0) \sim \sum_{n=0}^{\infty} \frac{1}{z^{n+1}} \mathscr{A}^{(n)}(0,0) \qquad (2.4.11)$$

and similarly for $\tilde{T}$; the operator coefficients $\mathscr{A}^{(n)}$ remain to be determined. Under an infinitesimal conformal transformation $z' = z + \epsilon v(z)$, a single pole in $v(z)T(z)\mathscr{A}(0,0)$ arises when the z^{-n-1} term of the $T\mathscr{A}$ OPE multiplies the term of order z^n in $v(z)$. Thus, the Ward identity in the form (2.3.11) implies that

$$\delta \mathscr{A}(z,\bar{z}) = -\epsilon \sum_{n=0}^{\infty} \frac{1}{n!} \left[\partial^n v(z) \mathscr{A}^{(n)}(z,\bar{z}) + \bar{\partial}^n v(z)^* \tilde{\mathscr{A}}^{(n)}(z,\bar{z}) \right] . \qquad (2.4.12)$$

Thus the operators $\mathscr{A}^{(n)}$ are determined by the conformal transformation of $\mathscr{A}$.

It is convenient to take a basis of local operators that are eigenstates under rigid transformation (2.4.9),

$$\mathscr{A}'(z',\bar{z}') = \zeta^{-h} \bar{\zeta}^{-\tilde{h}} \mathscr{A}(z,\bar{z}) . \qquad (2.4.13)$$

The $(h, \tilde{h})$ are known as the *weights* of $\mathscr{A}$. The sum $h + \tilde{h}$ is the dimension of $\mathscr{A}$, determining its behavior under scaling, while $h - \tilde{h}$ is the spin, determining its behavior under rotations. The derivative ∂_z increases h by one, and the derivative $\partial_{\bar{z}}$ increases $\tilde{h}$ by one. The Ward identities for the transformation (2.4.13) and for the translation $\delta \mathscr{A} = -\epsilon v^a \partial_a \mathscr{A}$ determine part of the OPE,

$$T(z) \mathscr{A}(0,0) = \ldots + \frac{h}{z^2} \mathscr{A}(0,0) + \frac{1}{z} \partial \mathscr{A}(0,0) + \ldots , \qquad (2.4.14)$$

and similarly for $\tilde{T}$.

An important special case is a *tensor operator* or *primary field*[3] $\mathcal{O}$, on which a general conformal transformation acts as

$$\mathcal{O}'(z',\bar{z}') = (\partial_z z')^{-h} (\partial_{\bar{z}} \bar{z}')^{-\tilde{h}} \mathcal{O}(z,\bar{z}) . \qquad (2.4.15)$$

The OPE (2.4.11) reduces to

$$T(z) \mathcal{O}(0,0) = \frac{h}{z^2} \mathcal{O}(0,0) + \frac{1}{z} \partial \mathcal{O}(0,0) + \ldots , \qquad (2.4.16)$$

the more singular terms in the general OPE (2.4.14) being absent.

Taking again the example of the free X^μ CFT, the weights of some typical operators are

$$
\left.
\begin{array}{llll}
X^\mu & (0,0) , & \partial X^\mu & (1,0) , \\
\bar{\partial} X^\mu & (0,1) , & \partial^2 X^\mu & (2,0) , \\
:e^{ik \cdot X}: & \left(\dfrac{\alpha' k^2}{4}, \dfrac{\alpha' k^2}{4} \right) . &&
\end{array}
\right\} \qquad (2.4.17)
$$

---

[3] In quantum field theory, one usually distinguishes the fundamental fields (the variables of integration in the path integral) from more general operators which may be composite. Actually, this distinction is primarily useful in weakly coupled field theories in three or more dimensions. It is of little use in CFT, in particular because of equivalences between different field theories, and so the term *field* is used for any local operator.

All transform as tensors except $\partial^2 X^\mu$. More generally, an exponential times a general product of derivatives,

$$: \left(\prod_i \partial^{m_i} X^{\mu_i} \right) \left(\prod_j \bar{\partial}^{n_j} X^{\nu_j} \right) e^{ik \cdot X} : , \tag{2.4.18}$$

has weight

$$\left(\frac{\alpha' k^2}{4} + \sum_i m_i \, , \, \frac{\alpha' k^2}{4} + \sum_j n_j \right) . \tag{2.4.19}$$

For any pair of operators, applying rigid translations, scale transformations, and rotations to both sides of the OPE determines the z-dependence of the coefficient functions completely,

$$\mathscr{A}_i(z_1, \bar{z}_1) \mathscr{A}_j(z_2, \bar{z}_2) = \sum_k z_{12}^{h_k - h_i - h_j} \bar{z}_{12}^{\tilde{h}_k - \tilde{h}_i - \tilde{h}_j} c^k{}_{ij} \, \mathscr{A}_k(z_2, \bar{z}_2) , \tag{2.4.20}$$

where the $c^k{}_{ij}$ are now constants. In all cases of interest the weights appearing on the right-hand side of the OPE (2.4.20) are bounded below, and so the degree of singularity in the operator product is bounded. More general conformal transformations put further constraints on the OPE: they determine the OPEs of *all* fields in terms of those of the primary fields. We will develop this in chapter 15.

Notice that the conformal transformation properties of normal ordered products are not in general given by the naive transformations of the product. For example, the transformation law (2.4.7) for X^μ would naively imply

$$\delta e^{ik \cdot X} = -\epsilon v(z) \partial e^{ik \cdot X} - \epsilon v(z)^* \bar{\partial} e^{ik \cdot X} \quad \text{(naive)} \tag{2.4.21}$$

for the exponential, making it a tensor of weight $(0,0)$. The modification is a quantum effect, due to the renormalization needed to define the product of operators at a point. Specifically it enters here because the subtraction $\ln|z_{12}|^2$ in $:\ :$ makes explicit reference to the coordinate frame.

Conformal properties of the energy-momentum tensor

The OPE of the energy-momentum tensor with itself was obtained in eq. (2.2.11),

$$T(z) T(0) = \frac{\eta^\mu{}_\mu}{2z^4} - \frac{2}{\alpha' z^2} : \partial X^\mu(z) \partial X_\mu(0) : + : T(z) T(0) :$$

$$\sim \frac{D}{2z^4} + \frac{2}{z^2} T(0) + \frac{1}{z} \partial T(0) . \tag{2.4.22}$$

A similar result holds for $\tilde{T}$. The $T(z) \tilde{T}(\bar{z}')$ OPE must be nonsingular. It cannot have poles in $(z - z')$ because it is antiholomorphic in z' at nonzero separation; similarly it cannot have poles in $(\bar{z} - \bar{z}')$ because it is holomorphic in z at nonzero separation.[4] The same holds for any OPE between a holomorphic and an antiholomorphic operator.

Thus T is *not* a tensor. Rather, the OPE (2.4.22) implies the transformation law

$$\epsilon^{-1}\delta T(z) = -\frac{D}{12}\partial_z^3 v(z) - 2\partial_z v(z)T(z) - v(z)\partial_z T(z) \ . \tag{2.4.23}$$

In a general CFT, $T(z)$ transforms as

$$\epsilon^{-1}\delta T(z) = -\frac{c}{12}\partial_z^3 v(z) - 2\partial_z v(z)T(z) - v(z)\partial_z T(z) \ , \tag{2.4.24}$$

with c a constant known as the *central charge*. The central charge of a free scalar is 1; for D free scalars it is D. The transformation (2.4.24) is the most general form that is linear in v, is consistent with the symmetry of the TT OPE, and has three lower z indices as required by rigid scale and rotation invariance. The scale, rotation, and translation symmetries determine the coefficients of the second and third terms. Further, by considering the commutator of two such transformations one can show that $\partial_a c = 0$, and it is a general result in quantum field theory that an operator that is independent of position must be a c-number.[5] The corresponding TT OPE is

$$T(z)T(0) \sim \frac{c}{2z^4} + \frac{2}{z^2}T(0) + \frac{1}{z}\partial T(0) \ . \tag{2.4.25}$$

The finite form of the transformation law (2.4.24) is

$$(\partial_z z')^2 T'(z') = T(z) - \frac{c}{12}\{z',z\} \ , \tag{2.4.26}$$

where $\{f,z\}$ denotes the *Schwarzian derivative*

$$\{f,z\} = \frac{2\partial_z^3 f \partial_z f - 3\partial_z^2 f \partial_z^2 f}{2\partial_z f \partial_z f} \ . \tag{2.4.27}$$

One can check this by verifying that it has the correct infinitesimal form and that it composes correctly under successive transformations (so that one can integrate the infinitesimal transformation). The corresponding forms hold for $\tilde{T}$, possibly with a different central charge $\tilde{c}$ in the general CFT.

The nontensor behavior of the energy-momentum tensor has a number of important physical consequences, as we will see. We should empha-

---

[4] Unless otherwise stated OPEs hold only at *nonzero* separation, ignoring possible delta functions. For our applications the latter will not matter. Occasionally it is useful to include the delta functions, but in general these depend partly on definitions so one must be careful.

[5] The argument is this. Suppose that $\mathscr{A}(\sigma)$ is independent of position, and consider its equal time commutator with any other local operator $\mathscr{B}(\sigma')$. These commute at spacelike separation, by locality. Since $\mathscr{A}(\sigma)$ is actually independent of position it also commutes at zero separation. Therefore $\mathscr{A}(\sigma)$ commutes with all local operators, and so must be a c-number.

size that 'nontensor' refers to conformal transformations. The energy-momentum tensor will have its usual tensor property under coordinate transformations.

2.5   Free CFTs

In this section we discuss three families of free-field CFTs — the linear dilaton, bc, and $\beta\gamma$ theories. The bc theory is the one of most immediate interest, as it will appear in the next chapter when we gauge-fix the Polyakov string, but all have a variety of applications in string theory.

Linear dilaton CFT

This family of CFTs is based on the same action (2.1.10) but with energy-momentum tensor

$$T(z) = -\frac{1}{\alpha'} :\partial X^\mu \partial X_\mu: +V_\mu \partial^2 X^\mu \,, \qquad (2.5.1a)$$

$$\tilde{T}(\bar{z}) = -\frac{1}{\alpha'} :\bar{\partial} X^\mu \bar{\partial} X_\mu: +V_\mu \bar{\partial}^2 X^\mu \,, \qquad (2.5.1b)$$

where V_μ is some fixed D-vector. Working out the TT OPE, one finds that it is of the standard form (2.4.25), but with central charge

$$c = \tilde{c} = D + 6\alpha' V_\mu V^\mu \,. \qquad (2.5.2)$$

The TX^μ OPE and the Ward identity (2.4.12) imply the conformal transformation

$$\delta X^\mu = -\epsilon v \partial X^\mu - \epsilon v^* \bar{\partial} X^\mu - \frac{\epsilon}{2} \alpha' V^\mu [\partial v + (\partial v)^*] \,. \qquad (2.5.3)$$

This is a *different* conformal symmetry of the same action. The field X^μ no longer transforms as a tensor, its variation now having an inhomogeneous piece. Incidentally, the free massless scalar in two dimensions has a remarkably large amount of symmetry — much more than we will have occasion to mention.

The energy-momentum tensor plays a special role in string theory — in particular, it tells us how to couple to a curved metric — so different values of V^μ are to be regarded as different CFTs. The vector V^μ picks out a direction in spacetime. This CFT is therefore not Lorentz-invariant and not of immediate interest to us. We will see one physical interpretation of this *linear dilaton* CFT in section 3.7, and encounter it in some technical applications later.

A different variation of the free scalar CFT is to take some of the X^μ to be periodic; we will take this up in chapter 8.

bc CFT

The second family of CFTs has *anticommuting* fields b and c with action

$$S = \frac{1}{2\pi} \int d^2z\, b\bar{\partial}c \, . \tag{2.5.4}$$

This is conformally invariant for b and c transforming as tensors of weights $(h_b, 0)$ and $(h_c, 0)$ such that

$$h_b = \lambda \, , \quad h_c = 1 - \lambda \tag{2.5.5}$$

for any given constant λ. Thus we have another family of CFTs (which is secretly the same as the linear dilaton family, as we will learn in chapter 10). The operator equations of motion, obtained by the same method (2.1.15), (2.1.18) as before, are

$$\bar{\partial}c(z) = \bar{\partial}b(z) = 0 \, , \tag{2.5.6a}$$

$$\bar{\partial}b(z)c(0) = 2\pi\delta^2(z, \bar{z}) \, . \tag{2.5.6b}$$

The bb and cc OPEs satisfy the equation of motion without source. The normal ordered bc product is

$$:b(z_1)c(z_2): = \ b(z_1)c(z_2) - \frac{1}{z_{12}} \, . \tag{2.5.7}$$

This satisfies the naive equations of motion as a consequence of

$$\bar{\partial}\frac{1}{z} = \partial\frac{1}{\bar{z}} = 2\pi\delta^2(z, \bar{z}) \, , \tag{2.5.8}$$

which can be verified by integrating over a region containing the origin and integrating the derivative by parts. Normal ordering of a general product of fields is combinatorially the same as for the X^μ CFT, a sum over contractions or subtractions. One must be careful because b and c are anticommuting so that interchange of fields flips the sign: one should anticommute the fields being paired until they are next to each other before replacing them with the subtraction (2.5.7).

The operator products are

$$b(z_1)c(z_2) \sim \frac{1}{z_{12}} \, , \quad c(z_1)b(z_2) \sim \frac{1}{z_{12}} \, , \tag{2.5.9}$$

where in the second OPE there have been two sign flips, one from anticommutation and one from $z_1 \leftrightarrow z_2$. Other operator products are nonsingular:

$$b(z_1)b(z_2) = O(z_{12}) \, , \quad c(z_1)c(z_2) = O(z_{12}) \, . \tag{2.5.10}$$

These are not only holomorphic but have a zero due to antisymmetry.

Noether's theorem gives the energy-momentum tensor

$$T(z) = :(\partial b)c: -\lambda\partial(:bc:) \, , \tag{2.5.11a}$$

$$\tilde{T}(\bar{z}) = 0 \, . \tag{2.5.11b}$$

One can also verify (2.5.11) by working out the OPE of T with b and c; it has the standard tensor form (2.4.16) with the given weights. The TT OPE is of standard form (2.4.25) with

$$c = -3(2\lambda - 1)^2 + 1 , \quad \tilde{c} = 0 . \tag{2.5.12}$$

This is a purely holomorphic CFT, and is an example where $c \neq \tilde{c}$. There is of course a corresponding antiholomorphic theory

$$S = \frac{1}{2\pi} \int d^2z \, \tilde{b}\partial\tilde{c} , \tag{2.5.13}$$

which is the same as the above with $z \leftrightarrow \bar{z}$.

The bc theory has a *ghost number* symmetry $\delta b = -i\epsilon b$, $\delta c = i\epsilon c$. The corresponding Noether current is

$$j = - :bc: . \tag{2.5.14}$$

Again the components are separately holomorphic and antiholomorphic, the latter vanishing. When there are both holomorphic and antiholomorphic bc fields, the ghost numbers are separately conserved.

This current is not a tensor,

$$T(z)j(0) \sim \frac{1-2\lambda}{z^3} + \frac{1}{z^2}j(0) + \frac{1}{z}\partial j(0) . \tag{2.5.15}$$

This implies the transformation law

$$\epsilon^{-1}\delta j = -v\partial j - j\partial v + \frac{2\lambda - 1}{2}\partial^2 v , \tag{2.5.16}$$

whose finite form is

$$(\partial_z z')j_{z'}(z') = j_z(z) + \frac{2\lambda - 1}{2}\frac{\partial_z^2 z'}{\partial_z z'} . \tag{2.5.17}$$

The one case where b and c have equal weight is $h_b = h_c = \frac{1}{2}$, for which the central charge $c = 1$. Here we will often use the notation $b \to \psi$, $c \to \bar{\psi}$. For this case the bc CFT can be split in two in a conformally invariant way,

$$\psi = 2^{-1/2}(\psi_1 + i\psi_2) , \quad \bar{\psi} = 2^{-1/2}(\psi_1 - i\psi_2) , \tag{2.5.18a}$$

$$S = \frac{1}{4\pi} \int d^2z \left(\psi_1\bar{\partial}\psi_1 + \psi_2\bar{\partial}\psi_2 \right) , \tag{2.5.18b}$$

$$T = -\frac{1}{2}\psi_1\partial\psi_1 - \frac{1}{2}\psi_2\partial\psi_2 . \tag{2.5.18c}$$

Each ψ theory has central charge $\frac{1}{2}$.

The bc theory for $\lambda = 2$, weights $(h_b, h_c) = (2, -1)$, will arise in the next chapter as the Faddeev–Popov ghosts from gauge-fixing the Polyakov string. The ψ theory will appear extensively in the more general string theories of volume two.

$\beta\gamma$ CFT

The third family of CFTs is much like the bc theory but with *commuting* fields; β is an $(h_\beta, 0)$ tensor and γ an $(h_\gamma, 0)$ tensor, where

$$h_\beta = \lambda, \quad h_\gamma = 1 - \lambda. \tag{2.5.19}$$

The action is

$$S = \frac{1}{2\pi} \int d^2z\, \beta \bar{\partial} \gamma. \tag{2.5.20}$$

These fields are again holomorphic by the equations of motion,

$$\bar{\partial}\gamma(z) = \bar{\partial}\beta(z) = 0. \tag{2.5.21}$$

The equations of motion and operator products are derived in the standard way. Because the statistics are changed, some signs in operator products are different,

$$\beta(z_1)\gamma(z_2) \sim -\frac{1}{z_{12}}, \quad \gamma(z_1)\beta(z_2) \sim \frac{1}{z_{12}}. \tag{2.5.22}$$

The energy-momentum tensor is

$$T = \,:(\partial\beta)\gamma: -\lambda\partial(:\beta\gamma:), \tag{2.5.23a}$$

$$\tilde{T} = 0. \tag{2.5.23b}$$

The central charge is simply reversed in sign,

$$c = 3(2\lambda - 1)^2 - 1, \quad \tilde{c} = 0. \tag{2.5.24}$$

The $\beta\gamma$ theory for $\lambda = \frac{3}{2}$, weights $(h_\beta, h_\gamma) = (\frac{3}{2}, -\frac{1}{2})$, will arise in chapter 10 as the Faddeev–Popov ghosts from gauge-fixing the superstring.

2.6　The Virasoro algebra

Thus far in this chapter we have studied local properties of the two-dimensional field theory. We now are interested in the spectrum of the theory. The spatial coordinate will be periodic, as in the closed string, or bounded, as in the open string.

For the periodic case let

$$\sigma^1 \sim \sigma^1 + 2\pi. \tag{2.6.1}$$

Let the Euclidean time coordinate run

$$-\infty < \sigma^2 < \infty \tag{2.6.2}$$

so that the two dimensions form an infinite cylinder. It is again useful to form a complex coordinate, and there are two natural choices. The first is

$$w = \sigma^1 + i\sigma^2, \tag{2.6.3}$$

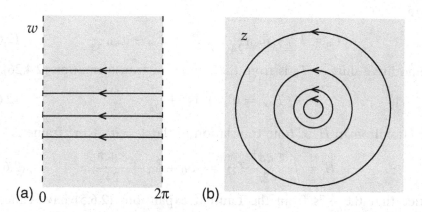

Fig. 2.3. Closed string coordinates. (a) Equal time contours in the w-plane. The dashed lines are identified. (b) The same contours in the z-plane.

so that $w \sim w + 2\pi$. The second is

$$z = \exp(-iw) = \exp(-i\sigma^1 + \sigma^2) . \tag{2.6.4}$$

These two coordinate systems are shown in figure 2.3. In terms of the w coordinate, time corresponds to translation of $\sigma^2 = \text{Im}\, w$. In terms of z, time runs radially, the origin being the infinite past. These coordinates are related by a conformal transformation. The w coordinate is natural for the canonical interpretation of the theory, but the z coordinate is also quite useful and most expressions are written in this frame.

For a holomorphic or antiholomorphic operator we can make a Laurent expansion,

$$T_{zz}(z) = \sum_{m=-\infty}^{\infty} \frac{L_m}{z^{m+2}} , \quad \tilde{T}_{\bar{z}\bar{z}}(\bar{z}) = \sum_{m=-\infty}^{\infty} \frac{\tilde{L}_m}{\bar{z}^{m+2}} . \tag{2.6.5}$$

The Laurent coefficients, known as the *Virasoro generators*, are given by the contour integrals

$$L_m = \oint_C \frac{dz}{2\pi i z} z^{m+2} T_{zz}(z) , \tag{2.6.6}$$

where C is any contour encircling the origin counterclockwise. The Laurent expansion is just the same as an ordinary Fourier transformation in the w frame at time $\sigma^2 = 0$:

$$T_{ww}(w) = -\sum_{m=-\infty}^{\infty} \exp(im\sigma^1 - m\sigma^2) T_m , \tag{2.6.7a}$$

$$T_{\bar{w}\bar{w}}(\bar{w}) = -\sum_{m=-\infty}^{\infty} \exp(-im\sigma^1 - m\sigma^2) \tilde{T}_m , \tag{2.6.7b}$$

where

$$T_m = L_m - \delta_{m,0}\frac{c}{24}\,, \quad \tilde{T}_m = \tilde{L}_m - \delta_{m,0}\frac{\tilde{c}}{24}\,. \tag{2.6.8}$$

The additive shift of T_0 is from the nontensor transformation (2.4.26),

$$T_{ww} = (\partial_w z)^2 T_{zz} + \frac{c}{24}\,. \tag{2.6.9}$$

The Hamiltonian H of time translation in the $w = \sigma^1 + i\sigma^2$ frame is

$$H = \int_0^{2\pi} \frac{d\sigma^1}{2\pi}\, T_{22} = L_0 + \tilde{L}_0 - \frac{c + \tilde{c}}{24}\,. \tag{2.6.10}$$

Notice that the $+2$s from the Laurent expansions (2.6.5) have canceled in the Fourier expansions (2.6.7) due to the conformal transformation of T. Similarly the Laurent expansion for a holomorphic field of weight h would include $+h$ in the exponent.

Cutting open the path integral on circles of constant time $\text{Im}\,w = \ln|z|$, the Virasoro generators become operators in the ordinary sense. (This idea of cutting open a path integral is developed in the appendix.) By holomorphicity the integrals (2.6.6) are independent of C and so in particular are invariant under time translation (radial rescaling). That is, they are conserved charges, the charges associated with the conformal transformations.

It is an important fact that the OPE of currents determines the algebra of the corresponding charges. Consider general charges Q_i, $i = 1, 2$, given as contour integrals of holomorphic currents,

$$Q_i\{C\} = \oint_C \frac{dz}{2\pi i}\, j_i\,. \tag{2.6.11}$$

Consider the combination

$$Q_1\{C_1\}Q_2\{C_2\} - Q_1\{C_3\}Q_2\{C_2\}\,, \tag{2.6.12}$$

where the contours are shown in figure 2.4(a). The order in which the factors are written is irrelevant, as these are just variables of integration in a path integral (unless both charges are anticommuting, in which case there is an additional sign and all the commutators become anticommutators). As discussed in the appendix, when we slice open the path integral to make an operator interpretation, what determines the operator ordering is the time ordering, which here is $t_1 > t_2 > t_3$. The path integral with the combination (2.6.12) thus corresponds to a matrix element of

$$\hat{Q}_1\hat{Q}_2 - \hat{Q}_2\hat{Q}_1 \equiv [\hat{Q}_1, \hat{Q}_2]\,. \tag{2.6.13}$$

Now, for a given point z_2 on the contour C_2, we can deform the difference of the C_1 and C_3 contours as shown in figure 2.4(b), so the commutator

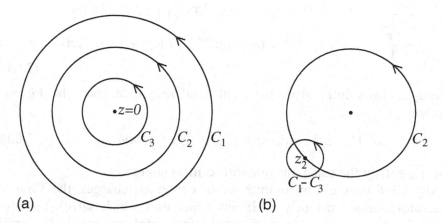

Fig. 2.4. (a) Contours centered on $z = 0$. (b) For given z_2 on contour C_2, contour $C_1 - C_3$ is contracted.

is given by the residue of the OPE,

$$[Q_1, Q_2]\{C_2\} = \oint_{C_2} \frac{dz_2}{2\pi i} \operatorname{Res}_{z_1 \to z_2} j_1(z_1) j_2(z_2) . \qquad (2.6.14)$$

This contour argument allows us to pass back and forth between OPEs and commutation relations. Let us emphasize this: for conserved currents, knowing the singular terms in the OPE is equivalent to knowing the commutator algebra of the corresponding charges. The calculation of figure 2.4 also applies with the conserved charge $Q_2\{C_2\}$ replaced by any operator,

$$[Q, \mathscr{A}(z_2, \bar{z}_2)] = \operatorname{Res}_{z_1 \to z_2} j(z_1) \mathscr{A}(z_2, \bar{z}_2) = \frac{1}{i\epsilon} \delta \mathscr{A}(z_2, \bar{z}_2) . \qquad (2.6.15)$$

This is just the familiar statement that a charge Q generates the corresponding transformation δ. Similarly for the contour integral of an antiholomorphic current

$$\tilde{Q}\{C\} = -\oint_C \frac{d\bar{z}}{2\pi i} \tilde{\jmath} , \qquad (2.6.16)$$

the Ward identity and contour argument imply

$$[\tilde{Q}, \mathscr{A}(z_2, \bar{z}_2)] = \overline{\operatorname{Res}}_{\bar{z}_1 \to \bar{z}_2} \tilde{\jmath}(\bar{z}_1) \mathscr{A}(z_2, \bar{z}_2) = \frac{1}{i\epsilon} \delta \mathscr{A}(z_2, \bar{z}_2) . \qquad (2.6.17)$$

Apply this to the Virasoro generators (2.6.6), where $j_m(z) = z^{m+1} T(z)$:

$$\operatorname{Res}_{z_1 \to z_2} z_1^{m+1} T(z_1) z_2^{n+1} T(z_2)$$

$$= \operatorname{Res}_{z_1 \to z_2} z_1^{m+1} z_2^{n+1} \left(\frac{c}{2 z_{12}^4} + \frac{2}{z_{12}^2} T(z_2) + \frac{1}{z_{12}} \partial T(z_2) \right)$$

$$= \frac{c}{12}(\partial^3 z_2^{m+1})z_2^{n+1} + 2(\partial z_2^{m+1})z_2^{n+1}T(z_2) + z_2^{m+n+2}\partial T(z_2)$$
$$= \frac{c}{12}(m^3 - m)z_2^{m+n-1} + (m - n)z_2^{m+n+1}T(z_2) + \text{total derivative} .$$

$$(2.6.18)$$

The z_2 contour integral of the right-hand side then gives the *Virasoro algebra*,

$$[L_m, L_n] = (m - n)L_{m+n} + \frac{c}{12}(m^3 - m)\delta_{m,-n} . \qquad (2.6.19)$$

The $\tilde{L}_m$ satisfy the same algebra with central charge $\tilde{c}$.

Any CFT thus has an infinite set of conserved charges, the Virasoro generators, which act in the Hilbert space and which satisfy the algebra (2.6.19). Let us for now notice just a few simple properties. Generally we work with eigenstates of L_0 and $\tilde{L}_0$. The generator L_0 satisfies

$$[L_0, L_n] = -nL_n . \qquad (2.6.20)$$

If $|\psi\rangle$ is an eigenstate of L_0 with eigenvalue h, then

$$L_0 L_n|\psi\rangle = L_n(L_0 - n)|\psi\rangle = (h - n)L_n|\psi\rangle , \qquad (2.6.21)$$

so that $L_n|\psi\rangle$ is an eigenstate with eigenvalue $h - n$. The generators with $n < 0$ raise the L_0 eigenvalue and those with $n > 0$ lower it.

The three generators L_0 and $L_{\pm 1}$ form a closed algebra without central charge,

$$[L_0, L_1] = -L_1 , \quad [L_0, L_{-1}] = L_{-1} , \quad [L_1, L_{-1}] = 2L_0 . \qquad (2.6.22)$$

This is the algebra $SL(2, \mathbf{R})$, which differs from $SU(2)$ by signs. For the Laurent coefficients of a holomorphic tensor field $\mathcal{O}$ of weight $(h, 0)$,

$$\mathcal{O}(z) = \sum_{m=-\infty}^{\infty} \frac{\mathcal{O}_m}{z^{m+h}} , \qquad (2.6.23)$$

one finds from the OPE (2.4.16) the commutator

$$[L_m, \mathcal{O}_n] = [(h - 1)m - n]\mathcal{O}_{m+n} . \qquad (2.6.24)$$

Again modes with $n > 0$ reduce L_0, while modes with $n < 0$ increase it.

In the open string, let

$$0 \leq \text{Re}\, w \leq \pi \quad \Leftrightarrow \quad \text{Im}\, z \geq 0 , \qquad (2.6.25)$$

where $z = -\exp(-iw)$. These coordinate regions are shown in figure 2.5. At a boundary, the energy-momentum tensor satisfies

$$T_{ab}n^a t^b = 0 , \qquad (2.6.26)$$

where n^a and t^b are again normal and tangent vectors. To see this, consider a coordinate system in which the boundary is straight. The presence of

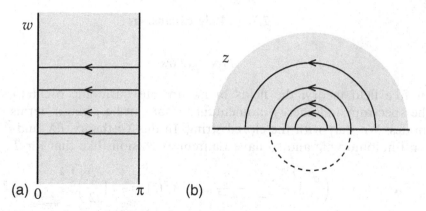

Fig. 2.5. Open string coordinates. (a) Equal time contours in the w-plane. (b) The same contours in the z-plane. The dashed line shows the extension of one contour, as used in the doubling trick.

the boundary breaks translation invariance in the normal direction but not the tangential, so that the current $T_{ab}t^b$ is still conserved. Then the boundary condition (2.6.26) is just the statement that the flow of this current out of the boundary is zero. In the present case, this becomes

$$T_{ww} = T_{\bar{w}\bar{w}} \,, \ \mathrm{Re}\, w = 0, \pi \quad \Leftrightarrow \quad T_{zz} = T_{\bar{z}\bar{z}} \,, \ \mathrm{Im}\, z = 0 \,. \qquad (2.6.27)$$

It is convenient to use the *doubling trick*. Define T_{zz} in the lower half-z-plane as the value of $T_{\bar{z}\bar{z}}$ at its image in the upper half-z-plane, $z' = \bar{z}$:

$$T_{zz}(z) \equiv T_{\bar{z}\bar{z}}(\bar{z}') \,, \quad \mathrm{Im}\, z < 0 \,. \qquad (2.6.28)$$

The equation of motion and boundary condition are then summarized by the statement that T_{zz} is holomorphic in the *whole* complex plane. There is only one set of Virasoro generators, because the boundary condition couples T and $\tilde{T}$,

$$L_m = \frac{1}{2\pi i} \int_C \left(dz\, z^{m+1} T_{zz} - d\bar{z}\, \bar{z}^{m+1} T_{\bar{z}\bar{z}} \right)$$

$$= \frac{1}{2\pi i} \oint dz\, z^{m+1} T_{zz}(z) \,. \qquad (2.6.29)$$

In the first line, the contour C is a semi-circle centered on the origin; in the second line, we have used the doubling trick to write L_m in terms of a closed contour. Again, these satisfy the Virasoro algebra

$$[L_m, L_n] = (m - n)L_{m+n} + \frac{c}{12}(m^3 - m)\delta_{m,-n} \,. \qquad (2.6.30)$$

2.7 Mode expansions

Free scalars

In free field theory, the fields break up into harmonic oscillators, and the spectrum and energy-momentum tensor can be given in terms of the modes. We start with the closed string. In the X^μ theory, ∂X and $\bar\partial X$ are (anti)holomorphic and so have Laurent expansions like that for T,

$$\partial X^\mu(z) = -i\left(\frac{\alpha'}{2}\right)^{1/2} \sum_{m=-\infty}^{\infty} \frac{\alpha_m^\mu}{z^{m+1}} \,, \quad \bar\partial X^\mu(\bar z) = -i\left(\frac{\alpha'}{2}\right)^{1/2} \sum_{m=-\infty}^{\infty} \frac{\tilde\alpha_m^\mu}{\bar z^{m+1}} \,.$$

(2.7.1)

Equivalently,

$$\alpha_m^\mu = \left(\frac{2}{\alpha'}\right)^{1/2} \oint \frac{dz}{2\pi} z^m \partial X^\mu(z) \,,$$

(2.7.2a)

$$\tilde\alpha_m^\mu = -\left(\frac{2}{\alpha'}\right)^{1/2} \oint \frac{d\bar z}{2\pi} \bar z^m \bar\partial X^\mu(\bar z) \,.$$

(2.7.2b)

Single-valuedness of X^μ implies that $\alpha_0^\mu = \tilde\alpha_0^\mu$. Moreover, the Noether current for spacetime translations is $i\partial_a X^\mu / \alpha'$, so the spacetime momentum is

$$p^\mu = \frac{1}{2\pi i} \oint_C (dz\, j^\mu - d\bar z\, \bar{\jmath}^\mu) = \left(\frac{2}{\alpha'}\right)^{1/2} \alpha_0^\mu = \left(\frac{2}{\alpha'}\right)^{1/2} \tilde\alpha_0^\mu \,.$$

(2.7.3)

Integrating the expansions (2.7.1) gives

$$X^\mu(z,\bar z) = x^\mu - i\frac{\alpha'}{2} p^\mu \ln|z|^2 + i\left(\frac{\alpha'}{2}\right)^{1/2} \sum_{\substack{m=-\infty\\m\neq 0}}^{\infty} \frac{1}{m}\left(\frac{\alpha_m^\mu}{z^m} + \frac{\tilde\alpha_m^\mu}{\bar z^m}\right) \,.$$

(2.7.4)

Either from standard canonical commutation, or from the contour argument and the XX OPE, one derives

$$[\alpha_m^\mu, \alpha_n^\nu] = [\tilde\alpha_m^\mu, \tilde\alpha_n^\nu] = m\delta_{m,-n}\eta^{\mu\nu} \,,$$

(2.7.5a)

$$[x^\mu, p^\nu] = i\eta^{\mu\nu} \,,$$

(2.7.5b)

with other commutators vanishing. The spectrum is given by starting with a state $|0; k\rangle$ that has momentum k^μ and is annihilated by all of the lowering modes, α_n^μ for $n > 0$, and acting in all possible ways with the raising ($n < 0$) modes.

We now wish to expand the Virasoro generators in terms of the mode operators. Insert the Laurent expansion for X^μ into the energy-momentum tensor (2.4.4) and collect terms with a given power of z, giving

$$L_m \sim \frac{1}{2} \sum_{n=-\infty}^{\infty} \alpha_{m-n}^{\mu} \alpha_{\mu n} \, . \tag{2.7.6}$$

The $\sim$ indicates that we have ignored the ordering of operators. For $m \neq 0$, the expansion (2.7.6) is well defined and correct as it stands — the mode operators in each term commute and so the ordering does not matter. For $m = 0$, put the lowering operators on the right and introduce a normal ordering constant,

$$L_0 = \frac{\alpha' p^2}{4} + \sum_{n=1}^{\infty} (\alpha_{-n}^{\mu} \alpha_{\mu n}) + a^X \, . \tag{2.7.7}$$

We encountered this same issue in section 1.3, where we treated it in a heuristic way. Now the left-hand side has a finite and unambiguous definition in terms of the Laurent coefficients of the : :-ordered energy-momentum tensor, so the normal ordering constant is finite and calculable. There are several ways to determine it. The simplest uses the Virasoro algebra,

$$2L_0|0;0\rangle = (L_1 L_{-1} - L_{-1}L_1)|0;0\rangle = 0 \, , \tag{2.7.8}$$

and so

$$a^X = 0 \, . \tag{2.7.9}$$

Here we have used the known form of L_1 and L_{-1}: every term of each contains either a lowering operator or p^{μ} and so annihilates $|0;0\rangle$.

We determined the central charge in the Virasoro algebra from the OPE. It can also be determined directly from the expression for the Virasoro generators in terms of the mode operators, though some care is needed. This is left as exercise 2.11.

Let us introduce a new notation. The symbol $^{\circ}_{\circ}$ $^{\circ}_{\circ}$ will denote *creation–annihilation normal ordering*, placing all lowering operators to the right of all raising operators, with a minus sign whenever anticommuting operators are switched. For the purposes of this definition, we will include p^{μ} with the lowering operators and x^{μ} with the raising operators. In this notation we can write

$$L_m = \frac{1}{2} \sum_{n=-\infty}^{\infty} {}^{\circ}_{\circ} \alpha_{m-n}^{\mu} \alpha_{\mu n} {}^{\circ}_{\circ} \tag{2.7.10}$$

since $a^X = 0$.

We have now introduced two forms of normal ordering, *conformal normal ordering* : : (which is what we will mean if we just refer to 'normal ordering') and creation–annihilation normal ordering $^{\circ}_{\circ}$ $^{\circ}_{\circ}$. The former is useful because it produces operators whose OPEs and conformal transformation properties are simple. The latter, which is probably more familiar to the reader, is useful for working out the matrix elements of the operators. Let us work out the relation between them. We start by comparing the time-ordered and creation–annihilation-ordered products. For the product $X^\mu(z,\bar{z})X^\nu(z',\bar{z}')$ with $|z| > |z'|$, insert the mode expansions and move the lowering operators in $X^\mu(z,\bar{z})$ to the right. Keeping track of the commutators gives

$$
X^\mu(z,\bar{z})X^\nu(z',\bar{z}') = {}^\circ_\circ X^\mu(z,\bar{z})X^\nu(z',\bar{z}') {}^\circ_\circ
$$
$$
+ \frac{\alpha'}{2}\eta^{\mu\nu}\left[-\ln|z|^2 + \sum_{m=1}^{\infty}\frac{1}{m}\left(\frac{z'^m}{z^m} + \frac{\bar{z}'^m}{\bar{z}^m}\right)\right]
$$
$$
= {}^\circ_\circ X^\mu(z,\bar{z})X^\nu(z',\bar{z}') {}^\circ_\circ - \frac{\alpha'}{2}\eta^{\mu\nu}\ln|z-z'|^2 . \qquad (2.7.11)
$$

Since $|z| > |z'|$, the left-hand side here is time-ordered and the sum converges. The definition (2.1.21) gave the relation between the time-ordered product and the conformal-normal-ordered product (eq. (2.1.21) was in path integral language, so the product on its right becomes time-ordered in operator formalism). It is the same as the relation (2.7.11), so that

$$
{}^\circ_\circ X^\mu(z,\bar{z})X^\nu(z',\bar{z}') {}^\circ_\circ = {:}X^\mu(z,\bar{z})X^\nu(z',\bar{z}'){:} . \qquad (2.7.12)
$$

Such a relation does not hold in general CFTs, and in fact the somewhat arbitrary grouping of p^μ with the lowering operators was done in part to give this simple result. It also does not hold for operators conformal-normal-ordered in terms of w rather than z, for example. From eq. (2.7.12) one can write the mode expansion (2.7.10) at once, giving a second derivation of $a^X = 0$.

Creation–annihilation normal-ordered products of more than two X^μs have the same combinatoric properties as for : : ordering. That is, they are obtained from the time-ordered product by summing over all subtractions — this is *Wick's theorem* from field theory. Also, to convert operators normal-ordered in one form into a different normal ordering, one sums over all subtractions using the *difference* of the two-point functions. That is, if we have two kinds of ordering,

$$
[X^\mu(z,\bar{z})X^\nu(z',\bar{z}')]_1 = [X^\mu(z,\bar{z})X^\nu(z',\bar{z}')]_2 + \eta^{\mu\nu}\Delta(z,\bar{z},z',\bar{z}') , \qquad (2.7.13)
$$

then for a general operator $\mathscr{F}$,

$$
[\mathscr{F}]_1 = \exp\left(\frac{1}{2}\int d^2z\, d^2z'\, \Delta(z,\bar{z},z',\bar{z}')\frac{\delta}{\delta X^\mu(z,\bar{z})}\frac{\delta}{\delta X_\mu(z',\bar{z}')}\right)[\mathscr{F}]_2 .
$$
$$
(2.7.14)
$$

For the linear dilaton CFT, the Laurent expansion and commutators are unchanged, while the Virasoro generators contain the extra term

$$L_m = \frac{1}{2} \sum_{n=-\infty}^{\infty} {}^{\circ}_{\circ}\alpha^{\mu}_{m-n}\alpha_{\mu n}{}^{\circ}_{\circ} + i \left(\frac{\alpha'}{2}\right)^{1/2} (m+1)V^{\mu}\alpha_{\mu m} . \tag{2.7.15}$$

bc CFT

The fields b and c have the Laurent expansions

$$b(z) = \sum_{m=-\infty}^{\infty} \frac{b_m}{z^{m+\lambda}} , \quad c(z) = \sum_{m=-\infty}^{\infty} \frac{c_m}{z^{m+1-\lambda}} . \tag{2.7.16}$$

To be precise, these are only Laurent expansions if λ is an integer, which we will assume for now. The half-integer case is also of interest, but will be dealt with in detail in chapter 10. The OPE gives the anticommutators

$$\{b_m, c_n\} = \delta_{m,-n} . \tag{2.7.17}$$

Consider first the states that are annihilated by all of the $n > 0$ operators. The b_0, c_0 oscillator algebra generates two such ground states $|\downarrow\rangle$ and $|\uparrow\rangle$, with the properties

$$b_0|\downarrow\rangle = 0 , \quad b_0|\uparrow\rangle = |\downarrow\rangle , \tag{2.7.18a}$$

$$c_0|\downarrow\rangle = |\uparrow\rangle , \quad c_0|\uparrow\rangle = 0 , \tag{2.7.18b}$$

$$b_n|\downarrow\rangle = b_n|\uparrow\rangle = c_n|\downarrow\rangle = c_n|\uparrow\rangle = 0 , \quad n > 0 . \tag{2.7.18c}$$

The general state is obtained by acting on these states with the $n < 0$ modes at most once each (because these anticommute). For reasons to appear later it is conventional to group b_0 with the lowering operators and c_0 with the raising operators, so we will single out $|\downarrow\rangle$ as the ghost vacuum $|0\rangle$. In string theory we will have a holomorphic bc theory and an antiholomorphic $\tilde{b}\tilde{c}$ theory, each with $\lambda = 2$. The closed string spectrum thus includes a product of two copies of the above.

The Virasoro generators are

$$L_m = \sum_{n=-\infty}^{\infty} (m\lambda - n){}^{\circ}_{\circ}b_n c_{m-n}{}^{\circ}_{\circ} + \delta_{m,0}a^{\mathrm{g}} . \tag{2.7.19}$$

The ordering constant can be determined as in eq. (2.7.8), which gives

$$\begin{aligned}
2L_0|\downarrow\rangle &= (L_1 L_{-1} - L_{-1}L_1)|\downarrow\rangle \\
&= (\lambda b_0 c_1)[(1-\lambda)b_{-1}c_0]|\downarrow\rangle = \lambda(1-\lambda)|\downarrow\rangle .
\end{aligned} \tag{2.7.20}$$

Thus, $a^{\mathrm{g}} = \frac{1}{2}\lambda(1-\lambda)$ and

$$L_m = \sum_{n=-\infty}^{\infty} (m\lambda - n){}^{\circ}_{\circ}b_n c_{m-n}{}^{\circ}_{\circ} + \frac{\lambda(1-\lambda)}{2}\delta_{m,0} . \tag{2.7.21}$$

The constant can also be obtained by working out the relation between $: :$ and $\overset{\circ}{\circ}\,\overset{\circ}{\circ}$ for the CFT (exercises 2.13, 2.14).

For the ghost number current (2.5.14), $j = - :bc:$, the charge is

$$N^{\mathrm{g}} = -\frac{1}{2\pi i} \int_0^{2\pi} dw\, j_w$$

$$= \sum_{n=1}^{\infty} (c_{-n}b_n - b_{-n}c_n) + c_0 b_0 - \frac{1}{2} \, . \tag{2.7.22}$$

It satisfies

$$[N^{\mathrm{g}}, b_m] = -b_m \, , \quad [N^{\mathrm{g}}, c_m] = c_m \, , \tag{2.7.23}$$

and so counts the number of c minus the number of b excitations. The ground states have ghost number $\pm\frac{1}{2}$:

$$N^{\mathrm{g}}|\!\downarrow\rangle = -\frac{1}{2}|\!\downarrow\rangle \, , \quad N^{\mathrm{g}}|\!\uparrow\rangle = \frac{1}{2}|\!\uparrow\rangle \, . \tag{2.7.24}$$

This depends on the value of the ordering constant, determined in exercise 2.13, but one might guess it on the grounds that the average ghost number of the ground states should be zero: the ghost number changes sign under $b \leftrightarrow c$.

The $\beta\gamma$ theory is similar; we leave the details until we need them in chapter 10.

Open strings

In the open string, the Neumann boundary condition becomes $\partial_z X^\mu = \partial_{\bar{z}} X^\mu$ on the real axis. There is only one set of modes, the boundary condition requiring $\alpha_m^\mu = \tilde{\alpha}_m^\mu$ in the expansions (2.7.1). The spacetime momentum integral (2.7.3) runs only over a semi-circle, so the normalization is now

$$\alpha_0^\mu = (2\alpha')^{1/2} p^\mu \, . \tag{2.7.25}$$

The expansion for X^μ is then

$$X^\mu(z, \bar{z}) = x^\mu - i\alpha' p^\mu \ln|z|^2 + i\left(\frac{\alpha'}{2}\right)^{1/2} \sum_{\substack{m=-\infty \\ m \neq 0}}^{\infty} \frac{\alpha_m^\mu}{m}(z^{-m} + \bar{z}^{-m}) \, . \tag{2.7.26}$$

Also,

$$L_0 = \alpha' p^2 + \sum_{n=1}^{\infty} \alpha_{-n}^\mu \alpha_{\mu n} \, . \tag{2.7.27}$$

The commutators are as before,

$$[\alpha_m^\mu, \alpha_n^\nu] = m\delta_{m,-n}\eta^{\mu\nu} \, , \quad [x^\mu, p^\nu] = i\eta^{\mu\nu} \, . \tag{2.7.28}$$

For the *bc* theory, the boundary conditions that will be relevant to the string are

$$c(z) = \tilde{c}(\bar{z}) , \quad b(z) = \tilde{b}(\bar{z}) , \quad \text{Im}\, z = 0 , \qquad (2.7.29)$$

written in terms of the z-coordinate where the boundary is the real axis. We can then use the doubling trick to write the holomorphic and antiholomorphic fields in the upper half-plane in terms of holomorphic fields in the whole plane,

$$c(z) \equiv \tilde{c}(\bar{z}') , \quad b(z) \equiv \tilde{b}(\bar{z}') , \quad \text{Im}(z) \leq 0 , \; z' = \bar{z} . \qquad (2.7.30)$$

The open string thus has a single set of Laurent modes for each of b and c.

2.8 Vertex operators

In quantum field theory, there is on the one hand the space of states of the theory and on the other hand the set of local operators. In conformal field theory there is a simple and useful isomorphism between these, with the CFT quantized on a circle. Consider the semi-infinite cylinder in the w-coordinate,

$$0 \leq \text{Re}\, w \leq 2\pi , \quad w \sim w + 2\pi , \quad \text{Im}\, w \leq 0 , \qquad (2.8.1)$$

which maps into the unit disk in the coordinate $z = \exp(-iw)$. This is shown in figure 2.6. To define the path integral in the w-coordinate one must in particular specify the boundary condition as $\text{Im}\, w \to -\infty$. That is, one must specify the initial state. In the z-coordinate, $\text{Im}\, w = -\infty$ maps to the origin, so this is equivalent to specifying the behavior of the fields at that point. In effect, this defines a local operator at the origin, known as the *vertex operator* associated with the state. Going in the other direction, the path integral on the disk with an operator $\mathscr{A}$ at the origin maps to the path integral on the cylinder with a specified initial state $|\mathscr{A}\rangle$.

For free field theories one can easily work out the detailed form of this isomorphism. Suppose one has a conserved charge Q acting on the state $|\mathscr{A}\rangle$ as in figure 2.6(a). One can find the corresponding local operator by using the OPE to evaluate the contour integral in figure 2.6(b). Let us use this to identify the state $|1\rangle$ corresponding to the unit operator. With no operator at the origin, ∂X^μ and $\bar{\partial} X^\mu$ are (anti)holomorphic inside the Q contour in figure 2.6(b). The contour integrals (2.7.2) defining α_m^μ and $\tilde{\alpha}_m^\mu$ for $m \geq 0$ then have no poles and so vanish. Thus $|1\rangle$ is annihilated by these modes. This identifies it as the ground state,

$$|1\rangle = |0; 0\rangle , \qquad (2.8.2)$$

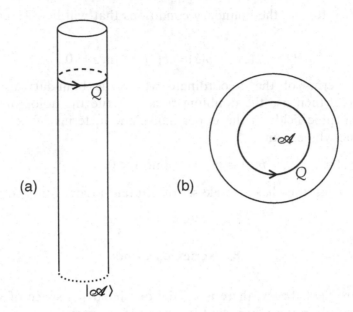

Fig. 2.6.   (a) A semi-infinite cylinder in w, with initial state $|\mathscr{A}\rangle$ and charge Q. (b) The conformally equivalent unit disk with local operator $\mathscr{A}$ and contour integral for Q.

with the normalization chosen for convenience. Now consider, for example, the state $\alpha^{\mu}_{-m}|1\rangle$ with m positive. Passing to figure 2.6(b) with $Q = \alpha^{\mu}_{-m}$ surrounding the unit operator, the fields are holomorphic inside the contour, and so we can evaluate

$$\alpha^{\mu}_{-m} = \left(\frac{2}{\alpha'}\right)^{1/2} \oint \frac{dz}{2\pi} z^{-m}\partial X^{\mu}(z) \;\rightarrow\; \left(\frac{2}{\alpha'}\right)^{1/2} \frac{i}{(m-1)!}\partial^{m}X^{\mu}(0) \qquad (2.8.3)$$

for $m \geq 1$. Thus

$$\alpha^{\mu}_{-m}|1\rangle \cong \left(\frac{2}{\alpha'}\right)^{1/2} \frac{i}{(m-1)!}\partial^{m}X^{\mu}(0) , \quad m \geq 1 . \qquad (2.8.4)$$

Similarly

$$\tilde{\alpha}^{\mu}_{-m}|1\rangle \cong \left(\frac{2}{\alpha'}\right)^{1/2} \frac{i}{(m-1)!}\bar{\partial}^{m}X^{\mu}(0) , \quad m \geq 1 . \qquad (2.8.5)$$

This correspondence continues to hold when α^{μ}_{-m} or $\tilde{\alpha}^{\mu}_{-m}$ acts on a general state $|\mathscr{A}\rangle$. If $:\mathscr{A}(0,0):$ is any normal-ordered operator, there may be singularities in the operator product of $\partial X^{\mu}(z)$ with $:\mathscr{A}(0,0):$, but it is not hard to check that

$$\alpha^{\mu}_{-m}:\mathscr{A}(0,0): = :\alpha^{\mu}_{-m}\mathscr{A}(0,0): \qquad (2.8.6)$$

for $m > 0$ because the contour integral of the contractions will never have a single pole. One can then carry out the same manipulation (2.8.3) inside the normal ordering, and so the state with several α oscillators excited comes out as the $:\ :$ normal-ordered product of the corresponding operators,

$$\alpha^\mu_{-m} \to i \left(\frac{2}{\alpha'}\right)^{1/2} \frac{1}{(m-1)!} \partial^m X^\mu(0) , \quad m \geq 1 , \tag{2.8.7a}$$

$$\tilde{\alpha}^\mu_{-m} \to i \left(\frac{2}{\alpha'}\right)^{1/2} \frac{1}{(m-1)!} \bar{\partial}^m X^\mu(0) , \quad m \geq 1 . \tag{2.8.7b}$$

Similarly,

$$x^\mu_0 \to X^\mu(0,0) . \tag{2.8.8}$$

Any state can be obtained from $|1\rangle$ by acting with the operators on the left-hand sides of eqs. (2.8.7) and (2.8.8). The operator corresponding to this state is then given by the $:\ :$ normal-ordered product of the corresponding local operators on the right-hand sides. For example,

$$|0;k\rangle \cong\ :e^{ik \cdot X(0,0)}:\ . \tag{2.8.9}$$

This is easy to understand: under the translation $X^\mu \to X^\mu + a^\mu$, both the state and the corresponding operator have the same transformation, being multiplied by $\exp(ik \cdot a)$.

The same method applies to the bc theory. For clarity we specialize to the case $\lambda = 2$ which is of interest for bosonic string theory, leaving the generalization to the reader. From the Laurent expansions (2.7.16) and the contour argument it follows that

$$b_m|1\rangle = 0 , \quad m \geq -1 , \qquad c_m|1\rangle = 0 , \quad m \geq 2 . \tag{2.8.10}$$

Note that due to the shifts in the exponents of the Laurent expansion coming from the conformal weights of b and c, the unit operator no longer maps to a ground state. Rather, relations (2.8.10) determine that

$$|1\rangle = b_{-1}|\downarrow\rangle . \tag{2.8.11}$$

The translation of the raising operators is straightforward,

$$b_{-m} \to \frac{1}{(m-2)!} \partial^{m-2} b(0) , \quad m \geq 2 , \tag{2.8.12a}$$

$$c_{-m} \to \frac{1}{(m+1)!} \partial^{m+1} c(0) , \quad m \geq -1 . \tag{2.8.12b}$$

Notice that the ghost number $-\frac{3}{2}$ state $b_{-1}|\downarrow\rangle$ maps to the ghost number 0 unit operator, and the ghost number $-\frac{1}{2}$ state $|\downarrow\rangle$ to the ghost number 1 operator c. The difference arises from the nontensor property (2.5.17) of the ghost number current,

$$(\partial_z w) j_w(w) = j_z(z) + q_0 \frac{\partial_z^2 w}{\partial_z w} = j_z(z) - \frac{q_0}{z} , \tag{2.8.13}$$

where $q_0 = \lambda - \frac{1}{2} = \frac{3}{2}$. The ghost number of states is conventionally defined by the cylindrical frame expression N^g, eq. (2.7.22), while the contour argument of figure 2.6 relates the ghost number of the *vertex operator* to the radial frame charge

$$Q^g \equiv \frac{1}{2\pi i} \oint dz\, j_z = N^g + q_0 . \tag{2.8.14}$$

This applies to other charges as well; for a tensor current the charges of a state and the corresponding operator are equal.

Most of the above can be extended to the $\beta\gamma$ theory, but there are certain complications in the superstring so we defer this to chapter 10.

All the ideas of this section extend to the open string. The semi-infinite strip

$$0 \leq \mathrm{Re}\, w \leq \pi , \quad \mathrm{Im}\, w \leq 0 , \tag{2.8.15}$$

maps to a half-disk, the intersection of the upper half-plane and the unit circle, under $z = -\exp(-iw)$. The initial state again maps to the origin, which is on the boundary, so there is an isomorphism

local operators on the boundary $\leftrightarrow$ states on the interval . (2.8.16)

The details, which are parallel to the above, are left for the reader. Again the doubling trick, extending the fields into the lower half-plane by holomorphicity, is useful in making the contour arguments.

Path integral derivation

The state–operator isomorphism is an important but unfamiliar idea and so it is useful to give also a more explicit path integral derivation. Consider a unit disk in the z-plane, with local operator $\mathscr{A}$ at the origin and with the path integral fields ϕ fixed to some specific boundary values ϕ_b on the unit circle. The path integral over the fields ϕ_i on the interior of the disk with ϕ_b held fixed produces some functional $\Psi_{\mathscr{A}}[\phi_b]$,

$$\Psi_{\mathscr{A}}[\phi_b] = \int [d\phi_i]_{\phi_b} \exp(-S[\phi_i]) \mathscr{A}(0) . \tag{2.8.17}$$

A functional of the fields is the Schrödinger representation of a state, so this is a mapping from operators to states. The Schrödinger representation, which assigns a complex amplitude to each field configuration, has many uses. It is often omitted from field theory texts, the Fock space representation being emphasized.

To go the other way, start with some state $\Psi[\phi_b]$. Consider a path integral over the annular region $1 \geq |z| \geq r$, with the fields ϕ_b on the outer circle fixed and the fields ϕ_b' integrated over the inner circle as follows:

$$\int [d\phi_b'][d\phi_i]_{\phi_b,\phi_b'} \exp(-S[\phi_i]) \, r^{-L_0-\tilde{L}_0} \Psi[\phi_b'] \, . \qquad (2.8.18)$$

That is, the integral is weighted by the state (functional) $r^{-L_0-\tilde{L}_0}\Psi[\phi_b']$. Now, the path integral over the annulus just corresponds to propagating from $|z| = r$ to $|z| = 1$, which is equivalent to acting with the operator $r^{+L_0+\tilde{L}_0}$. This undoes the operator acting on Ψ, so the path integral (2.8.18) is again $\Psi[\phi_b]$. Now take the limit as $r \to 0$. The annulus becomes a disk, and the limit of the path integral over the inner circle can be thought of as defining some local operator at the origin. By construction the path integral on the disk with this operator reproduces $\Psi[\phi_b]$ on the boundary.

Let us work this out explicitly for a single free scalar field X. On the unit circle expand

$$X_b(\theta) = \sum_{n=-\infty}^{\infty} X_n e^{in\theta} \, , \quad X_n^* = X_{-n} \, . \qquad (2.8.19)$$

The boundary state $\Psi[X_b]$ can thus be regarded as a function of all of the X_n. Let us first identify the state corresponding to the unit operator, given by the path integral without insertion,

$$\Psi_1[X_b] = \int [dX_i]_{X_b} \exp\left(-\frac{1}{2\pi\alpha'} \int d^2z \, \partial X \bar{\partial} X\right) \, . \qquad (2.8.20)$$

Evaluate this by the usual Gaussian method. Separate X_i into a classical part and a fluctuation,

$$X_i = X_{cl} + X_i' \, , \qquad (2.8.21a)$$

$$X_{cl}(z, \bar{z}) = X_0 + \sum_{n=1}^{\infty} (z^n X_n + \bar{z}^n X_{-n}) \, . \qquad (2.8.21b)$$

With this definition, X_{cl} satisfies the equation of motion and X_i' vanishes on the boundary. The path integral then separates,

$$\Psi_1[X_b] = \exp(-S_{cl}) \int [dX_i']_{X_b=0} \exp\left(-\frac{1}{2\pi\alpha'} \int d^2z \, \partial X' \bar{\partial} X'\right) \, , \qquad (2.8.22)$$

with

$$S_{cl} = \frac{1}{2\pi\alpha'} \sum_{m,n=1}^{\infty} mn X_m X_{-n} \int_{|z|<1} d^2z \, z^{m-1} \bar{z}^{n-1}$$

$$= \frac{1}{\alpha'} \sum_{m=1}^{\infty} m X_m X_{-m} \, . \qquad (2.8.23)$$

The X_i' integral is a constant, independent of the boundary condition, so

$$\Psi_1[X_b] \propto \exp\left(-\frac{1}{\alpha'} \sum_{m=1}^{\infty} m X_m X_{-m}\right) . \qquad (2.8.24)$$

This is a Gaussian, and is in fact the ground state. To see this, write the raising and lowering operators in the Schrödinger basis,

$$\alpha_n = -\frac{in}{(2\alpha')^{1/2}} X_{-n} - i\left(\frac{\alpha'}{2}\right)^{1/2} \frac{\partial}{\partial X_n} , \qquad (2.8.25a)$$

$$\tilde{\alpha}_n = -\frac{in}{(2\alpha')^{1/2}} X_n - i\left(\frac{\alpha'}{2}\right)^{1/2} \frac{\partial}{\partial X_{-n}} , \qquad (2.8.25b)$$

as follows from the Laurent expansion at $|z| = 1$ and the mode algebra. Acting on (2.8.24), we find

$$\alpha_n \Psi_1[X_b] = \tilde{\alpha}_n \Psi_1[X_b] = 0 , \quad n \geq 0 , \qquad (2.8.26)$$

so this is indeed the ground state. Thus,

$$|1\rangle \propto |0;0\rangle . \qquad (2.8.27)$$

Rather than try to keep track of the overall normalization of the path integral at this point, we define $|1\rangle = |0;0\rangle$.

Another easy calculation is the state corresponding to $\partial^k X$; this just adds a factor of $\partial^k X_{cl}(0) = k! X_k$ to the result, so

$$|\partial^k X\rangle = k! X_k \Psi_1 = -i\left(\frac{\alpha'}{2}\right)^{1/2} (k-1)! \alpha_{-k} |0;0\rangle , \qquad (2.8.28)$$

and similarly for $\bar{\partial}^k X$. This extends to all products of exponentials and derivatives of X. The conformal normal ordering just cancels the effect of the X_i' path integral.

2.9 More on states and operators

The OPE

In this section we make additional applications of the state–operator correspondence. The first is a simple and general derivation of the OPE, as shown in figure 2.7. Consider the product

$$\mathscr{A}_i(z, \bar{z}) \mathscr{A}_j(0, 0) , \qquad (2.9.1)$$

where $|z| < 1$. We can divide the path integral into an integral over fields ϕ_i on the interior of the unit disk, an integral over fields ϕ_b on the unit circle, and an integral over fields ϕ_e on the exterior of the unit disk. This sort of cutting open of path integrals is discussed in detail in the

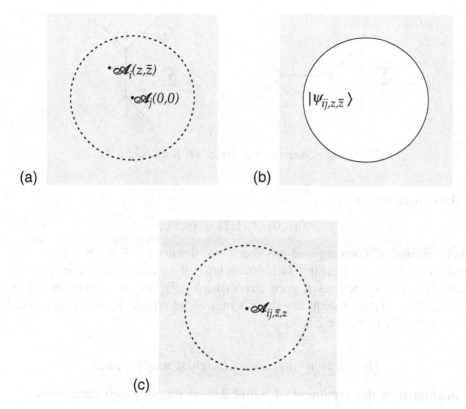

Fig. 2.7. (a) World-sheet with two local operators. (b) Integration over fields on the interior of the disk produces boundary state $|\psi_{ij,z,\bar{z}}\rangle$. (c) Sewing in a disk with the corresponding local operator. Expanding in operators of definite weight gives the OPE.

appendix. As in the discussion at the end of the previous section, the integral over ϕ_i leaves some functional of ϕ_b; call it $\Psi_{ij,z,\bar{z}}[\phi_b]$. By the state–operator correspondence this is equivalent to gluing back the disk with the appropriate local operator $\mathcal{A}_{ij,z,\bar{z}}(0,0)$ and so we are done! To put this in standard form expand in a complete set of L_0, $\tilde{L}_0$ eigenstates,

$$\mathcal{A}_{ij,z,\bar{z}} = \sum_k z^{h_k - h_i - h_j} \bar{z}^{\tilde{h}_k - \tilde{h}_i - \tilde{h}_j} c^k_{ij} \mathcal{A}_k \,, \qquad (2.9.2)$$

with the z- and $\bar{z}$-dependence determined by the conformal weights as in eq. (2.4.20). The convergence of the OPE is just the usual convergence of a complete set in quantum mechanics. The construction is possible as long as there are no other operators with $|z'| \leq |z|$, so that we can cut on a circle of radius $|z| + \epsilon$.

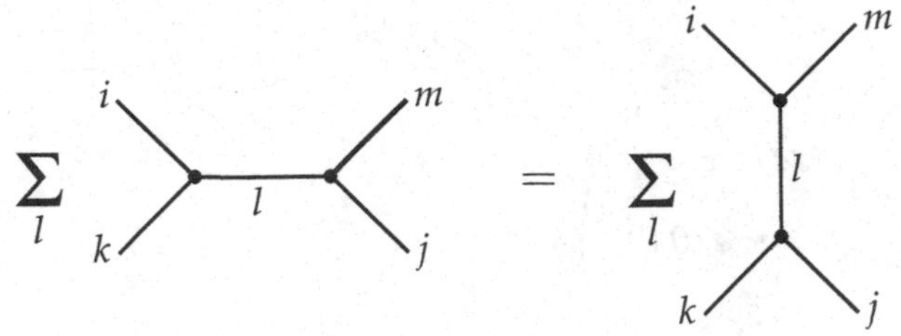

Fig. 2.8. Schematic picture of OPE associativity.

For three operators

$$\mathscr{A}_i(0,0)\mathscr{A}_j(1,1)\mathscr{A}_k(z,\bar{z}) \ , \tag{2.9.3}$$

the regions of convergence of the $z \to 0$ and $z \to 1$ OPEs, $|z| < 1$ and $|1 - z| < 1$, overlap. The coefficient of $\mathscr{A}_m$ in the triple product can then be written as a sum involving $c^l{}_{ik}c^m{}_{lj}$ or as a sum involving $c^l{}_{jk}c^m{}_{li}$. Associativity requires these sums to be equal. This is represented schematically in figure 2.8.

The Virasoro algebra and highest weight states

Consider now the argument of figure 2.6 for the Virasoro generators:

$$
\begin{aligned}
L_m|\mathscr{A}\rangle &\cong \oint \frac{dz}{2\pi i} z^{m+1} T(z)\mathscr{A}(0,0) \\
&\cong L_m \cdot \mathscr{A}(0,0) \ .
\end{aligned}
\tag{2.9.4}
$$

Here we have introduced a new idea and notation. Given the state–operator isomorphism, every operator acting on the Hilbert space has an image acting on the space of local operators. In other words, form the corresponding contour around the operator and evaluate the resulting local operator using the OPE. Corresponding to the L_m that map states to states are the $L_m \cdot$ that map local operators to local operators. In general there will be local operators at various positions z_i, and there will be a different copy of the Virasoro algebra for each one, obtained by a Laurent expansion centered on the position of the operator. In some geometries there may also be a standard set of generators, such as the ones centered on $z = 0$ for the cylinder. The '·' serves as a reminder that we are talking about the Laurent coefficients around that particular operator.

In this notation the $T\mathscr{A}$ OPE is

$$T(z)\mathscr{A}(0,0) = \sum_{m=-\infty}^{\infty} z^{-m-2} L_m \cdot \mathscr{A}(0,0) \ . \tag{2.9.5}$$

To relate to the earlier notation (2.4.11), $\mathscr{A}^{(n)} = L_{n-1} \cdot \mathscr{A}$ for $n \geq 0$, and the conformal transformation of $\mathscr{A}$ is

$$\delta\mathscr{A}(z,\bar{z}) = -\epsilon \sum_{n=0}^{\infty} \frac{1}{n!} \left[\partial^n v(z) L_{n-1} + (\partial^n v(z))^* \tilde{L}_{n-1} \right] \cdot \mathscr{A}(z,\bar{z}) . \qquad (2.9.6)$$

Using the general form (2.4.14) for the $T\mathscr{A}$ OPE, we have the very useful result

$$L_{-1} \cdot \mathscr{A} = \partial\mathscr{A} , \quad \tilde{L}_{-1} \cdot \mathscr{A} = \bar{\partial}\mathscr{A} , \qquad (2.9.7a)$$
$$L_0 \cdot \mathscr{A} = h\mathscr{A} , \quad \tilde{L}_0 \cdot \mathscr{A} = \tilde{h}\mathscr{A} . \qquad (2.9.7b)$$

The OPE (2.9.5) implies that a primary field $\mathcal{O}$ of weight $(h,\tilde{h})$ corresponds to a state satisfying

$$L_0|\mathcal{O}\rangle = h|\mathcal{O}\rangle , \quad \tilde{L}_0|\mathcal{O}\rangle = \tilde{h}|\mathcal{O}\rangle , \qquad (2.9.8a)$$
$$L_m|\mathcal{O}\rangle = \tilde{L}_m|\mathcal{O}\rangle = 0 , \quad m > 0 . \qquad (2.9.8b)$$

Such a state is conventionally known as a *highest weight state*. For the CFTs of interest L_0 and $\tilde{L}_0$ are bounded below. By acting repeatedly with lowering operators on any state we eventually reach a state annihilated by further lowering operators, a highest weight state.

An interesting special case is the unit operator. The operator product $T\,1$ is nonsingular, so it follows in any CFT that

$$L_m|1\rangle = \tilde{L}_m|1\rangle = 0 , \quad m \geq -1 . \qquad (2.9.9)$$

As noted in section 2.6, the operators L_0 and $L_{\pm 1}$ form a closed algebra, as do $\tilde{L}_0$ and $\tilde{L}_{\pm 1}$. The full algebra is

$$SL(2,\mathbf{C}) . \qquad (2.9.10)$$

Thus $|1\rangle$ is also called the $SL(2,\mathbf{C})$-invariant state. It is the only such state because the relations (2.9.7) imply that the operator $\mathscr{A}$ corresponding to any $SL(2,\mathbf{C})$-invariant state is independent of position. It must then be a c-number, as explained after eq. (2.4.24).

Unitary CFTs

Highest weight states play a special role in string theory, and also in the theory of representations of the Virasoro algebra. We will return to this at various points, but for now we derive a few important general results that hold in *unitary* CFTs. A unitary CFT is one that has a positive inner product $\langle\ |\ \rangle$ such that

$$L_m^\dagger = L_{-m} , \quad \tilde{L}_m^\dagger = \tilde{L}_{-m} . \qquad (2.9.11)$$

Recall that the inner product defines the adjoint via $\langle\alpha|A\beta\rangle = \langle A^\dagger\alpha|\beta\rangle$.

For example, the X^μ CFT is unitary for *spacelike* μ if we take the inner product

$$\langle 0; k | 0; k' \rangle = 2\pi \delta(k - k') \qquad (2.9.12)$$

for the ground state and define

$$\alpha_m^\dagger = \alpha_{-m} , \quad \tilde{\alpha}_m^\dagger = \tilde{\alpha}_{-m} . \qquad (2.9.13)$$

This implicitly defines the inner products of all higher states. This CFT is not unitary for $\mu = 0$ because of the opposite sign in the commutator there.

The first constraint is that any state in a unitary CFT must have $h, \tilde{h} \geq 0$. If this is true for highest weight states it is true for all states. For a highest weight state the Virasoro algebra gives

$$2h_{\mathcal{O}} \langle \mathcal{O} | \mathcal{O} \rangle = 2\langle \mathcal{O} | L_0 | \mathcal{O} \rangle = \langle \mathcal{O} | [L_1, L_{-1}] | \mathcal{O} \rangle = \| L_{-1} | \mathcal{O} \rangle \|^2 \geq 0 , \qquad (2.9.14)$$

so $h_{\mathcal{O}} \geq 0$. It also follows that if $h_{\mathcal{O}} = \tilde{h}_{\mathcal{O}} = 0$ then

$$L_{-1} \cdot \mathcal{O} = \tilde{L}_{-1} \cdot \mathcal{O} = 0 , \qquad (2.9.15)$$

and so $\mathcal{O}$ is independent of position. As noted before, $\mathcal{O}$ must then be a c-number. That is, the unit operator is the only $(0,0)$ operator in a unitary CFT. Curiously, X^μ itself is the one notable exception: the corresponding state $x^\mu | 0; 0 \rangle$ is nonnormalizable because of the infinite range of X^μ, and equation (2.9.14) no longer holds. The general theorems are of the most use for the CFTs corresponding to compactified dimensions, where this kind of exception cannot occur.

In a similar way, one finds that an operator in a unitary CFT is holomorphic if and only if $\tilde{h} = 0$, and antiholomorphic if and only if $h = 0$; this is an important result so we repeat it,

$$\partial \mathscr{A} = 0 \iff h = 0 , \quad \bar{\partial} \mathscr{A} = 0 \iff \tilde{h} = 0 . \qquad (2.9.16)$$

Finally, using the above argument with the commutator $[L_n, L_{-n}]$, one can show that $c, \tilde{c} \geq 0$ in a unitary CFT. In fact, the only unitary CFT with $c = 0$ is the trivial one, $L_n = 0$.

Zero-point energies

The state–operator mapping gives a simple alternative derivation of the various normal ordering constants. In any CFT, we know that $L_0 | 1 \rangle = 0$, and this determines the additive normalization of L_0. In the X CFT, $| 1 \rangle$ is the ground state $| 0; 0 \rangle$, so a^X vanishes. In the bc theory, $| 1 \rangle$ is the excited state $b_{-1} | \downarrow \rangle$, so the weight of $| \downarrow \rangle$ is -1 in agreement with the earlier result (2.7.21) for $\lambda = 2$.

This also provides one physical interpretation of the central charge. In a unitary CFT the ground state is $|1\rangle$ with $L_0 = \tilde{L}_0 = 0$. The conformal transformation (2.6.10) between the radial and time-translation generators then implies that

$$E = -\frac{c + \tilde{c}}{24} \qquad (2.9.17)$$

for the ground state. This is a Casimir energy, coming from the finite size of the system, and it depends only on the central charge. On dimensional grounds this energy is inverse to the size ℓ of the system, which is 2π here, so the general result would be

$$E = -\frac{\pi(c + \tilde{c})}{12\ell} \ . \qquad (2.9.18)$$

We have now given three honest ways of calculating normal ordering constants: from the Virasoro algebra as in eq. (2.7.8), by relating the two forms of normal ordering as in eq. (2.7.11), and from the state–operator mapping above. Nevertheless the idea of adding up zero-point energies is intuitive and a useful mnemonic, so we give a prescription that can be checked by one of the more honest methods:

1. Add the zero-point energies, $\frac{1}{2}\omega$ for each bosonic mode and $-\frac{1}{2}\omega$ for each fermionic (anticommuting) mode.

2. One encounters divergent sums of the form $\sum_{n=1}^{\infty}(n-\theta)$, the θ arising when one considers nontrivial periodicity conditions. Define this to be

$$\sum_{n=1}^{\infty}(n - \theta) = \frac{1}{24} - \frac{1}{8}(2\theta - 1)^2 \ . \qquad (2.9.19)$$

This is the value one obtains as in eq. (1.3.32) by regulating and discarding the quadratically divergent part.

3. The above gives the normal ordering constant for the w-frame generator T_0, eq. (2.6.8). For L_0 we must add the nontensor correction $\frac{1}{24}c$.

For the free boson, the modes are integer-valued so we get one-half of the sum (2.9.19) for $\theta = 0$ after step 2, which is $-\frac{1}{24}$. This is just offset by the correction in step 3, giving $a^X = 0$. For the ghosts we similarly get $\frac{2}{24} - \frac{26}{24} = -1$.

Exercises

2.1 Verify that

$$\partial\bar\partial \ln|z|^2 = \partial \frac{1}{\bar z} = \bar\partial \frac{1}{z} = 2\pi\delta^2(z,\bar z)$$

(a) by use of the divergence theorem (2.1.9);
(b) by regulating the singularity and then taking a limit.

2.2 Work out explicitly the expression (2.2.5) for the normal-ordered product of four X^μ fields. Show that it is a harmonic function of each of the positions.

2.3 The expectation value of a product of exponential operators on the plane is

$$\left\langle \prod_{i=1}^{n} :e^{ik_i \cdot X(z_i,\bar z_i)}: \right\rangle = iC^X(2\pi)^D\delta^D(\textstyle\sum_{i=1}^n k_i) \prod_{\substack{i,j=1\\i<j}}^{n} |z_{ij}|^{\alpha' k_i \cdot k_j} \;,$$

with C^X a constant. This can be obtained as a limit of the expectation value (6.2.17) on the sphere, which we will obtain by several methods in chapter 6.
(a) Show that the leading behavior as one vertex operator approaches another is in agreement with the OPE (2.2.14).
(b) As implied by eq. (2.2.13), the expectation value is $|z_{12}|^{\alpha' k_1 \cdot k_2}$ times a function that is smooth as $z_1 \to z_2$. For $n = 3$, work out the explicit expansion of this smooth function in powers of z_{12} and $\bar z_{12}$. From the ratio of the large-order terms find the radius of convergence.
(c) Give a general proof that the free-field OPE (2.2.4) is convergent inside the dashed circle in figure 2.1.

2.4 Extend the OPE (2.2.14) up to and including subleading terms of order z^2, $z\bar z$, and $\bar z^2$. Show that the $z,\bar z$-dependence of each term is in agreement with the general form (2.4.20). Give also the symmetric form as in eq. (2.2.15) and discuss the properties of the coefficients under interchange of the two operators.

2.5 Derive the classical Noether theorem in the form usually found in textbooks. That is, assume that

$$S[\phi] = \int d^d\sigma \, \mathcal{L}(\phi(\sigma), \partial_a\phi(\sigma))$$

and ignore the variation of the measure. Invariance of the action implies that the variation of the Lagrangian density is a total derivative,

$$\delta\mathcal{L} = \epsilon\partial_a\mathcal{K}^a$$

under a symmetry transformation (2.3.1). Then the classical result, with conventional string normalization, is

$$j^a = 2\pi i \left(\frac{\partial \mathscr{L}}{\partial \phi_{\alpha,a}} \epsilon^{-1} \delta\phi_\alpha - \mathscr{K}^a \right) .$$

Derive the Lagrangian equations of motion and show that they imply that this current is conserved. Show that it agrees with eq. (2.3.4).

2.6 Consider the flat Euclidean metric δ_{ab}, in d dimensions. An infinitesimal coordinate transformation $\delta\sigma^a = v^a(\sigma)$ changes the metric by $\delta g_{ab} = -\partial_a v_b - \partial_b v_a$. Determine the most general $v^a(\sigma)$ that leaves the metric invariant up to a local rescaling. There are $\frac{1}{2}(d+1)(d+2)$ independent solutions in $d > 2$, and an infinite number in $d = 2$. [Although the result is given in many places in the literature, the author has not found a simple derivation.]

2.7 (a) By computing the relevant OPEs, confirm the weights stated in eq. (2.4.17) and determine which operators are tensors.
(b) Do this for the same operators in the linear dilaton theory.

2.8 What is the weight of $f_{\mu\nu} : \partial X^\mu \bar{\partial} X^\nu e^{ik\cdot X} :$? What are the conditions on $f_{\mu\nu}$ and k_μ in order for it to be a tensor?

2.9 Derive the central charges for the linear dilaton, bc, and $\beta\gamma$ CFTs by working out the TT OPEs.

2.10 Consider now a patch of world-sheet with boundary. For convenience suppose that the patch lies in the upper half-z-plane, with the real axis being the boundary. Show that expectation values of a normal-ordered operator, say $: \partial X^\mu(z) \bar{\partial} X_\mu(\bar{z}) :$, although finite in the interior, diverge as z approaches the boundary (represent the effect of the boundary by an image charge). Define *boundary normal ordering* $\overset{\star}{\star}\ \overset{\star}{\star}$, which is the same as $: :$ except that the contraction includes the image charge piece as well. Operators on the boundary are finite if they are boundary normal ordered.

In a general CFT on a manifold with boundary, the interior operators and the boundary operators are independent. Label the bases $\mathscr{A}_i$ and $\mathscr{B}_r$ respectively, and define the former by $: :$ and the latter by $\overset{\star}{\star}\ \overset{\star}{\star}$. Each set has its own closed OPE: $\mathscr{A}\mathscr{A} \to \mathscr{A}$ and $\mathscr{B}\mathscr{B} \to \mathscr{B}$. In addition, the sets are related: $\mathscr{A}_i$, as it approaches the boundary, can be expanded in terms of the $\mathscr{B}_r$. Find the leading behaviors of

$$\overset{\star}{\star} e^{ik_1\cdot X(y_1)} \overset{\star}{\star}\ \overset{\star}{\star} e^{ik_2\cdot X(y_2)} \overset{\star}{\star} , \quad y_1 \to y_2 \ (y \ \text{real}) ,$$
$$: e^{ik\cdot X(z,\bar{z})} : , \quad \text{Im}(z) \to 0 ,$$

The identity (2.7.14) is useful for the latter.

2.11 Evaluate the central charge in the Virasoro algebra for X^μ by calculating

$$L_m(L_{-m}|0;0\rangle) - L_{-m}(L_m|0;0\rangle) \ .$$

2.12 Use the OPE and the contour results (2.6.14) and (2.6.15) to derive the commutators (2.7.5) and (2.7.17).

2.13 (a) Show that

$$:b(z)c(z'): \ - \ {}^{\circ}_{\circ}b(z)c(z'){}^{\circ}_{\circ} = \frac{(z/z')^{1-\lambda} - 1}{z - z'}$$

by the method of eq. (2.7.11).

(b) Use this to determine the ordering constant in N^g, eq. (2.7.22). You also need the conformal transformation (2.8.14).

(c) Show that one obtains the same value for N^g by a heuristic treatment of the ordering similar to that in section 2.9.

2.14 (a) Determine the ordering constant in L_0^g, eq. (2.7.21), using part (a) of the previous exercise.

(b) Determine it using the heuristic rules in section 2.9.

2.15 Work out the state–operator mapping for the open string X^μ CFT.

2.16 Let $\mathcal{O}$ be a tensor field of weight $(h, \tilde{h})$. Evaluate the commutators $[L_m, \mathcal{O}(z, \bar{z})]$ using the contour argument of figure 2.4. Show in particular that the commutator of a $(1,1)$ tensor is a total derivative.

2.17 Extend the argument of eq. (2.9.14) to show that $c \geq 0$.

3

The Polyakov path integral

We now begin a systematic study of string theory using the Polyakov path integral and conformal field theory. After an introduction to the path integral picture, we discuss gauge fixing, the Weyl anomaly, and vertex operators. We then generalize to strings propagating in curved spacetime.

3.1 Sums over world-sheets

The Feynman path integral is one way to represent a quantum theory, and it is a very natural method for describing interactions in string theory. In path integral quantization, amplitudes are given by summing over all possible histories interpolating between the initial and final states. Each history is weighted by

$$\exp(iS_{\text{cl}}/\hbar) , \qquad (3.1.1)$$

with S_{cl} the classical action for the given history. Thus, one would define an amplitude in string theory by summing over all world-sheets connecting given initial and final curves, as in figure 3.1(a) for the open string and figure 3.1(b) for the closed string. The analogous sum for the relativistic point particle produces the free propagator (exercise 5.1).

One can imagine a number of ways in which strings might interact. One would be a contact interaction, an energy when two strings intersect. Another would be a long-range force mediated by some quantum field. However, it is not possible to *add* such interactions to string theory in a way that is consistent with the symmetries; we will get some idea later of why this is so. Rather, the only interactions that are allowed are those that are already implicit in the sum over world-sheets. Consider, for example, the world-sheets shown in figure 3.2. Figure 3.2(a) looks like a quantum correction to the open string amplitude of figure 3.1(a), involving intermediate states with two open strings. Figure 3.2(b) has three external

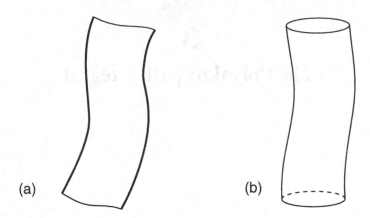

Fig. 3.1. (a) An open string world-sheet with the topology of a strip. The heavier curves are the world-lines of string endpoints. (b) A closed string world-sheet with the topology of a cylinder.

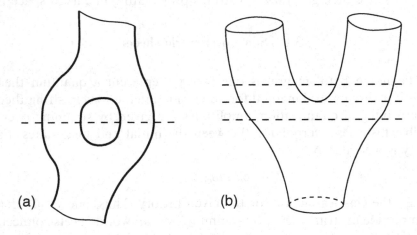

Fig. 3.2. (a) Quantum correction to open string propagation. (b) Decay of one closed string into two. The dashed lines are time-slices.

closed strings and represents one string decaying into two. It turns out that this is the correct way to introduce interactions into string theory. We will see that these interactions produce a consistent theory, finite and unitary, which includes gravity.

It is interesting to consider the process of figure 3.2(b) as seen at successive times as shown in figure 3.3. A closed string splits into two, or in the reverse process two join into one. This is the basic closed string interaction. In closed string theory, all particles are obtained as various states of excitation of the string, and all interactions (gauge, gravitational,

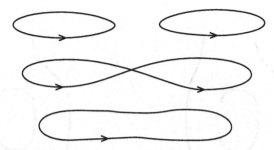

Fig. 3.3. Successive time-slices from figure 3.2(b). The arrows indicate the orientation of the string.

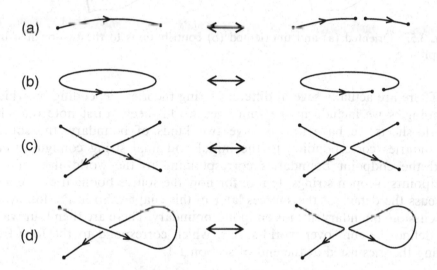

Fig. 3.4. Processes involving open strings: (a) open ↔ open + open; (b) closed ↔ open; (c) open + open ↔ open + open; (d) open ↔ open + closed.

and Yukawa, for example) arise from the single process of figure 3.3. For theories with open strings there are several additional processes that can occur, as shown in figure 3.4. With a given slicing, the interactions of figures 3.3 and 3.4 appear to occur at a definite spacetime point, but this is an illusion. A different slicing, in a boosted Lorentz frame, will put the apparent interaction at a different point; there is no distinguished point in spacetime. The interaction arises only from the global topology of the world-sheet, while the local properties of the world-sheet are the same as they were in the free case. As discussed in the introduction, it is this smearing out of the interaction that cuts off the short-distance divergences of gravity.

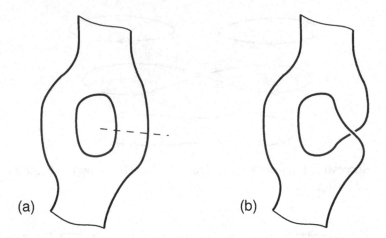

Fig. 3.5.   Oriented (a) and unoriented (b) contributions to the two-open-string amplitude.

There are actually several different string theories, depending on which topologies we include in the sum over world-sheets. First note that the world-sheets we have drawn have two kinds of boundary, the source boundaries corresponding to the initial and final string configurations, and the endpoint boundaries corresponding to the world-lines of the endpoints of open strings. Ignore for now the source boundaries; we will discuss the details of the sources later in this chapter. So in the following discussion 'boundary' means endpoint boundary. There are then four ways to define the sum over world-sheets, which correspond to the four free string theories listed at the end of section 1.4:

1. Closed oriented: All oriented world-sheets without boundary.

2. Closed unoriented: All world-sheets without boundary.

3. Closed plus open oriented: All oriented world-sheets with any number of boundaries.

4. Closed plus open unoriented: All world-sheets with any number of boundaries.

There are two things that we should expand upon here. The first is the connection between the inclusion of unoriented world-sheets and the $\Omega = +1$ projection described in section 1.4. Figure 3.5 shows two world-sheets that are included in the unoriented theory with open strings. The unoriented world-sheet of figure 3.5(b) is equivalent to cutting the oriented surface of figure 3.5(a) along the dashed line, which we will parameterize

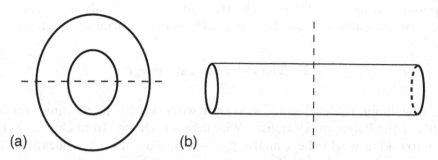

Fig. 3.6. (a) An annulus: the intermediate state is two open strings. (b) The same topology drawn as a cylinder: the intermediate state is one closed string.

by $0 \leq \sigma \leq \pi$, and requiring the fields on the cut edges to satisfy

$$X^{\mu}_{\text{upper}}(\sigma) = X^{\mu}_{\text{lower}}(\pi - \sigma) \,. \tag{3.1.2}$$

This in turn is equivalent to acting with the operator Ω on the open string state. Adding these two surfaces with appropriate weight then amounts to inserting the operator $\frac{1}{2}(1 + \Omega)$, which projects onto states of $\Omega = +1$. Summing over all oriented and unoriented surfaces inserts projection operators on all intermediate states, reducing the spectrum to the $\Omega = +1$ sector.

Second, the list above does not include any theory with open strings but not closed. Let us explain in more detail why this is the case. Consider a world-sheet with the topology of an annulus, figure 3.6(a). For convenience we have drawn only the vacuum amplitude, but by attaching appropriate sources the same argument will also apply to scattering amplitudes. We see from the dashed line that this is a process with an intermediate state of two open strings, and so world-sheets of this topology must be present in any theory with open strings. In figure 3.6(b) we have drawn the same topology as a cylinder and cut it open on an intermediate state of a single closed string. Thus, even if we start with open strings only, the sum over world-sheets will necessarily include processes in which scattering of open strings produces closed strings. As we develop the amplitudes we will see how this works in detail.

The need for closed strings in open string theory can also be seen in another way. Consider the interactions in figures 3.4(a) and (b), but time reversed: respectively two open strings joining into one, and an open to closed string transition. Near the interaction point these are the same, two endpoints coalescing. To have the first interaction without the second would require some nonlocal constraint on the dynamics of the string; this would surely be inconsistent. One can make the same argument with figures 3.4(c) and (d), where the interaction is locally the reconnection of a pair of strings. So if any open string interaction is allowed, then so is some process in which closed strings are produced from open strings.

3.2   The Polyakov path integral

We now begin to develop the sum over world-sheets, integrating over the fields of the Polyakov formalism. We make one change from chapter 1: the Minkowskian world-sheet metric γ_{ab} is replaced with a Euclidean world-sheet metric $g_{ab}(\sigma^1, \sigma^2)$, of signature $(+, +)$. The integral runs over all Euclidean metrics and over all embeddings $X^\mu(\sigma^1, \sigma^2)$ of the world-sheet in Minkowski spacetime:

$$\int [dX\, dg]\, \exp(-S) \,. \tag{3.2.1}$$

The Euclidean action is

$$S = S_X + \lambda \chi \,, \tag{3.2.2}$$

with

$$S_X = \frac{1}{4\pi\alpha'} \int_M d^2\sigma\, g^{1/2} g^{ab} \partial_a X^\mu \partial_b X_\mu \,, \tag{3.2.3a}$$

$$\chi = \frac{1}{4\pi} \int_M d^2\sigma\, g^{1/2} R + \frac{1}{2\pi} \int_{\partial M} ds\, k \,. \tag{3.2.3b}$$

The geodesic curvature k was defined in exercise 1.3; note that the boundary now is always spacelike.

The advantage of the Euclidean path integral is that the integral over metrics is better defined. The topologically nontrivial world-sheets we have been describing can have nonsingular Euclidean metrics but not Minkowskian ones. We will take the Euclidean theory (3.2.1) as our starting point, and we will show how to give it a precise definition and that it defines a consistent spacetime theory with the advertised properties. However, let us give a brief formal argument that it is equivalent to the Minkowski theory with which we began.

We start again with the example of the point particle, where the path integral

$$\int [d\eta\, dX] \exp\left[\frac{i}{2} \int d\tau \left(\eta^{-1} \dot{X}^\mu \dot{X}_\mu - \eta m^2\right)\right] \tag{3.2.4}$$

is oscillatory. The path integral over η and X^μ is a product of ordinary integrals, and so we can deform contours just as in ordinary integration. If we take

$$\eta(\tau) \to e^{-i\theta} \eta(\tau) \,, \quad X^0(\tau) \to e^{-i\theta} X^0(\tau) \,, \tag{3.2.5}$$

for infinitesimal θ, all terms in the exponent acquire a negative real part and so this acts as a convergence factor. Now we can perform a contour rotation in field space, all the way to $\eta = -ie$, $X^0 = -iX^D$. The integral becomes

$$\int [de\, dX] \exp\left[-\frac{1}{2}\int d\tau \left(e^{-1}\sum_{\mu=1}^{D} \dot{X}^{\mu}\dot{X}^{\mu} + em^2\right)\right]. \qquad (3.2.6)$$

This is the Euclidean point-particle analog of the path integral (3.2.1). We have just made a contour rotation, so the Euclidean path integral gives the same amplitude as the Minkowski one.

The same procedure works for the Polyakov action if we write the metric in terms of a tetrad, $\gamma_{ab} = -e_a^0 e_b^0 + e_a^1 e_b^1$, and make the same rotation on e_a^0. This provides a formal justification for the equivalence of the Minkowski and Euclidean path integrals.[1] It has been shown by explicit calculation that they define the same amplitudes, respectively in the light-cone and conformal gauges. We have in the action (3.2.3a) left the rotation of X^0 implicit, as in the previous chapter, so as to emphasize the Minkowski nature of spacetime. Written in terms of X^0, the equations of motion and the OPE are covariant, with metric $\eta^{\mu\nu}$.

We have noted that χ is locally a total derivative in two dimensions and therefore depends only on the topology of the world-sheet — it is the *Euler number* of the world-sheet. Thus, the $e^{-\lambda\chi}$ factor in the path integral affects only the relative weighting of different topologies in the sum over world-sheets. In particular, if one adds an extra strip to a world-sheet, as we have done in going from figure 3.1(a) to figure 3.2(a), the Euler number decreases by one and the path integral is weighted by an extra factor of e^{λ}. Since adding a strip corresponds to emitting and reabsorbing a virtual open string, the amplitude for emitting an open string from any process is proportional to $e^{\lambda/2}$. Adding a handle to any world-sheet reduces the Euler number by 2 and adds a factor $e^{2\lambda}$. Since this corresponds to emitting and reabsorbing a closed string, the amplitude for emitting a closed string is proportional to e^{λ}. The Euler term in the action thus controls the coupling constants in the string theory, with

$$g_o^2 \sim g_c \sim e^{\lambda}. \qquad (3.2.7)$$

By the way, λ might seem to be a free parameter in the theory, contradicting the statement in the introduction that there are no such parameters

---

[1] In more than two dimensions, things are not so simple because the Hilbert action behaves in a more complicated way under the rotation. No simple rotation damps the path integral. In particular, the meaning of the Euclidean path integral for four-dimensional gravity is very uncertain.

in string theory. This is an important point, and we will resolve it in section 3.7.

As an aside, the counting of couplings extends in a simple way to world-sheets with stringy sources. The sources for closed strings are closed boundary loops, while the open string sources are boundary segments. For convenience we will require the open string source boundaries to meet the endpoint boundaries at right angles in the world-sheet metric. We have to be careful because the boundary curvature k diverges at the corners: the Euler number is

$$\chi = \tilde{\chi} + \frac{1}{4}n_c \,, \tag{3.2.8}$$

where $\tilde{\chi}$ includes only the integral over the smooth segments of boundary and n_c is the number of corners. The correct weight for the path integral is

$$\exp(-\lambda\tilde{\chi}) = \exp(-\lambda\chi + \lambda n_c/4) \,. \tag{3.2.9}$$

This follows from unitarity: the λ-dependence must be unchanged if we cut through a world-sheet, leaving fewer internal lines and more external sources. This will be the case for the weight (3.2.9) because the surface integral in $\tilde{\chi}$ cancels on the cut edges. Adding a closed string source amounts to cutting a hole in the world-sheet and decreases χ by 1. Adding an open string source leaves χ unchanged and increases n_c by 2. Thus we reach the same conclusion (3.2.7) about the amplitude for emission of closed or open strings.

3.3   Gauge fixing

The path integral (3.2.1) is not quite right. It contains an enormous overcounting, because configurations (X, g) and (X', g') that are related to one another by the local diff$\times$Weyl symmetry represent the same physical configuration. We need, in effect, to divide by the volume of this local symmetry group,

$$\int \frac{[dX\, dg]}{V_{\text{diff}\times\text{Weyl}}} \exp(-S) \equiv Z \,. \tag{3.3.1}$$

We will carry this out by gauge-fixing, integrating over a slice that cuts through each gauge equivalence class once and obtaining the correct measure on the slice by the Faddeev–Popov method.

In chapter 1 we imposed light-cone gauge. This is useful for some purposes but hides some of the symmetry of the theory, so we now make a different choice. Note that the metric, being symmetric, has three independent components, and that there are three gauge functions, the two coordinates and the local scale of the metric. Thus there is just enough gauge freedom to eliminate the integration over the metric, fixing it at some specific functional form which we will call the *fiducial* metric,

$$g_{ab}(\sigma) \to \hat{g}_{ab}(\sigma) \, . \tag{3.3.2}$$

A simple choice is the flat or *unit gauge* metric

$$\hat{g}_{ab}(\sigma) = \delta_{ab} \, . \tag{3.3.3}$$

One sometimes wishes to consider the effect of the diff group alone. In this case, one can bring an arbitrary metric to within a Weyl transformation of the unit form. This is the *conformal gauge*,

$$\hat{g}_{ab}(\sigma) = \exp[2\omega(\sigma)]\delta_{ab} \, . \tag{3.3.4}$$

Let us see explicitly that this counting of fields and gauge symmetries works — that any metric can be brought to the flat form at least locally, in a given neighborhood on the world-sheet. First, make a Weyl transformation to set to zero the Ricci scalar built from g_{ab}. The Weyl transformation of the Ricci scalar is

$$g'^{1/2}R' = g^{1/2}(R - 2\nabla^2\omega) \, . \tag{3.3.5}$$

To set $R' = 0$ requires solving $2\nabla^2\omega = R$ as an equation for ω. This is always possible, at least locally, by a standard argument from electrostatics. Now, in two dimensions the vanishing of the Ricci scalar implies the vanishing of the entire Riemann tensor, because the symmetries of the Riemann tensor imply that

$$R_{abcd} = \frac{1}{2}(g_{ac}g_{bd} - g_{ad}g_{bc})R \, . \tag{3.3.6}$$

So the metric is flat and coordinate-equivalent to the unit metric (3.3.3).

Actually, locally there is a little bit of extra gauge freedom. There are diff×Weyl transformations that leave the metric in unit gauge and so are not fixed by the choice of metric. It is convenient here to introduce a complex coordinate $z = \sigma^1 + i\sigma^2$ as in chapter 2, where the flat metric is $ds^2 = dzd\bar{z}$. Consider a coordinate transformation such that z' is a holomorphic function of z,

$$z' \equiv \sigma'^1 + i\sigma'^2 = f(z) \, , \tag{3.3.7}$$

combined with a Weyl transformation. The new metric is

$$ds'^2 = \exp(2\omega)|\partial_z f|^{-2}dz'd\bar{z}' \, . \tag{3.3.8}$$

Then for

$$\omega = \ln|\partial_z f| \tag{3.3.9}$$

the metric is invariant. This extra gauge freedom does not conflict with our earlier counting ($3 = 3$) because the mismatch, the set of holomorphic reparameterizations, is of measure zero compared to all reparameterizations.

When we consider the full world-sheet, most or all of this extra freedom will be eliminated by the boundary conditions. There are two issues here. First, in the discussion of Noether's theorem and the Ward identities, we were careful to emphasize below eq. (2.3.10) that these results depend only on the symmetry being defined locally on the world-sheet. The transformation (3.3.7), (3.3.9) will therefore give rise to a conserved current and to Ward identities. This is precisely the conformal invariance studied in chapter 2. We see that it arises as the subgroup of diff×Weyl that leaves invariant the unit metric.

The second issue is what happens to our counting of metric and gauge degrees of freedom when we do consider the full world-sheet. We will take this up in chapter 5, where we will see that globally there is a small mismatch between the two, a finite number of parameters. Until then, we will be concerned only with local properties of the world-sheet theory.

The Faddeev–Popov determinant

After fixing the metric, the functional integral runs along a slice parameterized by X^μ alone. In order to obtain the correct measure, we follow the Faddeev–Popov procedure. The steps are the same as were used to obtain the gauge-fixed measure in Yang-Mills theory, but we will give a self-contained discussion. The idea, illustrated in figure 3.7, is to separate the path integral into an integral over the gauge group times an integral along the gauge slice, and to divide out the former. The Faddeev–Popov determinant is the Jacobian for this change of variables.

Let ζ be shorthand for a combined coordinate and Weyl transformation,

$$\zeta : g \to g^\zeta , \quad g^\zeta_{ab}(\sigma') = \exp[2\omega(\sigma)]\frac{\partial\sigma^c}{\partial\sigma'^a}\frac{\partial\sigma^d}{\partial\sigma'^b}g_{cd}(\sigma) . \tag{3.3.10}$$

Following a standard route, we define the Faddeev–Popov measure Δ_{FP} by

$$1 = \Delta_{FP}(g)\int[d\zeta]\,\delta(g - \hat{g}^\zeta) , \tag{3.3.11}$$

where again $\hat{g}_{ab}$ is the fiducial metric. In (3.3.11), $[d\zeta]$ is a gauge-invariant measure on the diff×Weyl group. We will not need the explicit form of this measure; we will discuss the question of its existence in the next section. The delta function is actually a delta functional, requiring $g_{ab}(\sigma) = \hat{g}^\zeta_{ab}(\sigma)$ at every point.

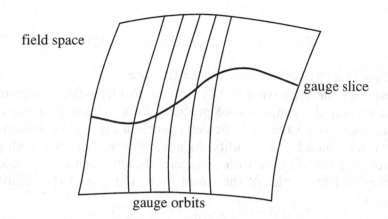

Fig. 3.7. Schematic picture of the Faddeev–Popov procedure. The gauge orbits are families of gauge-equivalent configurations. Integrating over the whole field space and dividing by the orbit volume is equivalent to integrating over a slice that intersects each orbit once, with appropriate Jacobian.

Inserting (3.3.11) into the functional integral (3.3.1), the latter becomes

$$Z[\hat{g}] = \int \frac{[d\zeta \, dX \, dg]}{V_{\text{diff} \times \text{Weyl}}} \, \Delta_{\text{FP}}(g)\delta(g - \hat{g}^{\zeta}) \exp(-S[X, g]) \,. \qquad (3.3.12)$$

For later reference, we denote explicitly the dependence of Z on the choice of fiducial metric. Carry out the integration over g_{ab}, and also rename the dummy variable $X \rightarrow X^{\zeta}$, to obtain

$$Z[\hat{g}] = \int \frac{[d\zeta \, dX^{\zeta}]}{V_{\text{diff} \times \text{Weyl}}} \Delta_{\text{FP}}(\hat{g}^{\zeta}) \exp(-S[X^{\zeta}, \hat{g}^{\zeta}]) \,. \qquad (3.3.13)$$

Now use the gauge invariance[2] of Δ_{FP}, of the functional integral measure $[dX]$, and of the action to obtain

$$Z[\hat{g}] = \int \frac{[d\zeta \, dX]}{V_{\text{diff} \times \text{Weyl}}} \, \Delta_{\text{FP}}(\hat{g}) \exp(-S[X, \hat{g}]) \,. \qquad (3.3.14)$$

Finally, nothing in the integrand depends on ζ, so the integral over ζ just produces the volume of the gauge group and cancels the denominator,

---

[2] To show this invariance:

$$\Delta_{\text{FP}}(g^{\zeta})^{-1} = \int [d\zeta'] \, \delta(g^{\zeta} - \hat{g}^{\zeta'}) = \int [d\zeta'] \, \delta(g - \hat{g}^{\zeta^{-1} \cdot \zeta'})$$

$$= \int [d\zeta''] \, \delta(g - \hat{g}^{\zeta''}) = \Delta_{\text{FP}}(g)^{-1} \,,$$

where $\zeta'' = \zeta^{-1} \cdot \zeta'$. In the second equality we have used the gauge invariance of the delta function, and in the third the invariance of the measure.

leaving

$$Z[\hat{g}] = \int [dX] \, \Delta_{\text{FP}}(\hat{g}) \exp(-S[X, \hat{g})) . \qquad (3.3.15)$$

Thus, $\Delta_{\text{FP}}(\hat{g})$ is the correct measure on the slice.

To evaluate the expression (3.3.11) for the Faddeev–Popov measure, let us pretend that the gauge choice precisely fixes the gauge symmetry, so that for exactly one value of ζ the delta function $\delta(g - \hat{g}^\zeta)$ is nonzero; for $\delta(\hat{g} - \hat{g}^\zeta)$ this would be the identity. As noted above, there is a small global mismatch, but we will deal with this when the time comes — it does not affect the local properties of the world-sheet. For ζ near the identity, we can expand

$$\delta g_{ab} = 2\delta\omega \, g_{ab} - \nabla_a \delta\sigma_b - \nabla_b \delta\sigma_a$$
$$= (2\delta\omega - \nabla_c \delta\sigma^c) g_{ab} - 2(P_1 \delta\sigma)_{ab} . \qquad (3.3.16)$$

We have defined a differential operator P_1 that takes vectors into traceless symmetric 2-tensors,

$$(P_1 \delta\sigma)_{ab} = \frac{1}{2}(\nabla_a \delta\sigma_b + \nabla_b \delta\sigma_a - g_{ab}\nabla_c \delta\sigma^c) . \qquad (3.3.17)$$

Near the identity, the inverse determinant (3.3.11) becomes

$$\Delta_{\text{FP}}(\hat{g})^{-1}$$
$$= \int [d\delta\omega \, d\delta\sigma] \, \delta\left[-(2\delta\omega - \hat{\nabla}\cdot\delta\sigma)\hat{g} + 2\hat{P}_1\delta\sigma\right]$$
$$= \int [d\delta\omega \, d\beta \, d\delta\sigma] \exp\left\{2\pi i \int d^2\sigma \, \hat{g}^{1/2}\beta^{ab}\left[-(2\delta\omega - \hat{\nabla}\cdot\delta\sigma)\hat{g} + 2\hat{P}_1\delta\sigma\right]_{ab}\right\}$$
$$= \int [d\beta' \, d\delta\sigma] \exp\left\{4\pi i \int d^2\sigma \, \hat{g}^{1/2}\beta'^{ab}(\hat{P}_1\delta\sigma)_{ab}\right\} . \qquad (3.3.18)$$

A hat on a differential operator indicates that it contains the fiducial metric $\hat{g}$. In the second equality we have used the integral representation for the functional delta function, introducing the symmetric tensor field β^{ab}. In the third equality we have integrated over $\delta\omega$, which produces a delta functional forcing β^{ab} to be traceless; the functional integral $[d\beta']$ is then over *traceless* symmetric tensors. We now have a representation for $\Delta_{\text{FP}}(\hat{g})^{-1}$ as a functional integral over a vector field $\delta\sigma^a$ and over a traceless symmetric tensor β'^{ab}.

As discussed in the appendix, we can invert this path integral by replacing each bosonic field with a corresponding Grassmann *ghost* field.

$$\delta\sigma^a \rightarrow c^a , \qquad (3.3.19a)$$
$$\beta'_{ab} \rightarrow b_{ab} , \qquad (3.3.19b)$$

with b^{ab}, like β'^{ab}, being traceless. Thus,

$$\Delta_{\mathrm{FP}}(\hat{g}) = \int [db\, dc]\, \exp(-S_g)\,, \qquad (3.3.20)$$

where the ghost action S_g, with a convenient normalization for the fields, is

$$S_g = \frac{1}{2\pi} \int d^2\sigma\, \hat{g}^{1/2} b_{ab} \hat{\nabla}^a c^b = \frac{1}{2\pi} \int d^2\sigma\, \hat{g}^{1/2} b_{ab} (\hat{P}_1 c)^{ab}\,. \qquad (3.3.21)$$

Locally on the world-sheet, the Polyakov path integral is now

$$Z[\hat{g}] = \int [dX\, db\, dc]\, \exp(-S_X - S_g)\,. \qquad (3.3.22)$$

The action is quadratic in the fields, so we can evaluate the path integral in terms of determinants as discussed in the appendix. This step will be carried out in more detail in the chapter 5; the result here is

$$Z[\hat{g}] = (\det \hat{\nabla}^2)^{-D/2} \det \hat{P}_1\,, \qquad (3.3.23)$$

the first determinant coming from the X integration and the second from the ghost integration.

In conformal gauge, $\hat{g}_{ab}(\sigma) = \exp[2\omega(\sigma)]\delta_{ab}$, the ghost action (3.3.21) is

$$\begin{aligned}
S_g &= \frac{1}{2\pi} \int d^2z \left(b_{zz} \nabla_{\bar{z}} c^z + b_{\bar{z}\bar{z}} \nabla_z c^{\bar{z}} \right) \\
&= \frac{1}{2\pi} \int d^2z \left(b_{zz} \partial_{\bar{z}} c^z + b_{\bar{z}\bar{z}} \partial_z c^{\bar{z}} \right)\,.
\end{aligned} \qquad (3.3.24)$$

Notice that $\omega(\sigma)$ does not appear in the final form. This is because the covariant $\bar{z}$-derivative of a tensor that has only z indices reduces to the simple derivative, and vice versa, as the reader can check by working out the connection tensor. We have positioned the indices, lower on b and upper on c, to take advantage of this. Since the action (3.3.24) is Weyl-invariant, it follows that b_{ab} and c^a must be neutral under the Weyl transformation (in contrast to b^{ab} and c_a, for example, which contain additional factors of the metric). Since the conformal transformation is a combination of a coordinate transformation (3.3.7) and a Weyl transformation (3.3.9), we can immediately read off the conformal transformations of b_{ab} and c^a from the tensor indices: this is a bc CFT with $(h_b, h_c) = (2, -1)$ and a $\tilde{b}\tilde{c}$ CFT with $(\tilde{h}_b, \tilde{h}_c) = (2, -1)$.

The derivation that we have given of the gauge-fixed path integral is sometimes referred to as the 'heuristic' derivation. The problem is that the starting point, the Polyakov functional integral (3.3.1), is ill defined due to the enormous gauge overcounting. The gauge-fixed functional integral (3.3.22), on the other hand, is quite well defined, and can be calculated explicitly. For practical purposes we should perhaps regard the gauge-fixed form as our true starting point, with the gauge-invariant one giving an intuitive interpretation. In the gauge-fixed form, the essential content of the gauge symmetry is still present as *BRST invariance*, which we will take up in the next chapter.

In the open string we need also to consider a patch containing a segment of the boundary. It is convenient to regard functional integrals as being over fields on a fixed coordinate region, so the diff invariance is limited to coordinate changes that take the boundary into itself. The variation $\delta\sigma^a$ therefore has vanishing normal component,

$$n_a \delta\sigma^a = 0 \ . \tag{3.3.25}$$

This boundary condition is inherited by the corresponding ghost field,

$$n_a c^a = 0 \ , \tag{3.3.26}$$

so c^a is proportional to the tangent vector t^a. The equations of motion then provide a boundary condition on b_{ab}. They have a surface term

$$\int_{\partial M} ds \, n^a b_{ab} \delta c^b = 0 \ . \tag{3.3.27}$$

With the boundary condition on c^a, this implies

$$n_a t_b b^{ab} = 0 \ . \tag{3.3.28}$$

These are the boundary conditions that we used in section 2.7.

3.4   The Weyl anomaly

A key feature of string theory is that it is not consistent in all spacetime backgrounds, but only in those satisfying certain conditions. For the bosonic string theory in flat spacetime, the condition is $D = 26$. This was first discovered as a pathology of the scattering amplitudes. In the light-cone analysis, it arises as a loss of Lorentz invariance, as we briefly and rather heuristically explained in section 1.3. The underlying source of the restriction is an *anomaly* in the local world-sheet symmetries. In our present formalism, the anomaly is in the Weyl symmetry: $T^a{}_a$ vanishes classically, but not in the quantum theory.

The reader who has studied quantum field theory has already encountered a related anomaly, although it was likely not described in the same language. A global Weyl transformation, $\omega(\sigma) = $ constant, is equivalent to an overall rescaling of lengths. A number of familiar field theories in four dimensions are classically invariant under such a rescaling. These include massless scalar field theory with a ϕ^4 interaction, and non-Abelian gauge theory. The scale invariance is evident because the Lagrangian contains no dimensionful parameters — the coupling constant in each case is dimensionless. However, we know that due to divergences in the quantum theory there is a nonzero renormalization group *beta function* in each theory, implying that the effective coupling constant is in fact a function of the length scale. Correspondingly, the trace of the energy-momentum tensor vanishes in the classical theory but is nonzero and proportional to the beta function when quantum effects are taken into account.

In the previous section, we ignored the possibility of anomalies. We need to check this: is the gauge-fixed path integral $Z[g]$ really independent of the choice of fiducial metric,

$$Z[g^\zeta] = Z[g] \ ? \tag{3.4.1}$$

In section 4.2 we will extend this to more general changes of gauge. For convenience, we omit the hat on g henceforth. We will also be interested in path integrals with additional insertions,

$$\langle \ldots \rangle_g \equiv \int [dX\, db\, dc]\, \exp(-S[X, b, c, g]) \ldots . \tag{3.4.2}$$

Then we also require

$$\langle \ldots \rangle_{g^\zeta} = \langle \ldots \rangle_g . \tag{3.4.3}$$

We are not at the present time interested in the details of the insertions themselves — that will be the subject of the next two sections — so we will restrict attention to gauge transformations ζ vanishing near the positions of the fields in '...'. That is, we are asking that Weyl invariance hold as an operator equation.

It is easy to preserve the diff and Poincaré invariances in the quantum theory. For example, one may define the gauge-fixed path integral using a Pauli–Villars regulator as is done for the harmonic oscillator in section A.1, dividing by the path integral for a massive regulator field. The massive regulator field can be coupled to the metric in a diff- and Poincaré-invariant way. However, the diff- and Poincaré-invariant mass term $\mu^2 \int d^2\sigma\, g^{1/2}\, Y^\mu Y_\mu$ for a regulator field Y^μ is not Weyl-invariant. We must therefore check Weyl invariance by hand: does the invariance (3.4.3) hold for Weyl transformations? Any Weyl transformation can be obtained as the result of repeated infinitesimal transformations, so it is sufficient to study the latter.

The energy-momentum tensor T^{ab} is the infinitesimal variation of the path integral with respect to the metric,

$$\delta \langle \ldots \rangle_g = -\frac{1}{4\pi} \int d^2\sigma\, g(\sigma)^{1/2} \delta g_{ab}(\sigma) \left\langle T^{ab}(\sigma) \ldots \right\rangle_g . \tag{3.4.4}$$

Classically, T^{ab} comes entirely from the variation of the action, and this coincides with the Noether definition

$$T^{ab}(\sigma) \stackrel{\text{classical}}{\equiv} \frac{4\pi}{g(\sigma)^{1/2}} \frac{\delta}{\delta g_{ab}(\sigma)} S . \tag{3.4.5}$$

It also reduces to the earlier definition in the limit of a flat world-sheet. If one takes δg_{ab} to have the form of a coordinate transformation, the diff invariance of the path integral implies the conservation of T^{ab}.

In the following analysis, we will at first neglect possible boundary terms. For a Weyl transformation, the definition (3.4.4) of T^{ab} means that

$$\delta_W \langle \dots \rangle_g = -\frac{1}{2\pi} \int d^2\sigma \, g(\sigma)^{1/2} \delta\omega(\sigma) \langle T^a{}_a(\sigma) \dots \rangle_g \,, \qquad (3.4.6)$$

so Weyl invariance with the general insertions '...' can be phrased as the operator statement that the energy-momentum tensor is traceless:

$$T^a{}_a \overset{?}{=} 0 \,. \qquad (3.4.7)$$

The classical action is Weyl-invariant, but in the quantum theory the trace might be nonzero because we have not managed to find a fully gauge-invariant regulator. The trace must be diff- and Poincaré-invariant, because we have preserved these symmetries, and it must vanish in the flat case because we know from the previous chapter that that theory is conformally invariant. This leaves only one possibility,

$$T^a{}_a = a_1 R \qquad (3.4.8)$$

for some constant a_1, with R again the world-sheet Ricci scalar. Terms with more than two derivatives are forbidden for dimensional reasons. In units of world-sheet length, g_{ab} and X^μ are both dimensionless, so the constant a_1 in eq. (3.4.8) is dimensionless as well. Terms with more derivatives would have a coefficient with positive powers of the cutoff length used to define the path integral, and so vanish in the limit where the cutoff is taken to zero.

Calculation of the Weyl anomaly

The possible obstruction to a gauge-invariant theory has been reduced to the single number a_1. In fact, this number is proportional to something we have already calculated, the *central charge* c of the CFT on a flat world-sheet. Both a_1 and c are related to the two-point function of the energy-momentum tensor, though to different components. To get the precise relation between them we will need to use the diff invariance, as follows.

In conformal gauge,

$$T_{z\bar{z}} = \frac{a_1}{2} g_{z\bar{z}} R \,. \qquad (3.4.9)$$

Taking the covariant derivative,

$$\nabla^{\bar{z}} T_{\bar{z}z} = \frac{a_1}{2} \nabla^{\bar{z}}(g_{z\bar{z}} R) = \frac{a_1}{2} \partial_z R \,, \qquad (3.4.10)$$

where we have used the fact that the metric is covariantly constant. Then by conservation of T_{ab},

$$\nabla^z T_{zz} = -\nabla^{\bar{z}} T_{\bar{z}z} = -\frac{a_1}{2} \partial_z R \ . \tag{3.4.11}$$

To fix a_1 we compare the Weyl transformations on the left and right. The Weyl transformation (3.3.5) of the right is

$$a_1 \partial_z \nabla^2 \delta\omega \approx 4a_1 \partial_z^2 \partial_{\bar{z}} \delta\omega \ , \tag{3.4.12}$$

where we have expanded around a flat world-sheet. To get the Weyl transformation of T_{zz} we use first the conformal transformation (2.4.24),

$$\epsilon^{-1}\delta T_{zz}(z) = -\frac{c}{12}\partial_z^3 v^z(z) - 2\partial_z v^z(z) T_{zz}(z) - v^z(z)\partial_z T_{zz}(z) \ . \tag{3.4.13}$$

From the discussion in section 3.3, this conformal transformation consists of a coordinate transformation $\delta z = \epsilon v$ plus a Weyl transformation $2\delta\omega = \epsilon \partial v + \epsilon(\partial v)^*$. The last two terms in the variation are the coordinate transformation of the tensor, so the Weyl transformation, to leading order around flat space, is

$$\delta_{\mathrm{W}} T_{zz} = -\frac{c}{6}\partial_z^2 \delta\omega \ . \tag{3.4.14}$$

Acting with $\partial^z = 2\partial_{\bar{z}}$ and comparing with the transformation (3.4.12) gives

$$c = -12a_1 \ , \qquad T^a{}_a = -\frac{c}{12}R \ . \tag{3.4.15}$$

Let us derive this again by a slightly longer route, along which we obtain a useful intermediate result. In conformal gauge,

$$R = -2\exp(-2\omega)\partial_a\partial_a\omega \ , \tag{3.4.16a}$$
$$\nabla^2 = \exp(-2\omega)\partial_a\partial_a \ . \tag{3.4.16b}$$

By the contraction with two lowered indices we mean $\delta^{ab}\partial_a\partial_b$. The Weyl variation (3.4.6) of $Z[g]$ becomes

$$\delta_{\mathrm{W}} Z[\exp(2\omega)\delta] = \frac{a_1}{\pi} Z[\exp(2\omega)\delta] \int d^2\sigma \, \delta\omega\partial_a\partial_a\omega \ , \tag{3.4.17}$$

where 'exp$(2\omega)\delta$' denotes conformal gauge. This integrates immediately to

$$Z[\exp(2\omega)\delta] = Z[\delta] \exp\left(-\frac{a_1}{2\pi}\int d^2\sigma \, \partial_a\omega\partial_a\omega\right) \ . \tag{3.4.18}$$

Since every metric is diff-equivalent to a conformal metric, eq. (3.4.18) actually determines the complete metric-dependence of $Z[g]$. We need to find the diff-invariant expression that reduces to eq. (3.4.18) in conformal gauge. Using the relations (3.4.16), one verifies that the desired expression is

$$Z[g] = Z[\delta] \exp\left[\frac{a_1}{8\pi} \int d^2\sigma \int d^2\sigma' g^{1/2} R(\sigma) G(\sigma, \sigma') g^{1/2} R(\sigma')\right] , \quad (3.4.19)$$

where

$$g(\sigma)^{1/2} \nabla^2 G(\sigma, \sigma') = \delta^2(\sigma - \sigma') \quad (3.4.20)$$

defines the scalar Green's function. This is an interesting result: the path integral is completely determined by the anomaly equation.

Now expand $Z[g]$ around a flat background, $g_{ab} = \delta_{ab} + h_{ab}$, and keep terms of second order in $h_{\bar{z}\bar{z}}$. The Ricci scalar is $4\partial_z^2 h_{\bar{z}\bar{z}}$ to first order in $h_{\bar{z}\bar{z}}$, so (ignoring contact terms, where $z = z'$)

$$\ln \frac{Z[\delta + h]}{Z[\delta]} \approx \frac{a_1}{8\pi^2} \int d^2z \int d^2z' \, (\partial_z^2 \ln|z - z'|^2) h_{\bar{z}\bar{z}}(z, \bar{z}) \partial_{z'}^2 h_{\bar{z}\bar{z}}(z', \bar{z}')$$

$$= -\frac{3a_1}{4\pi^2} \int d^2z \int d^2z' \, \frac{h_{\bar{z}\bar{z}}(z, \bar{z}) h_{\bar{z}\bar{z}}(z', \bar{z}')}{(z - z')^4} . \quad (3.4.21)$$

We can also calculate this by using second order perturbation theory in the metric, which by eq. (3.4.4) gives

$$\ln \frac{Z[\delta + h]}{Z[\delta]} \approx \frac{1}{8\pi^2} \int d^2z \int d^2z' \, h_{\bar{z}\bar{z}}(z, \bar{z}) h_{\bar{z}\bar{z}}(z', \bar{z}') \langle T_{zz}(z) T_{zz}(z') \rangle_\delta . \quad (3.4.22)$$

Now use the standard TT OPE (2.4.25). All terms except the first involve operators of nonzero spin and so vanish by rotational invariance, leaving $\langle T_{zz}(z) T_{zz}(z') \rangle_\delta = \frac{1}{2} c(z - z')^{-4}$. Comparison with the result (3.4.21) from the Weyl anomaly gives again eq. (3.4.15).

Discussion

The world-sheet theory consists of the X^μ, with central charge D, and the ghosts, whose central charge (2.5.12) is -26 for $\lambda = 2$, giving

$$c = c^X + c^g = D - 26 . \quad (3.4.23)$$

The theory is Weyl-invariant only for $D = 26$, the same as the condition for Lorentz invariance found in chapter 1. When the Weyl anomaly is nonvanishing, different gauge choices are inequivalent, and as with anomalous gauge theories any choice leads to a pathology such as loss of covariance or loss of unitarity. In light-cone gauge, for example, choosing different spacetime axes requires a change of gauge, so the underlying Weyl anomaly is translated into a Lorentz anomaly.

There is another thing that one might try: ignore the Weyl invariance and treat only the diff invariance as a gauge symmetry. The metric would then have one real degree of freedom $\omega(\sigma)$, to be integrated over rather than gauge-fixed. This is particularly plausible because the quantum effects in eq. (3.4.18) have generated an action for ω which looks just like that for one of the X^μ; for $D < 26$ the sign is that of a spacelike dimension. However, the physics is a little exotic because the diff transformation (and therefore the energy-momentum tensor) of ω is different from that of the X^μ. In fact the theory is the linear dilaton CFT. So there is no $(D + 1)$-dimensional Lorentz invariance and this theory is not useful for our immediate purpose, which is to understand physics in our part of the universe. It is an interesting model, however, and we will return to this point briefly in section 9.9.

We have found above that a flat world-sheet CFT can be coupled to a curved metric in a Weyl-invariant way if and only if the total central charge c is zero. There is a paradox here, however, because the same argument applies to the antiholomorphic central charge $\tilde c$, and we have seen from the example of the bc CFT that c need not equal $\tilde c$. The point is that we have implicitly assumed world-sheet diff invariance in eqs. (3.4.11) and (3.4.19). If $c \neq \tilde c$ there is no diff-invariant expression that agrees with both the $O(h_{zz}^2)$ and $O(h_{\bar z \bar z}^2)$ calculations. The CFT then *cannot* be coupled to a curved metric in a diff-invariant way; there is a *diff* or *gravitational* anomaly. The paradox shows that $c = \tilde c$ is necessary for a CFT to be free of gravitational anomalies, and it can also be shown that this is sufficient.

Inclusion of boundaries

Let us first return to eq. (3.4.8) and consider the more general possibility

$$T^a{}_a = a_1 R + a_2 . \tag{3.4.24}$$

This is allowed by diff invariance but would be nonvanishing in the flat limit. Earlier we set a_2 to zero by the choice (2.3.15b) of T_{ab}, explained in the footnote below that equation. Equivalently, the existence of more than one conserved T_{ab} on the flat world-sheet means that there is some freedom in how we couple to a curved metric. In particular, we can add to the action

$$S_{ct} = b \int d^2\sigma \, g^{1/2} , \tag{3.4.25}$$

which is a two-dimensional cosmological constant. This is not Weyl-invariant. It adds $2\pi b g_{ab}$ to T_{ab} and so the trace (3.4.24) becomes

$$T^a{}_a = a_1 R + (a_2 + 4\pi b) . \tag{3.4.26}$$

We can therefore make the second term vanish by setting $b = -a_2/4\pi$; this is what we have done implicitly in eq. (2.3.15b). The separate values of a_2 and b depend on definitions and are uninteresting.

No such local counterterm can remove a_1. For example, $\int d^2\sigma\, g^{1/2} R$ is Weyl-invariant, as we have already noted. So if a_1 is nonzero, there is a real anomaly in the Weyl symmetry.

We now consider the effects of boundaries. Including boundary terms, the most general possible variation is[3]

$$
\delta_{\mathrm{W}} \ln \langle \ldots \rangle_g = -\frac{1}{2\pi} \int_M d^2\sigma\, g^{1/2} (a_1 R + a_2) \delta\omega
$$
$$
- \frac{1}{2\pi} \int_{\partial M} ds\, (a_3 + a_4 k + a_5 n^a \partial_a) \delta\omega \, , \qquad (3.4.27)
$$

where again $k = -t^a n_b \nabla_a t^b$ is the geodesic curvature of the boundary. The possible counterterms are

$$
S_{\mathrm{ct}} = \int_M d^2\sigma\, g^{1/2} b_1 + \int_{\partial M} ds\, (b_2 + b_3 k) \, . \qquad (3.4.28)
$$

The b_3 term is in addition to the geodesic curvature term in the Euler number, in eq. (3.2.3b). The Weyl variation of S_{ct}, being careful to include the dependence of ds and of the unit vectors t and n on the metric, is

$$
\delta_{\mathrm{W}} S_{\mathrm{ct}} = 2 \int_M d^2\sigma\, g^{1/2} b_1 \delta\omega + \int_{\partial M} ds\, (b_2 + b_3 n^a \partial_a) \delta\omega \, . \qquad (3.4.29)
$$

We see that b_1, b_2, and b_3 can be chosen to set a_2, a_3, and a_5 to zero, leaving a_1 and a_4 as potential anomalies. There is one more tool at our disposal, the *Wess–Zumino consistency condition*. Take the second functional variation

$$
\delta_{\mathrm{W}_1} (\delta_{\mathrm{W}_2} \ln \langle \ldots \rangle_g)
$$
$$
= \frac{a_1}{\pi} \int_M d^2\sigma\, g^{1/2}\, \delta\omega_2 \nabla^2 \delta\omega_1 - \frac{a_4}{2\pi} \int_{\partial M} ds\, \delta\omega_2\, n^a \partial_a \delta\omega_1
$$
$$
= -\frac{a_1}{\pi} \int_M d^2\sigma\, g^{1/2}\, \partial_a \delta\omega_2 \partial^a \delta\omega_1 + \frac{2a_1 - a_4}{2\pi} \int_{\partial M} ds\, n^a \delta\omega_2\, \partial_a \delta\omega_1 \, .
$$
$$
(3.4.30)
$$

This is a second (functional) derivative and so by definition must be symmetric in $\delta\omega_1$ and $\delta\omega_2$. The first term is symmetric but the second one is not, and so $a_4 = 2a_1$. It follows that boundaries, with open string boundary conditions, give rise to no new Weyl anomalies. Weyl invariance holds in the quantum theory if and only if $a_1 = -c/12$ vanishes. This is determined by the local physics in the *interior* of the world-sheet.

The Wess–Zumino consistency condition is a very powerful restriction on the form of possible anomalies. A second application enables us to show that the central charge is a constant. This followed above from

---

[3] We require all terms to be independent of the fields X^μ, as can be shown to hold in free field theory. When we consider the interacting case in section 3.7, the Weyl anomaly will take a more general form.

Lorentz invariance and dimensional analysis, but in more general CFTs one might imagine that

$$T^a{}_a(\sigma) = -\frac{\mathscr{C}(\sigma)}{12} R(\sigma) \,, \tag{3.4.31}$$

where $\mathscr{C}(\sigma)$ is some local operator. In particular, the trace would still vanish on a flat world-sheet. Repeating the calculation (3.4.30) with $-\mathscr{C}(\sigma)/12$ in place of a_1, there is an additional asymmetric term proportional to $\partial_a \mathscr{C}(\sigma)$ from the integration by parts. The consistency condition thus implies that expectation values involving $\mathscr{C}(\sigma)$ are position-independent, implying that $\mathscr{C}(\sigma)$ itself is just a numerical constant c.

3.5 Scattering amplitudes

The idea of a sum over all world-sheets bounded by given initial and final curves seems like a natural one. However, it is difficult to define this in a way that is consistent with the local world-sheet symmetries, and the resulting amplitudes are rather complicated. There is one special case where the amplitudes simplify. This is the limit where the string sources are taken to infinity. This corresponds to a scattering amplitude, an S-matrix element, with the incoming and outgoing strings specified. For most of this book we will confine ourselves to the S-matrix and similar 'on-shell' questions. This is not completely satisfying, because one would like to be able to discuss the state of the system at finite times, not just asymptotic processes. For now we focus on the S-matrix because it is well understood, while it is not clear what is the correct way to think about string theory off-shell. We will return to this issue briefly in the next section, and again in chapter 9.

So we are now considering a process like that shown in figure 3.8(a), where the sources are pulled off to infinity. We will give a heuristic argument for how these sources should be represented in the path integral. Away from the scattering process the strings propagate freely. Each of the incoming and outgoing legs is a long cylinder, which can be described with a complex coordinate w,

$$-2\pi t \leq \text{Im}\, w \leq 0 \,, \quad w \cong w + 2\pi \,. \tag{3.5.1}$$

The lower end of the cylinder, $\text{Im}\, w = -2\pi t$, is the end with the external source, the upper end fits onto the rest of the world-sheet, and the circumference is the periodic $\text{Re}\, w$ direction. The limit corresponding to the scattering process is $t \to \infty$. It may seem that we are confusing a long distance in spacetime with a long cylinder in the world-sheet coordinate, but this will turn out to be correct: we will see later that propagation

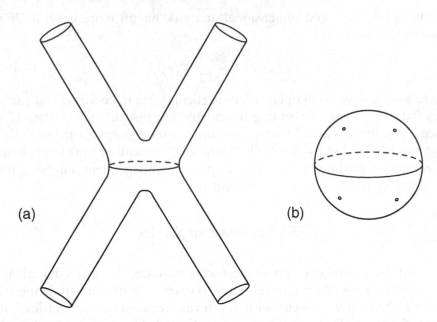

Fig. 3.8. (a) Scattering of four closed strings, with sources approaching $X^0 = \pm\infty$. (b) Conformally equivalent picture of the four-closed-string scattering: a sphere with small holes.

over long spacetime distances comes precisely from world-sheets where a cylinder is growing long in the sense above.

As we know from chapter 2, the cylinder has a conformally equivalent description in terms of the coordinate z,

$$z = \exp(-iw) , \quad \exp(-2\pi t) \leq |z| \leq 1 . \tag{3.5.2}$$

In this picture, the long cylinder is mapped into the unit disk, where the external string state is a tiny circle in the center. In the geometry of the internal metric, the process of figure 3.8(a) now looks like figure 3.8(b). In the limit $t \to \infty$, the tiny circles shrink to points and the world-sheet reduces to a sphere with a point-like insertion for each external state.

The same idea holds for external open string states. The long strips in the process of figure 3.9(a) can be described as the region

$$-2\pi t \leq \text{Im}\, w \leq 0 , \quad 0 \leq \text{Re}\, w \leq \pi , \tag{3.5.3}$$

where $\text{Im}\, w = -2\pi t$ is the source and $\text{Re}\, w = 0, \pi$ are endpoint boundaries. Under $z = -\exp(-iw)$, this maps into the intersection of the unit disk with the upper half-plane, with a tiny semi-circle cut out at the origin,

$$\exp(-2\pi t) \leq |z| \leq 1 , \quad \text{Im}\, z \geq 0 . \tag{3.5.4}$$

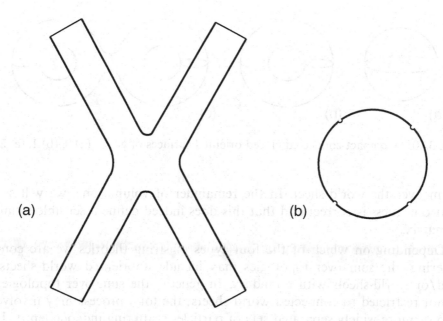

Fig. 3.9. (a) Scattering of four open strings, with sources approaching $X^0 = \pm\infty$. (b) Conformally equivalent picture: a disk with small dents.

The scattering process now looks like figure 3.9(b). In the limit $t \to \infty$, the sources shrink to points located on the (endpoint) boundary.

This is the state–operator mapping, which we have already seen. Each string source becomes a *local* disturbance on the world-sheet. To a given incoming or outgoing string, with D-momentum k^μ and internal state j, there corresponds a local *vertex operator* $\mathscr{V}_j(k)$ determined by the limiting process. Incoming and outgoing states are distinguished by the sign of k^0; for an incoming state $k^\mu = (E, \mathbf{k})$ and for an outgoing state $k^\mu = -(E, \mathbf{k})$. Figures 3.8 and 3.9 depict only the lowest order amplitudes, but the construction is clearly general: we may restrict attention to compact world-sheets, with no tubes going off to infinity but with a pointlike insertion in the interior for each external closed string and a pointlike insertion on the boundary for each external open string. An n-particle S-matrix element is then given by

$$S_{j_1\ldots j_n}(k_1, \ldots, k_n)$$

$$= \sum_{\substack{\text{compact} \\ \text{topologies}}} \int \frac{[dX\, dg]}{V_{\text{diff} \times \text{Weyl}}} \exp(-S_X - \lambda\chi) \prod_{i=1}^{n} \int d^2\sigma_i \, g(\sigma_i)^{1/2} \mathscr{V}_{j_i}(k_i, \sigma_i) \, .$$

$$(3.5.5)$$

To make the vertex operator insertions diff-invariant, we have integrated

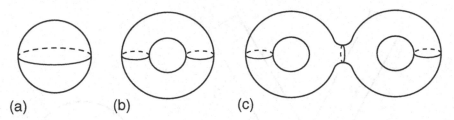

Fig. 3.10. Compact connected closed oriented surfaces of genus (a) 0, (b) 1, (c) 2.

them over the world-sheet. In the remainder of volume one, we will see that our guess is correct and that this does indeed define a sensible string S-matrix.

Depending on which of the four types of string theories we are considering, the sum over topologies may include unoriented world-sheets, and/or world-sheets with boundary. In general, the sum over topologies is not restricted to connected world-sheets; the total process may involve two or more widely separated sets of particles scattering independently. It is convenient to focus on the connected S-matrix, given by restricting the sum to connected surfaces.

To obtain the connected S-matrix, then, we must sum over all compact, connected topologies for the world-sheet. In two dimensions, the classification of these topologies is well known. Any compact, connected, oriented two-dimensional surface without boundary is topologically equivalent to a sphere with g handles; g is known as the *genus* of the surface. In figure 3.10 we show $g = 0$, 1, and 2. Boundaries can be added by cutting holes in a closed surface. Any compact, connected, oriented two-dimensional surface is topologically equivalent to a sphere with g handles and b holes. For example, $(g, b) = (0, 1)$ is a disk, $(0, 2)$ is an annulus, and $(0, 3)$ is a pair of pants.

To describe unoriented surfaces, it is useful to introduce the cross-cap: cut a hole in the surface and identify diametrically opposite points. In more detail, take a complex coordinate z, cut out a disk of slightly less than unit radius, and glue the opposite points together by defining pairs of points with $z' = -1/\bar{z}$ to be equivalent. The resulting surface is unoriented because the gluing is antiholomorphic. Also, unlike the case of a boundary, the cross-cap introduces no edge or other local feature: the edge introduced by the cutting is only the boundary of a coordinate patch. Any compact connected closed surface, oriented or unoriented, can be obtained by adding g handles and c cross-caps to the sphere; any compact connected surface can be obtained by adding g handles, b boundaries, and c cross-caps to the sphere. Actually, these descriptions are somewhat redundant: in either the case with boundary or the case without, one obtains every topology exactly once by restricting the sum to let only one of g and c be nonzero. For example, $(g, b, c) = (0, 0, 1)$ is the projective plane, $(g, b, c) = (0, 1, 1)$ is the Möbius strip, $(g, b, c) = (0, 0, 2)$ is the Klein bottle, and a torus with cross-cap can be obtained either as $(g, b, c) = (0, 0, 3)$ or as $(g, b, c) = (1, 0, 1)$, trading two cross-caps for a handle. The Euler number is given by

$$\chi = 2 - 2g - b - c . \tag{3.5.6}$$

The number of distinct topologies is very small compared to the number of distinct Feynman graphs at a given order in field theory. For example, in the closed oriented theory, there is exactly one topology at each order of perturbation theory. In a field theory, the number of graphs grows factorially with the order. The single string graph contains all the field theory graphs. In various limits, where handles become long compared to their circumferences, we can approximate the handles by lines and the string graph in that limit is approximated by the corresponding Feynman graph.

3.6   Vertex operators

Using the state–operator mapping, the vertex operator for the closed string tachyon is

$$V_0 = 2g_c \int d^2\sigma \, g^{1/2} \, e^{ik \cdot X}$$

$$\rightarrow g_c \int d^2z \; :e^{ik \cdot X}: \; . \tag{3.6.1}$$

We have now included a factor g_c in the mapping. This is the closed string coupling constant, from adding an extra string to the process: we use the normalization of the vertex operator as the definition of the coupling. In the second line we have gone to the flat world-sheet. The vertex operator must be (diff×Weyl)-invariant, so in particular it must be conformally invariant on the flat world-sheet. To offset the transformation of d^2z, the operator must be a tensor of weight $(1, 1)$. By a straightforward OPE calculation, $e^{ik \cdot X}$ is a tensor of weight $h = \tilde{h} = \alpha' k^2/4$, so the condition is

$$m^2 = -k^2 = -\frac{4}{\alpha'} . \tag{3.6.2}$$

This is precisely the mass found in the light-cone quantization.

Similarly, the tensor states at the first excited level of the closed string have the flat world-sheet vertex operators

$$\frac{2g_c}{\alpha'} \int d^2z \; :\partial X^\mu \bar\partial X^\nu e^{ik\cdot X}: \; . \tag{3.6.3}$$

The normalization relative to the tachyon vertex operator follows from the state–operator mapping, and we will see in chapter 6 that the same relative normalization is obtained from unitarity of the S-matrix. The weight is

$$h = \tilde h = 1 + \frac{\alpha' k^2}{4} \; , \tag{3.6.4}$$

so these are massless, again agreeing with the light-cone quantization. There are further conditions in order that this be a tensor, as we will work out below.

Vertex operators in the Polyakov approach

We have in the state–operator mapping a systematic means of writing the flat world-sheet vertex operator for any state. Since the world-sheet can always be made locally flat, this is in principle all we need. For perspective, however, it is useful to study the curved world-sheet vertex operators in the Polyakov formalism. We will do this in the remainder of the section. The method we use will be somewhat clumsy and not as systematic as the state–operator mapping, but some of the results are useful. The reader need not work out all the details.

Any operator to be included in the Polyakov path integral (3.5.5) must respect the local diff×Weyl symmetry of the theory. We have not been specific in the first line of eq. (3.6.1) as to how the operator is to be defined on the curved world-sheet. As in the earlier discussion of the Weyl anomaly, it is convenient to use a method that preserves diff invariance automatically, and to check the Weyl transformation by hand. For most purposes *dimensional regularization* is used in the literature. However, we will not have extensive need for this and so will not introduce it; rather, we will generalize the normal ordering introduced earlier. Define a renormalized operator $[\mathscr{F}]_r$ by

$$[\mathscr{F}]_r = \exp\left(\frac{1}{2}\int d^2\sigma d^2\sigma' \Delta(\sigma,\sigma')\frac{\delta}{\delta X^\mu(\sigma)}\frac{\delta}{\delta X_\mu(\sigma')}\right)\mathscr{F} \; . \tag{3.6.5}$$

Here

$$\Delta(\sigma,\sigma') = \frac{\alpha'}{2}\ln d^2(\sigma,\sigma') \; , \tag{3.6.6}$$

where $d(\sigma,\sigma')$ is the geodesic distance between points σ and σ'. As with normal ordering, the expression (3.6.5) instructs us to sum over all ways of contracting pairs in $\mathscr{F}$ using $\Delta(\sigma, \sigma')$. On the flat world-sheet, $d^2(\sigma, \sigma') = |z - z'|^2$ and this reduces to conformal normal ordering, which we have studied in detail. On a curved world-sheet it cancels the singular part of the self-contractions of fields in $\mathscr{F}$. Diff invariance is obvious, but the contraction depends on the metric and so introduces the Weyl dependence

$$\delta_W [\mathscr{F}]_r = [\delta_W \mathscr{F}]_r + \frac{1}{2} \int d^2\sigma d^2\sigma' \delta_W \Delta(\sigma, \sigma') \frac{\delta}{\delta X^\mu(\sigma)} \frac{\delta}{\delta X_\mu(\sigma')} [\mathscr{F}]_r \,,$$

(3.6.7)

the first term being the explicit Weyl dependence in the operator.

A vertex operator for a state of momentum k^μ must transform under translations in the same way as the state, and so will be of the form $e^{ik \cdot X}$ times derivatives of X^μ. Operators with different numbers of derivatives will not mix under Weyl transformations, so we can start by considering zero derivatives, the operator (3.6.1). The Weyl variation comes from the explicit factor of $g^{1/2}$ as well as from the renormalization as given in eq. (3.6.7),

$$\delta_W V_0 = 2g_c \int d^2\sigma \, g^{1/2} \left(2\delta\omega(\sigma) - \frac{k^2}{2} \delta_W \Delta(\sigma, \sigma) \right) [e^{ik \cdot X(\sigma)}]_r \,.$$

(3.6.8)

At short distance,

$$d^2(\sigma, \sigma') \approx (\sigma - \sigma')^2 \exp(2\omega(\sigma))$$

(3.6.9)

and so

$$\Delta(\sigma, \sigma') \approx \alpha'\omega(\sigma) + \frac{\alpha'}{2} \ln(\sigma - \sigma')^2 \,.$$

(3.6.10)

The Weyl variation is nonsingular in the limit $\sigma' \to \sigma$,

$$\delta_W \Delta(\sigma, \sigma) = \alpha' \delta\omega(\sigma) \,.$$

(3.6.11)

The condition for the Weyl variation (3.6.8) to vanish is then $k^2 = 4/\alpha'$, as already deduced above.

Off-shell amplitudes?

Observe that a naive attempt to define off-shell amplitudes, by taking k^μ off the mass shell, is inconsistent with the local world-sheet symmetries. In position space this means that a local probe of the string world-sheet, given by inserting

$$\delta^D(X(\sigma) - x_0) = \int \frac{d^D k}{(2\pi)^D} \exp[ik \cdot (X(\sigma) - x_0)]$$

(3.6.12)

into the path integral, is inconsistent because it involves all momenta.

From several points of view this is not surprising. First, the argument given above that we could use pointlike sources (vertex operators) for string states required a limiting process that essentially restricts us to on-shell questions. Second, string theory contains gravity, and in general relativity simple off-shell amplitudes do not exist. This is because we have to specify the location of the probe, but the coordinates are unphysical and coordinate-invariant off-shell quantities are much more complicated. Third, in string theory we do not have the freedom to introduce additional fields to measure local observables (analogous to the way electroweak processes are used to probe strongly interacting systems): we must use the strings themselves, or other objects that we will discuss (D-branes and solitons) that are intrinsic to the theory.

One can define off-shell amplitudes, at least in perturbation theory, if one fixes a gauge (light-cone gauge being simplest for this) and uses stringy sources. The S-matrix formula (3.5.5) can then be derived in a way more akin to the reduction formula from quantum field theory, where it is possible to define finite-time transition amplitudes and then take the infinite-time limit. Incidentally, the reader who is trying to relate the present discussion to field theory should note that the path integral expression (3.5.5) is not analogous to a Green's function but will in fact have the properties of an S-matrix element. The analogous object in field theory is the Green's function with its external propagators amputated and its external momenta restricted to the mass shell.

The discussion of the local probe (3.6.12) also shows the problem with a contact interaction between strings. The simplest form for such an interaction would be

$$\int_M d^2\sigma \, g(\sigma)^{1/2} \int_M d^2\sigma' \, g(\sigma')^{1/2} \, \delta^D(X(\sigma) - X(\sigma')) \,, \qquad (3.6.13)$$

which is nonvanishing whenever the world-sheet self-intersects. However, the delta function involves all momenta, as in eq. (3.6.12), so this is not Weyl-invariant. The structure of string theory is quite rigid: strings can couple to strings only in the way we have described at the beginning of this chapter.

Massless closed string vertex operators

It is interesting, if a little messy, to carry the Polyakov treatment of vertex operators one level further. There is no way to make a world-sheet scalar with exactly one derivative, so the next case is two derivatives. The diff-invariant possibilities are

$$V_1 = \frac{g_c}{\alpha'} \int d^2\sigma \, g^{1/2} \Big\{ (g^{ab} s_{\mu\nu} + i\epsilon^{ab} a_{\mu\nu}) [\, \partial_a X^\mu \partial_b X^\nu e^{ik\cdot X} \,]_{\rm r}$$

$$+ \alpha' \phi R \, [\, e^{ik\cdot X} \,]_{\rm r} \Big\} \,, \qquad (3.6.14)$$

where $s_{\mu\nu}$, $a_{\mu\nu}$, and ϕ are respectively a symmetric matrix, an antisymmetric matrix, and a constant. The antisymmetric *tensor* ϵ^{ab} is normalized $g^{1/2}\epsilon^{12} = 1$, or $g^{1/2}\epsilon^{z\bar{z}} = -i$. The i accompanying the antisymmetric tensor in the vertex operator can be understood as arising from the Euclidean continuation, because this term necessarily has an odd number of time derivatives (one) . This result applies generally to Euclidean actions and vertex operators.

To extend the analysis of Weyl invariance we need to work out the Weyl dependence (3.6.11) of the geodesic distance to higher order. One finds

$$\partial_a\delta_W\Delta(\sigma,\sigma')\Big|_{\sigma'=\sigma} = \frac{1}{2}\alpha'\partial_a\delta\omega(\sigma) , \tag{3.6.15a}$$

$$\partial_a\partial_b'\delta_W\Delta(\sigma,\sigma')\Big|_{\sigma'=\sigma} = \frac{1+\gamma}{2}\alpha'\nabla_a\partial_b\delta\omega(\sigma) , \tag{3.6.15b}$$

$$\nabla_a\partial_b\delta_W\Delta(\sigma,\sigma')\Big|_{\sigma'=\sigma} = -\frac{\gamma}{2}\alpha'\nabla_a\partial_b\delta\omega(\sigma) . \tag{3.6.15c}$$

Here, $\gamma = -\frac{2}{3}$, but we have left it as a parameter for later reference; the third equation can be obtained from the second and the gradient of the first. Using eq. (3.3.5) for the variation of the curvature and eqs. (3.6.7) and (3.6.15) for the variation of the renormalization gives

$$\delta_W V_1 = \frac{g_c}{2}\int d^2\sigma\, g^{1/2}\delta\omega\Big\{(g^{ab}S_{\mu\nu} + i\epsilon^{ab}A_{\mu\nu})[\partial_a X^\mu\partial_b X^\nu e^{ik\cdot X}]_r$$
$$+\alpha'FR[e^{ik\cdot X}]_r\Big\} , \tag{3.6.16}$$

where

$$S_{\mu\nu} = -k^2 s_{\mu\nu} + k_\nu k^\omega s_{\mu\omega} + k_\mu k^\omega s_{\nu\omega} - (1+\gamma)k_\mu k_\nu s^\omega{}_\omega + 4k_\mu k_\nu\phi , \tag{3.6.17a}$$

$$A_{\mu\nu} = -k^2 a_{\mu\nu} + k_\nu k^\omega a_{\mu\omega} - k_\mu k^\omega a_{\nu\omega} , \tag{3.6.17b}$$

$$F = (\gamma-1)k^2\phi + \frac{1}{2}\gamma k^\mu k^\nu s_{\mu\nu} - \frac{1}{4}\gamma(1+\gamma)k^2 s^\nu{}_\nu . \tag{3.6.17c}$$

In deriving the Weyl variation (3.6.16) we have integrated by parts and also used the relation

$$[\nabla^2 X^\mu e^{ik\cdot X}]_r = i\frac{\alpha'\gamma}{4}k^\mu R[e^{ik\cdot X}]_r . \tag{3.6.18}$$

The left-hand side would vanish by the naive equation of motion, but here the equation of motion multiplies another operator at the same point. The general principle that holds in this case is that the operator need not vanish but it is not independent — it can be expanded in terms of the other local operators in the theory. In general the coefficients of these operators depend on the renormalization scheme; the result (3.6.18) can be obtained by taking the Weyl variation of each side.

Since $S_{\mu\nu}$, $A_{\mu\nu}$, and F multiply independent operators, the condition for Weyl invariance is

$$S_{\mu\nu} = A_{\mu\nu} = F = 0 . \qquad (3.6.19)$$

In order to get an accurate count of the number of operators, we must note that not all V_1 of the form (3.6.14) are independent. Rather, under

$$s_{\mu\nu} \rightarrow s_{\mu\nu} + \xi_\mu k_\nu + k_\mu \xi_\nu , \qquad (3.6.20a)$$

$$a_{\mu\nu} \rightarrow a_{\mu\nu} + \zeta_\mu k_\nu - k_\mu \zeta_\nu , \qquad (3.6.20b)$$

$$\phi \rightarrow \phi + \frac{\gamma}{2} k \cdot \xi , \qquad (3.6.20c)$$

the change in V_1 integrates to zero using the equation of motion (3.6.18). Choose now a vector n^μ satisfying $n \cdot k = 1$ and $n^2 = 0$. A complete set of independent operators is obtained by restricting to

$$n^\mu s_{\mu\nu} = n^\mu a_{\mu\nu} = 0 . \qquad (3.6.21)$$

This is $2D$ equations for the $2D$ parameters ξ_μ and ζ_μ; one can show that one equation and one parameter are trivial, but that eqs. (3.6.21) are otherwise nondegenerate and so define an independent set of vertex operators.

By using condition (3.6.21) and solving first $S_{\mu\nu} n^\mu n^\nu = 0$, then $S_{\mu\nu} n^\mu = A_{\mu\nu} n^\mu = 0$, and finally the full $S_{\mu\nu} = A_{\mu\nu} = F = 0$, one finds that

$$k^2 = 0 , \qquad (3.6.22a)$$

$$k^\nu s_{\mu\nu} = k^\mu a_{\mu\nu} = 0 , \qquad (3.6.22b)$$

$$\phi = \frac{1+\gamma}{4} s^\mu_{\ \mu} . \qquad (3.6.22c)$$

Again we find a mass-shell condition (3.6.22a), this time corresponding to the first excited states of the closed string. There is also the condition (3.6.22b) that the polarization be transverse to the momentum, as required for a physical polarization of a massless tensor field. In the frame in which

$$k^\mu = (1, 1, 0, 0, \ldots, 0) , \quad n^\mu = \frac{1}{2}(-1, 1, 0, 0, \ldots, 0) , \qquad (3.6.23)$$

conditions (3.6.21) and (3.6.22b) imply that all 0- and 1-components vanish, while eq. (3.6.22c) fixes ϕ. This leaves precisely $(D - 2)^2$ independent operators, the same number of massless closed string states as found in the light-cone gauge.

The condition $n^\mu s_{\mu\nu} = n^\mu a_{\mu\nu} = 0$ is needed to remove zero-norm light-like polarizations. However, the introduction of the vector n^μ spoils manifest Lorentz invariance, and the theory is Lorentz-invariant only because different choices of n^μ are equivalent due to (3.6.20). The equivalence relations (3.6.20) are therefore essential in order for the theory to be

Lorentz-invariant with a positive Hilbert space norm. These relations are the signatures of local spacetime symmetries. In particular, ξ_μ is an infinitesimal *spacetime* coordinate transformation, and ζ_μ is the local spacetime symmetry of the antisymmetric tensor field. We now see in the interacting theory the existence of local spacetime symmetries, which we argued in section 1.4 had to be present because of the existence of massless spin-1 and spin-2 fields.

There is a technical point we should mention. Different renormalization schemes assign different names to the renormalized operators. The most common scheme is dimensional regularization; the operators in this scheme are related to those in the scheme above by

$$[e^{ik\cdot X}]_{\mathrm{DR}} = [e^{ik\cdot X}]_{\mathrm{r}}\,, \tag{3.6.24a}$$

$$[\partial_a X^\mu e^{ik\cdot X}]_{\mathrm{DR}} = [\partial_a X^\mu e^{ik\cdot X}]_{\mathrm{r}}\,, \tag{3.6.24b}$$

$$[\partial_a X^\mu \partial_b X^\nu e^{ik\cdot X}]_{\mathrm{DR}} = [\partial_a X^\mu \partial_b X^\nu e^{ik\cdot X}]_{\mathrm{r}} - \frac{\alpha'}{12} g_{ab} \eta^{\mu\nu} R[e^{ik\cdot X}]_{\mathrm{r}}\,, \tag{3.6.24c}$$

$$[\nabla_a \partial_b X^\mu e^{ik\cdot X}]_{\mathrm{DR}} = [\nabla_a \partial_b X^\mu e^{ik\cdot X}]_{\mathrm{r}} + i\frac{\alpha'}{12} g_{ab} k^\mu R[e^{ik\cdot X}]_{\mathrm{r}}\,. \tag{3.6.24d}$$

Comparing with eq. (3.6.18), we see that in dimensional regularization $\gamma = 0$, so the equations of motion hold inside the regulated operator. This is a convenience — many equations simplify. In particular, in the spacetime gauge transformation (3.6.20), ϕ is invariant.

Open string vertex operators

The extension to the open string is straightforward. We leave the details for the reader as exercise 3.9, giving only some results. The tachyon vertex operator is

$$g_0 \int_{\partial M} ds \, [e^{ik\cdot X}]_{\mathrm{r}} \tag{3.6.25}$$

and is Weyl-invariant for $k^2 = 1/\alpha'$. The photon vertex operator is

$$-i\frac{g_0}{(2\alpha')^{1/2}} e_\mu \int_{\partial M} ds \, [\dot{X}^\mu e^{ik\cdot X}]_{\mathrm{r}}\,, \tag{3.6.26}$$

where the normalization relative to the tachyon is obtained either from the state–operator mapping or unitarity, the i is as discussed for the closed string vertex operator (3.6.14), and the sign is a matter of convention. This operator is Weyl-invariant if $k^2 = 0$ and $k \cdot e = 0$. There is an equivalence $e_\mu \cong e_\mu + \lambda k_\mu$, which is a spacetime gauge transformation. This leaves $D - 2$ transverse polarizations.

3.7 Strings in curved spacetime

Now we are ready to consider an important new issue, strings moving in a curved spacetime. Let us recall first that the point-particle action has a natural extension to curved spacetime. Replacing the flat metric $\eta_{\mu\nu}$ with a general metric $G_{\mu\nu}(x)$ gives the action

$$S_{\rm pp} = \frac{1}{2} \int d\tau \left(\eta^{-1} G_{\mu\nu}(X)\dot{X}^\mu \dot{X}^\nu - \eta m^2 \right) . \tag{3.7.1}$$

After eliminating η this becomes the invariant proper time along the particle world-line. Its variation is well known to give the geodesic equation, representing the motion of a particle in a gravitational field.

Making the same replacement in the Polyakov action gives

$$S_\sigma = \frac{1}{4\pi\alpha'} \int_M d^2\sigma\, g^{1/2} g^{ab} G_{\mu\nu}(X)\partial_a X^\mu \partial_b X^\nu . \tag{3.7.2}$$

This is a natural guess, but the reader might make the following objection. We have learned that the graviton itself is one state of the string. A curved spacetime is, roughly speaking, a coherent background of gravitons, and therefore in string theory it is a coherent state of strings. Merely putting a curved metric in the action (3.7.2) doesn't seem stringy enough. To see why it is justified, consider a spacetime that is close to flat,

$$G_{\mu\nu}(X) = \eta_{\mu\nu} + \chi_{\mu\nu}(X) \tag{3.7.3}$$

with $\chi_{\mu\nu}$ small. The integrand in the world-sheet path integral is

$$\exp(-S_\sigma) = \exp(-S_{\rm P})\left(1 - \frac{1}{4\pi\alpha'} \int_M d^2\sigma\, g^{1/2} g^{ab} \chi_{\mu\nu}(X)\partial_a X^\mu \partial_b X^\nu + \dots \right) . \tag{3.7.4}$$

The term of order χ is precisely the vertex operator for the graviton state of the string, eq. (3.6.14) with

$$\chi_{\mu\nu}(X) = -4\pi g_c e^{ik \cdot X} s_{\mu\nu}. \tag{3.7.5}$$

So indeed, inserting the curved metric does seem to make contact with what we already know. The action (3.7.2) can be thought of as describing a coherent state of gravitons by exponentiating the graviton vertex operator.

This suggests a natural generalization: include backgrounds of the other massless string states as well. From the form of the vertex operators, eq. (3.6.14), one has

$$S_\sigma = \frac{1}{4\pi\alpha'} \int_M d^2\sigma\, g^{1/2} \left[\left(g^{ab} G_{\mu\nu}(X) + i\epsilon^{ab} B_{\mu\nu}(X) \right) \partial_a X^\mu \partial_b X^\nu + \alpha' R\Phi(X) \right]. \tag{3.7.6}$$

The field $B_{\mu\nu}(X)$ is the antisymmetric tensor, and the dilaton involves both Φ and the diagonal part of $G_{\mu\nu}$, as implied by the equation of motion (3.6.22c).

As a check, let us see that the spacetime gauge invariances are respected. Under a change of variables in the path integral corresponding to a field redefinition $X'^{\mu}(X)$, the action (3.7.6) is invariant with $G_{\mu\nu}$ and $B_{\mu\nu}$ transforming as tensors and Φ as a scalar. From the spacetime point of view, this is a coordinate transformation. The action is also invariant under

$$\delta B_{\mu\nu}(X) = \partial_\mu \zeta_\nu(X) - \partial_\nu \zeta_\mu(X) , \qquad (3.7.7)$$

which adds a total derivative to the Lagrangian density. This is a generalization of the electromagnetic gauge transformation to a potential with two antisymmetric indices. The three-index field strength

$$H_{\omega\mu\nu} = \partial_\omega B_{\mu\nu} + \partial_\mu B_{\nu\omega} + \partial_\nu B_{\omega\mu} \qquad (3.7.8)$$

is invariant. There is a similar generalization to an antisymmetric n-tensor potential, which plays an important role in the superstring and will be developed in volume two.

The theory thus depends only on gauge-invariant objects built out of the metric and other fields. One can think of the fields X^μ as being coordinates on a manifold. This is called the *target space*, because the X^μ define an embedding,

$$\text{world-sheet} \rightarrow \text{target} . \qquad (3.7.9)$$

In string theory, the target space is spacetime itself. A field theory such as (3.7.2), in which the kinetic term is field-dependent and so field space is effectively a curved manifold, is known for historic reasons as a *nonlinear sigma model*. Nonlinear sigma models have had many applications in particle physics and quantum field theory. For example, the neutral and charged pion fields can be regarded to good approximation as coordinates on the group manifold $SU(2)$.

The nonlinear sigma model action is no longer quadratic in X^μ, so the path integral is now an *interacting* two-dimensional quantum field theory. Expand the path integral around the classical solution $X^\mu(\sigma) = x_0^\mu$ for any chosen point x_0^μ. With $X^\mu(\sigma) = x_0^\mu + Y^\mu(\sigma)$,

$$G_{\mu\nu}(X)\partial_a X^\mu \partial_b X^\nu = \Big[G_{\mu\nu}(x_0) + G_{\mu\nu,\omega}(x_0)Y^\omega$$
$$+ \tfrac{1}{2} G_{\mu\nu,\omega\rho}(x_0)Y^\omega Y^\rho + \dots \Big]\partial_a Y^\mu \partial_b Y^\nu . \qquad (3.7.10)$$

The first term in this expansion is a quadratic kinetic term for the field Y^μ. The next term is a cubic interaction, and so forth. The coupling constants in the expansion, $G_{\mu\nu,\omega}(x_0)$ and so on, involve derivatives of the metric at x_0. In a target space with characteristic radius of curvature R_c, derivatives of the metric are of order R_c^{-1}. The effective dimensionless coupling in the theory is therefore $\alpha'^{1/2}R_c^{-1}$. If the radius of curvature R_c is much greater than the characteristic length scale of the string, then this coupling is small and perturbation theory in the two-dimensional field theory is a useful tool. Notice that in this same regime we can employ another tool as well. Since the characteristic length scale is long compared to the string, we can ignore the internal structure of the string and use *low energy effective field theory*. String theory enters in determining the effective low energy action for the fields, as we will soon see. We have also implicitly used $\alpha'^{1/2}R_c^{-1} \ll 1$ when restricting our attention to *massless* backgrounds: when wavelengths are long compared to the string scale, massive string states are not created.

The nonlinear sigma model (3.7.2) is a renormalizable theory: the fields Y^μ have dimension zero and so the interactions all have dimension two. Nevertheless the couplings, which are the coefficients in the expansion of the metric, are infinite in number (unless some symmetry restricts the form of the metric). In fact it is more useful to think of the metric function itself, rather than its expansion coefficients, as defining the couplings of the theory.

Weyl invariance

We have emphasized that Weyl invariance is essential to the consistency of string theory. The action (3.7.6) will define a consistent string theory only if the two-dimensional quantum field theory is Weyl-invariant. This action happens to be the most general action that is classically invariant under a *rigid* Weyl transformation, with $\delta\omega$ independent of σ. This is easy to see: under a rigid Weyl transformation, the scaling of a term with n derivatives will be proportional to $2 - n$. Of course, it is necessary also to consider local Weyl transformations, under which the $G_{\mu\nu}$ and $B_{\mu\nu}$ terms in the action (3.7.6) are classically invariant while the Φ term is not, and to include the quantum contributions to the Weyl transformation properties. In the limit that $B_{\mu\nu}$ and Φ are small, and $G_{\mu\nu}$ is close to $\eta_{\mu\nu}$, we can use the results of the previous section to find the Weyl transformation. Write $S_\sigma = S_P - V_1 + \ldots$, with S_P the flat-space action and V_1 the vertex operator (3.6.14). Then

$$G_{\mu\nu}(X) = \eta_{\mu\nu} - 4\pi g_c s_{\mu\nu} e^{ik\cdot X} , \tag{3.7.11a}$$

$$B_{\mu\nu}(X) = -4\pi g_c a_{\mu\nu} e^{ik\cdot X} , \tag{3.7.11b}$$

$$\Phi(X) = -4\pi g_c \phi e^{ik\cdot X} . \tag{3.7.11c}$$

We could of course take linear combinations of vertex operators with different momenta. To first order the Weyl variation is then given in eq. (3.6.16); we take for convenience a renormalization scheme in which $\gamma = 0$. Relating this Weyl transformation to T^a_a as in eq. (3.4.6) gives

$$T^a_a = -\frac{1}{2\alpha'}\beta^G_{\mu\nu}g^{ab}\partial_a X^\mu \partial_b X^\nu - \frac{i}{2\alpha'}\beta^B_{\mu\nu}\epsilon^{ab}\partial_a X^\mu \partial_b X^\nu - \frac{1}{2}\beta^\Phi R \,, \quad (3.7.12)$$

where, to linear order in $\chi_{\mu\nu}$, $B_{\mu\nu}$ and Φ,

$$\beta^G_{\mu\nu} \approx -\frac{\alpha'}{2}\left(\partial^2 \chi_{\mu\nu} - \partial_\nu \partial^\omega \chi_{\mu\omega} - \partial_\mu \partial^\omega \chi_{\omega\nu} + \partial_\mu \partial_\nu \chi^\omega_\omega \right) + 2\alpha'\partial_\mu \partial_\nu \Phi \,,$$
$$(3.7.13a)$$

$$\beta^B_{\mu\nu} \approx -\frac{\alpha'}{2}\partial^\omega H_{\omega\mu\nu} \,, \quad (3.7.13b)$$

$$\beta^\Phi \approx \frac{D-26}{6} - \frac{\alpha'}{2}\partial^2 \Phi \,. \quad (3.7.13c)$$

We have included in β^Φ the flat spacetime anomaly found in section 3.4, including the contribution proportional to 26 from the ghost fields. The symbol β is used for the coefficients in T^a_a because these are essentially the renormalization group beta functions, governing the dependence of the physics on world-sheet scale. We will discuss this connection further, though it does not come up in detail until chapter 15.

The Weyl anomalies (3.7.13) have further contributions from higher orders in the fields. For example, expanding the path integral to second order in the cubic interaction from the expansion (3.7.10) will give rise to divergences when the two interaction vertices approach one another. Their OPE includes a singularity of the form $|z|^{-2}\partial X^\mu \bar\partial X^\nu$ times the square of the derivative of $G_{\alpha\beta}$. Integrating d^2z produces a logarithmic divergence. Diff invariance requires that the integral be cut off in terms of the invariant distance, $|z|\exp(\omega) > a_0$. This introduces a dependence on the scale of the metric, $\ln z_{\min} = -\omega + \ln a_0$. The Weyl variation $\beta^G_{\mu\nu}$ thus gets a contribution proportional to $O(G_{\alpha\beta,\gamma})^2$ which combines with the linear second derivative term to form the spacetime Ricci tensor.

We will not work this out in detail, but quote the result keeping all terms with up to two spacetime derivatives

$$\beta^G_{\mu\nu} = \alpha' R_{\mu\nu} + 2\alpha'\nabla_\mu \nabla_\nu \Phi - \frac{\alpha'}{4}H_{\mu\lambda\omega}H_\nu{}^{\lambda\omega} + O(\alpha'^2) \,, \quad (3.7.14a)$$

$$\beta^B_{\mu\nu} = -\frac{\alpha'}{2}\nabla^\omega H_{\omega\mu\nu} + \alpha'\nabla^\omega \Phi H_{\omega\mu\nu} + O(\alpha'^2) \,, \quad (3.7.14b)$$

$$\beta^\Phi = \frac{D-26}{6} - \frac{\alpha'}{2}\nabla^2 \Phi + \alpha'\nabla_\omega \Phi \nabla^\omega \Phi - \frac{\alpha'}{24}H_{\mu\nu\lambda}H^{\mu\nu\lambda} + O(\alpha'^2)$$
$$(3.7.14c)$$

Several terms in (3.7.14) can be recognized from the linear approximation

(3.7.13), but now made covariant under a change of spacetime coordinates. The spacetime Ricci tensor is $R_{\mu\nu}$, distinguished from the world-sheet Ricci tensor R_{ab}. Terms with more derivatives are higher order in the world-sheet $\alpha'^{1/2} R_c^{-1}$ expansion. Incidentally, the most efficient method to carry the calculation of the Weyl anomaly to higher order is not the one we have described, but rather dimensional regularization.

The condition that the world-sheet theory be Weyl-invariant is thus

$$\beta_{\mu\nu}^G = \beta_{\mu\nu}^B = \beta^\Phi = 0 \ . \tag{3.7.15}$$

These are sensible-looking equations of motion. The equation $\beta_{\mu\nu}^G = 0$ resembles Einstein's equation with source terms from the antisymmetric tensor field and the dilaton. The equation $\beta_{\mu\nu}^B = 0$ is the antisymmetric tensor generalization of Maxwell's equation, determining the divergence of the field strength.

Backgrounds

One other qualitative feature of the field equations is readily understood: the field Φ always occurs differentiated, so there is an invariance under X^μ-independent shifts of Φ. This is because such a shift only changes the world-sheet action (3.7.6) by a term proportional to the Euler number, and so does not affect local properties like Weyl invariance. In particular, the background

$$G_{\mu\nu}(X) = \eta_{\mu\nu} \ , \quad B_{\mu\nu}(X) = 0 \ , \quad \Phi(X) = \Phi_0 \tag{3.7.16}$$

is exactly Weyl-invariant for any constant Φ_0. This is just the flat spacetime action (3.2.2), with

$$\lambda = \Phi_0 \ . \tag{3.7.17}$$

The value of λ determines the coupling strength between strings, but we now see that this does not mean that there are different string theories with different values of this parameter. Different values of λ correspond not to different *theories*, but to different *backgrounds* in a single theory. The other parameter appearing in the Nambu–Goto and Polyakov actions, the Regge slope α', is also not really a free parameter because it is dimensionful. It simply defines the unit of length, and can be absorbed in the definition of X^μ.

Changing the fields $G_{\mu\nu}$, $B_{\mu\nu}$, and Φ in the world-sheet action would seem, from the two-dimensional point of view, to give a new theory. From the point of view of the full string theory, one is merely looking at a different background — a different state — in the same theory. This is one of the features that makes string theory attractive. In the Standard Model, and in most attempts to unify the Standard Model within quantum field theory, there are many constants that are not determined by the theory; any values for these constants give a consistent theory. In string theory there are no such free parameters — the coupling constants depend on the state and are determined in principle by the dynamics. Of course, for now this only moves the difficulty elsewhere, because we do not understand the dynamics well enough to know the values of the background fields.

It is striking that Einstein's equation turns up in what seems to be a rather out-of-the-way place, as the condition for Weyl invariance of a two-dimensional field theory. String theory provides the physical connection between these two ideas. It is also notable that, because of the need for Weyl invariance, strings can propagate consistently only in a background that satisfies appropriate field equations. This parallels our earlier discovery that only on-shell vertex operators make sense.

Another observation: the condition $D = 26$ came from the R term in $T^a{}_a$. In the nonlinear sigma model this is generalized to $\beta^\Phi = 0$. We see in eq. (3.7.14c) that the leading term in β^Φ is proportional to $D - 26$ but that there are corrections involving the gradients of the fields. Evidently we can have other values of D if the fields are not constant. This is true, although we cannot really conclude it from eq. (3.7.14c) because this has been derived in the approximation $\alpha'^{1/2} R_c^{-1} \ll 1$ where the correction terms in β^Φ are small. In fact, exact solutions with $D \neq 26$ are known. For now we give one simple example, saving others for later. This is

$$G_{\mu\nu}(X) = \eta_{\mu\nu} \,, \quad B_{\mu\nu}(X) = 0 \,, \quad \Phi(X) = V_\mu X^\mu \,. \tag{3.7.18}$$

The beta functions (3.7.14) vanish if

$$V_\mu V^\mu = \frac{26 - D}{6\alpha'} \,. \tag{3.7.19}$$

This result is actually *exact*, because the fields (3.7.18) are constant or linear in X^μ, so the world-sheet path integral remains Gaussian. Varying g_{ab} to determine T_{ab} for this theory, one finds that it is none other than the linear dilaton CFT (hence the name), whose central charge $c = D + 6\alpha' V_\mu V^\mu$ is indeed 26 under the condition (3.7.19). Because Φ must have a large gradient in order to change D, this CFT does not describe our roughly static and homogeneous four-dimensional spacetime, but it may have cosmological applications. This CFT is also important because the cases $D = 1$ and $D = 2$ are simple enough for the string theory to be exactly solvable, as we will discuss in section 9.9.

The spacetime action

The field equations (3.7.15) can be derived from the spacetime action

$$S = \frac{1}{2\kappa_0^2} \int d^D x\, (-G)^{1/2} e^{-2\Phi} \left[-\frac{2(D-26)}{3\alpha'} + R - \frac{1}{12} H_{\mu\nu\lambda} H^{\mu\nu\lambda} \right.$$
$$\left. + 4\partial_\mu \Phi \partial^\mu \Phi + O(\alpha') \right] . \qquad (3.7.20)$$

The normalization constant κ_0 is not determined by the field equations and has no physical significance since it can be changed by a redefinition of Φ. One can verify that

$$\delta S = -\frac{1}{2\kappa_0^2 \alpha'} \int d^D x\, (-G)^{1/2} e^{-2\Phi} \left[\delta G_{\mu\nu} \beta^{G\mu\nu} + \delta B_{\mu\nu} \beta^{B\mu\nu} \right.$$
$$\left. + \left(2\delta\Phi - \tfrac{1}{2} G^{\mu\nu} \delta G_{\mu\nu} \right) (\beta^{G\omega}{}_\omega - 4\beta^\Phi) \right] . \qquad (3.7.21)$$

The action is written in bold as a reminder that this is the effective action governing the low energy spacetime fields, unlike the world-sheet actions that have appeared everywhere previously in this book.

It is often useful to make a field redefinition of the form

$$\tilde{G}_{\mu\nu}(x) = \exp(2\omega(x)) G_{\mu\nu}(x) , \qquad (3.7.22)$$

which is a spacetime Weyl transformation. The Ricci scalar constructed from $\tilde{G}_{\mu\nu}$ is

$$\tilde{R} = \exp(-2\omega) \left[R - 2(D-1)\nabla^2\omega - (D-2)(D-1)\partial_\mu\omega\partial^\mu\omega \right] . \qquad (3.7.23)$$

For the special case $D = 2$, this is the Weyl transformation (3.3.5). Let $\omega = 2(\Phi_0 - \Phi)/(D-2)$ and define

$$\tilde{\Phi} = \Phi - \Phi_0 , \qquad (3.7.24)$$

which has vanishing expectation value. The action becomes

$$S = \frac{1}{2\kappa^2} \int d^D X\, (-\tilde{G})^{1/2} \left[-\frac{2(D-26)}{3\alpha'} e^{4\tilde{\Phi}/(D-2)} + \tilde{R} \right.$$
$$\left. - \frac{1}{12} e^{-8\tilde{\Phi}/(D-2)} H_{\mu\nu\lambda} \tilde{H}^{\mu\nu\lambda} - \frac{4}{D-2} \partial_\mu\tilde{\Phi}\tilde{\partial}^\mu\tilde{\Phi} + O(\alpha') \right] , \qquad (3.7.25)$$

where tildes have been inserted as a reminder that indices here are raised with $\tilde{G}^{\mu\nu}$. In terms of $\tilde{G}_{\mu\nu}$, the gravitational Lagrangian density takes the standard Hilbert form $(-\tilde{G})^{1/2} \tilde{R}/2\kappa^2$. The constant $\kappa = \kappa_0 e^{\Phi_0}$ is the gravitational coupling, which in four-dimensional gravity has the value

$$\kappa = (8\pi G_{\mathrm{N}})^{1/2} = \frac{(8\pi)^{1/2}}{M_{\mathrm{P}}} = (2.43 \times 10^{18}\ \mathrm{GeV})^{-1} . \qquad (3.7.26)$$

Commonly, $G_{\mu\nu}$ is called the *sigma model metric* or *string metric*, and $\tilde{G}_{\mu\nu}$ the *Einstein metric*. Because of the force from dilaton exchange, there is no equivalence principle and so no way to single out a preferred definition of the metric. Looking ahead, there are higher order effects in string theory that can give the dilaton a mass and so a finite range — a good thing, because the equivalence principle does of course hold to good accuracy.

Note that the dilaton appears in the action (3.7.20) in an overall factor $e^{-2\Phi}$ and otherwise is always differentiated. This is consistent with the fact that adding an overall constant to the dilaton has no effect on the Weyl anomaly. From the spacetime point of view this multiplies the action by a constant, which does not change the equations of motion. It also reflects the fact that in quantum field theory it is possible to rescale the fields in such a way that the coupling g appears only as an overall factor of g^{-2}, which in string theory is $e^{-2\Phi}$. For example, the Yang–Mills action is usually introduced in the form

$$S = -\frac{1}{4} \int d^D x \, \mathrm{Tr} \left(F_{\mu\nu} F^{\mu\nu} \right) , \qquad (3.7.27a)$$

$$F_{\mu\nu} = \partial_\mu A_\nu - \partial_\nu A_\mu - ig[A_\mu, A_\nu] \qquad (3.7.27b)$$

(matrix notation). By defining

$$gA_\mu = A'_\mu , \quad gF_{\mu\nu} = F'_{\mu\nu} , \qquad (3.7.28)$$

g is removed from the field strength, covariant derivative, and gauge transformation and appears only in an overall g^{-2} in the action. In string theory this holds when the action is written in terms of the fields appearing on the string world-sheet. If we make a dilaton-dependent field redefinition as in going to the form (3.7.25) then it no longer holds.

When g appears only in the normalization of the action, every propagator contains a factor of g^2 and every interaction a factor of g^{-2}. It is not hard to show that an L-loop contribution to the effective action has exactly $L-1$ more propagators than vertices and so scales as $g^{2(L-1)}$. In string theory this is $e^{-\chi\Phi}$, as we would expect.

The discussion of backgrounds extends to open strings. We have seen that open string vertex operators are integrated along the boundary, so open string fields appear as boundary terms in the sigma model action.

Compactification and CFT

The four string theories studied thus far (with or without boundaries, oriented or unoriented) have a number of features in common. One good feature is the automatic inclusion of general relativity. There are also several bad features: the need for 26 dimensions, the existence of the tachyon, and the absence of fermions in the spectrum. The elimination of the tachyon and the introduction of fermions will have to wait until volume two. However, the presence of general relativity provides a natural way to account for the extra dimensions. In general relativity the geometry of spacetime is dynamical, not fixed. Flat spacetime is only one of many solutions to the field equations. There may well be solutions in which some dimensions are large and flat and others are small and highly curved. The metric would be

$$g_{MN} = \begin{bmatrix} \eta_{\mu\nu} & 0 \\ 0 & g_{mn}(x^p) \end{bmatrix}. \qquad (3.7.29)$$

We have divided the 26 coordinates $M, N = 0, \ldots, 25$ into four 'spacetime' coordinates $\mu, \nu = 0, \ldots, 3$ and 22 'internal' coordinates $m, n, p = 4, \ldots, 25$. The metric (3.7.29) is flat in the four spacetime dimensions and curved in the 22 internal dimensions. The internal dimensions are assumed to be compact. As one example, the background field equations (3.7.15) are satisfied to order α' if the dilaton field is constant, the antisymmetric tensor is zero, and the metric for the internal space is Ricci-flat, $R_{mn} = 0$.

In other words, spacetime is of the form $M^d \times K$, where M^d is d-dimensional Minkowski space and K is some $(26-d)$-dimensional compact Riemannian space.[4] In such a spacetime, the physics on length scales much longer than the size of K is the same as in a d-dimensional Minkowski space. One says that $26 - d$ dimensions have been *compactified,* though cosmologically it seems more appropriate to say that the large dimensions have decompactified. As we discussed briefly in the introduction, there are good reasons, independent of string theory, to consider the possibility that our spacetime might be of this form. The compact dimensions must be small enough not to have been observed. In fact, we will see in chapter 8, and in more detail in chapter 18, that the size of any extra dimensions must be within a few orders of magnitude of the Planck length, 1.6×10^{-33} cm.

We can also think about generalizing this idea further. The basic physical consistency conditions are few in number. We require Lorentz invariance, and we require that quantum mechanical probabilities be positive and conserved — to date, nothing in string theory seems to require that the standard quantum mechanical framework be modified, though it would be interesting to learn otherwise. Although these conditions are simple there is a tension between them, in gauge theory and gravitation as well as in string theory. In a manifestly covariant gauge there are negative norm timelike excitations that must decouple from physical amplitudes. In the light-cone gauge the inner product is positive but covariance is no longer manifest.

---

[4] We will generally use D to refer to the total dimension of a string theory and d when we may be talking about a subset of the world-sheet fields.

One necessary condition is therefore two-dimensional diff invariance, so that the unphysical oscillations are merely oscillations of the coordinate system. Weyl invariance is a more technical requirement: the extra degree of freedom that appears when Weyl invariance is lost is less intuitive than an oscillation of the coordinate system, but we have argued in section 3.4 that it is incompatible with Lorentz invariance.

We will also make an additional technical assumption: that the embedding time X^0 appears in the world-sheet action only in the term

$$- \frac{1}{4\pi\alpha'} \int_M d^2\sigma \, g^{1/2} g^{ab} \partial_a X^0 \partial_b X^0 \, . \tag{3.7.30}$$

For example, in the sigma model this means that the fields are static with $G_{0\mu} = \eta_{0\mu}$ and $B_{0\mu} = 0$. This assumption is not much of a restriction. The form (3.7.30) is sufficient if one is interested in static states. Time-dependent backgrounds are certainly of interest for cosmological and other reasons, but they can be analyzed using low energy effective field theory except in the extreme case that the time scale is of the order of the string scale. The reason for making the assumption (3.7.30) is to put the main problem, the wrong-sign nature of X^0, in a rather explicit form.

The requirement that the local invariances be maintained leads to the following proposal for a more general string theory: *Replace the 25 spatial X^μ fields with any unitary CFT having $c = \tilde{c} = 25$.* This ensures that the full two-dimensional theory, with X^0 and the ghosts, can be coupled to a curved metric in a (diff×Weyl)-invariant way. The unitary (positive inner product) condition is necessary because there is only enough local symmetry to remove the unphysical X^0 excitation; the world-sheet inner product is relevant here because it becomes the spacetime inner product in the one-particle sector. We will see in volume two that further generalization is possible, enlarging simultaneously the local symmetry group and the set of unphysical excitations.

If we are interested in a bosonic string theory whose low energy physics looks four-dimensional, we keep the free fields X^μ for $\mu = 0, 1, 2, 3$, and replace the remaining X^μ with a compact unitary CFT having $c = \tilde{c} = 22$. By compact, we mean that the spectrum is discrete, just as for a quantum field theory in a finite volume. There have been some efforts to construct all CFTs, or all CFTs having various additional properties. We will describe some of these in chapter 15. Incidentally, all of these theories have tachyons, the product of the unit state $|1\rangle$ from the internal CFT with the ground state $|0; k\rangle$ from the four-dimensional CFT.

How large a generalization have we just made? If the local symmetries are the only essential restriction, we are free to introduce fields on the world-sheet with various world-sheet and spacetime quantum numbers. This seems to take us far beyond the original theory of 26 X^μ fields. However, in two dimensions there are many equivalences between seemingly different quantum field theories, so that the situation is not clear.

It is generally assumed that all string theories based on different CFTs but the same world-sheet gauge symmetries and topology are different ground states of the same theory. This causes a semantic problem: we often refer to strings with different world-sheet Lagrangians as different *theories*, though they are most likely just different vacua in one theory. We will take this even further in chapter 14, seeing that *all* string theories appear to be vacua of a single theory.

As we develop the techniques of string theory in volume one, we will usually take the simplest case of flat 26-dimensional spacetime in explicit examples. However, many results will be developed in such a way as to apply to the very general string theory we have described above.

Exercises

3.1 (a) Evaluate the Euler number χ for a flat disk and for a disk with the metric of a hemisphere.
(b) Show that the Euler number of a sphere with b holes is $2 - b$. You may take any convenient metric on the sphere.
(c) Extend this to the result $2 - b - 2g$ for the sphere with b holes and g handles.

3.2 (a) Show that an n-index traceless symmetric tensor in two dimensions has exactly two independent components. The traceless condition means that the tensor vanishes when any pair of indices is contracted with g^{ab}.
(b) Find a differential operator P_n that takes n-index traceless symmetric tensors into $(n + 1)$-index traceless symmetric tensors.
(c) Find a differential operator P_n^T that does the reverse.
(d) For u an $(n + 1)$-index traceless symmetric tensor and v an n-index traceless symmetric tensor, show that with appropriate normalization $(u, P_n v) = (P_n^T u, v)$. The inner product of traceless symmetric tensors of any rank is

$$(t, t') = \int d^2\sigma \, g^{1/2} t \cdot t' \,,$$

the dot denoting contraction of all indices.

3.3 (a) Work out the form of the covariant derivative in conformal gauge. Show in particular that ∇_z reduces to ∂_z when acting on a tensor component with all $\bar{z}$ indices, and vice versa. Since raising or lowering reverses z and $\bar{z}$ indices, this means that it is always possible to work with simple derivatives by appropriately positioning the indices.
(b) Use this to give simple expressions for P_n and P_n^T.

3.4 To gain familiarity with the Faddeev–Popov procedure, it is useful to consider it in a more general and abstract context. This is described in the first page of section 4.2 (which is independent of section 4.1). Derive the general result (4.2.3).

3.5 (A continuation of the previous problem.) Consider two different gauge choices, $F^A(\phi) = 0$ and $G^A(\phi) = 0$. The Faddeev–Popov procedure shows that the gauge-fixed path integrals (4.2.3) are both equal to the original gauge-invariant path integral and so equal to each other. However, this passes through the somewhat heuristically defined gauge-invariant path integral. Adapt the Faddeev–Popov procedure to show directly the equivalence of the two gauge-fixed path integrals. This is what is relevant in practice, showing for example that manifestly covariant and manifestly unitary gauges define the same theory, which therefore has both properties. In chapter 4 we will show this equivalence by a more abstract method, BRST symmetry.

3.6 On dimensional grounds, a current that is conserved on a flat worldsheet has the possible anomaly

$$\nabla_a j^a = aR \ .$$

By comparing the Weyl transformations of the two sides, show that the sum of the coefficients of z^{-3} in Tj and of $\bar{z}^{-3}$ in $\tilde{T}\tilde{\jmath}$ is $4a$.

3.7 In our second calculation of the Weyl anomaly, we have in a roundabout way calculated a Feynman graph with a loop of scalars or ghosts and with two external gravitational fields. Do the calculation for scalars directly, in momentum space. You will need to use your favorite invariant regulator: dimensional, Pauli–Villars, or zeta function. Show that the result agrees with eq. (3.4.19). The linearized expression for the curvature can be conveniently read from eqs. (3.7.13a) and (3.7.14a). (Warning: the full calculation is a bit lengthy.)

3.8 Verify the Weyl transformation (3.6.16) of the vertex operator V_1.

3.9 Find the Weyl invariance conditions for the open string tachyon and photon vertex operators. It is simplest to take a coordinate system in which the boundary is a straight line. Recall from exercise 2.10 that the propagator includes an image charge term, so it is effectively doubled for fields directly on the boundary.

3.10 Find the general vertex operator and the conditions for Weyl invariance at the level $m^2 = 1/\alpha'$ of the open string. Find a complete set of independent Weyl-invariant operators and compare with the states found in the light-cone quantization. [Answer: there appears to be one extra operator. In fact its amplitudes vanish, but we need the formalism of the next chapter to understand the reason, that it corresponds to a null state.]

3.11 Calculate the order H^2 term in $\beta^G_{\mu\nu}$ by the method suggested above eq. (3.7.14). You need precisely the term $(z\bar{z})^{-1}$ in the OPE; more singular terms require a shift in the tachyon background, but this is an artifact of the bosonic string.

3.12 If $\beta^G_{\mu\nu} = \beta^B_{\mu\nu} = 0$, the nonlinear sigma model will be conformally invariant on a flat world-sheet. According to the discussion at the end of section 3.4, in this case β^Φ must be independent of X. Show this for the form (3.7.14); you will need the Bianchi identities.

3.13 Consider a spacetime with d flat dimensions and 3 dimensions in the shape of a 3-sphere. Let H be proportional to the completely antisymmetric tensor on the 3-sphere and let the dilaton be constant. Using the form (3.7.14) for the equations of motion, show that there are solutions with $d + 3 \neq 26$. These solutions are outside the range of validity of eq. (3.7.14), but we will see in chapter 15 that there are exact solutions of this form, though with H limited to certain quantized values.

3.14 Verify that the background (3.7.18) gives the linear dilaton energy-momentum tensor (2.5.1).

4

The string spectrum

4.1 Old covariant quantization

In conformal gauge, the world-sheet fields are X^μ and the Faddeev–Popov ghosts b_{ab} and c^a. The Hilbert space is bigger than the actual physical spectrum of the string: the D sets of α^μ oscillators include unphysical oscillations of the coordinate system, and there are ghost oscillators. As is generally the case in covariant gauges, there are negative norm states from the timelike oscillators (because the commutator is proportional to the spacetime metric $\eta_{\mu\nu}$), and also from the ghosts.

The actual physical space is smaller. To see how to identify this smaller space, consider the amplitude on the infinite cylinder for some initial state $|i\rangle$ to propagate to some final state $|f\rangle$. Suppose that we have initially used the local symmetries to fix the metric to a form $g_{ab}(\sigma)$. Consider now a different gauge, with metric $g_{ab}(\sigma) + \delta g_{ab}(\sigma)$. A physical amplitude should not depend on this choice. Of course, for a change in the metric we know how the path integral changes. From the definition (3.4.4) of T^{ab},

$$\delta \langle f|i\rangle = -\frac{1}{4\pi} \int d^2\sigma\, g(\sigma)^{1/2} \delta g_{ab}(\sigma) \langle f|T^{ab}(\sigma)|i\rangle . \qquad (4.1.1)$$

In order that the variation vanish for arbitrary changes in the metric, we need

$$\langle \psi|T^{ab}(\sigma)|\psi'\rangle = 0 \qquad (4.1.2)$$

for arbitrary physical states $|\psi\rangle$, $|\psi'\rangle$.

There is another way to think about this. The original equation of motion from variation of g_{ab} was $T^{ab} = 0$. After gauge-fixing, this does not hold as an operator equation: we have a missing equation of motion because we do not vary g_{ab} in the gauge-fixed theory. The condition (4.1.2) says that the missing equation of motion must hold for matrix elements between physical states. When we vary the gauge we must take into account the change in the Faddeev–Popov determinant, so the energy-momentum tensor in the matrix element is the sum of the X^μ and ghost contributions,

$$T_{ab} = T_{ab}^X + T_{ab}^{\mathrm{g}} . \tag{4.1.3}$$

The X^μ may be replaced by a more general CFT (which we will refer to as the *matter* CFT), in which case

$$T_{ab} = T_{ab}^{\mathrm{m}} + T_{ab}^{\mathrm{g}} . \tag{4.1.4}$$

In the remainder of this section we will impose the condition (4.1.2) in a simple but somewhat *ad hoc* way. This is known as the *old covariant quantization (OCQ)*, and it is sufficient for many purposes. In the next section we will take a more systematic approach, BRST quantization. These are in fact equivalent, as will be shown in section 4.4.

In the *ad hoc* approach, we will simply ignore the ghosts and try to restrict the matter Hilbert space so that the missing equation of motion $T_{ab}^{\mathrm{m}} = 0$ holds for matrix elements. In terms of the Laurent coefficients this is $L_n^{\mathrm{m}} = 0$, and in the closed string also $\tilde{L}_n^{\mathrm{m}} = 0$. One might first try to require physical states to satisfy $L_n^{\mathrm{m}}|\psi\rangle = 0$ for all n, but this is too strong; acting on this equation with L_m^{m} and forming the commutator, one encounters an inconsistency due to the central charge in the Virasoro algebra. However, it is sufficient that only the Virasoro lowering and zero operator annihilate physical states,

$$(L_n^{\mathrm{m}} + A\delta_{n,0})|\psi\rangle = 0 \quad \text{for } n \geq 0 . \tag{4.1.5}$$

For $n < 0$ we then have

$$\langle\psi|L_n^{\mathrm{m}}|\psi'\rangle = \langle L_{-n}^{\mathrm{m}}\psi|\psi'\rangle = 0 . \tag{4.1.6}$$

We have used

$$L_n^{\mathrm{m}\dagger} = L_{-n}^{\mathrm{m}} \tag{4.1.7}$$

as follows from the Hermiticity of the energy-momentum tensor. The conditions (4.1.5) are consistent with the Virasoro algebra. At $n = 0$ we have as usual included the possibility of an ordering constant. A state satisfying (4.1.5) is called *physical*. In the terminology of eq. (2.9.8), a physical state is a highest weight state of weight $-A$. The condition (4.1.5) is similar to the Gupta–Bleuler quantization of electrodynamics.

Using the adjoint (4.1.7), one sees that a state of the form

$$|\chi\rangle = \sum_{n=1}^{\infty} L_{-n}^{\mathrm{m}}|\chi_n\rangle \tag{4.1.8}$$

is orthogonal to all physical states for any $|\chi_n\rangle$. Such a state is called *spurious*. A state that is both physical and spurious is called *null*. If $|\psi\rangle$

is physical and $|\chi\rangle$ is null, then $|\psi\rangle + |\chi\rangle$ is also physical and its inner product with any physical state is the same as that of $|\psi\rangle$. Therefore these two states are physically indistinguishable, and we identify

$$|\psi\rangle \cong |\psi\rangle + |\chi\rangle . \tag{4.1.9}$$

The real physical Hilbert space is then the set of equivalence classes,

$$\mathscr{H}_{\text{OCQ}} = \frac{\mathscr{H}_{\text{phys}}}{\mathscr{H}_{\text{null}}} . \tag{4.1.10}$$

Let us see how this works for the first two levels of the open string in flat spacetime, not necessarily assuming $D = 26$. The only relevant terms are

$$L_0^{\text{m}} = \alpha' p^2 + \alpha_{-1} \cdot \alpha_1 + \dots , \tag{4.1.11a}$$

$$L_{\pm 1}^{\text{m}} = (2\alpha')^{1/2} p \cdot \alpha_{\pm 1} + \dots . \tag{4.1.11b}$$

At the lowest mass level, the only state is $|0; k\rangle$. The only nontrivial condition at this level is $(L_0^{\text{m}} + A)|\psi\rangle = 0$, giving $m^2 = A/\alpha'$. There are no null states at this level — the Virasoro generators in the spurious state (4.1.8) all contain raising operators. Thus there is one equivalence class, corresponding to a scalar particle.

At the next level there are D states

$$|e; k\rangle = e \cdot \alpha_{-1} |0; k\rangle . \tag{4.1.12}$$

The norm is

$$\begin{aligned}
\langle e; k | e; k' \rangle &= \langle 0; k | e^* \cdot \alpha_1 \, e \cdot \alpha_{-1} | 0; k' \rangle \\
&= \langle 0; k | (e^* \cdot e + e^* \cdot \alpha_{-1} e \cdot \alpha_1) | 0; k' \rangle \\
&= e^{\mu*} e_\mu (2\pi)^D \delta^D (k - k') .
\end{aligned} \tag{4.1.13}$$

We have used

$$\alpha_n^{\mu\dagger} = \alpha_{-n}^\mu , \tag{4.1.14}$$

as follows from the Hermiticity of X^μ, and also

$$\langle 0; k | 0; k' \rangle = (2\pi)^D \delta^D (k - k') , \tag{4.1.15}$$

as follows from momentum conservation. The timelike excitation has a negative norm.

The L_0^{m} condition gives

$$m^2 = \frac{1 + A}{\alpha'} . \tag{4.1.16}$$

The other nontrivial physical state condition is

$$L_1^{\text{m}} |e; k\rangle \propto p \cdot \alpha_1 e \cdot \alpha_{-1} |0; k\rangle = e \cdot k |0; k\rangle = 0 \tag{4.1.17}$$

and so $k \cdot e = 0$. There is a spurious state at this level:

$$L^{m}_{-1}|0;k\rangle = (2\alpha')^{1/2}k \cdot \alpha_{-1}|0;k\rangle \ . \qquad (4.1.18)$$

That is, $e^{\mu} \propto k^{\mu}$ is spurious. There are now three cases:

(i) If $A > -1$, the mass-squared is positive. Going to the rest frame, the physical state condition removes the negative norm timelike polarization. The spurious state (4.1.18) is not physical, $k \cdot k \neq 0$, so there are no null states and the spectrum consists of the $D - 1$ positive-norm states of a massive vector particle.

(ii) If $A = -1$, the mass-squared is zero. Now $k \cdot k = 0$ so the spurious state is physical and null. Thus,

$$k \cdot e = 0 \ , \quad e_{\mu} \cong e_{\mu} + \gamma k_{\mu} \ . \qquad (4.1.19)$$

This describes the $D - 2$ positive-norm states of a massless vector particle.

(iii) If $A < -1$, the mass-squared is negative. The momentum is spacelike, so the physical state condition removes a positive-norm *spacelike* polarization. The spurious state is not physical, so we are left with a tachyonic vector particle with $D - 2$ positive-norm states and one negative-norm state.

Case *(iii)* is unacceptable. Case *(ii)* is the same as the light-cone quantization. Case *(i)* is not the same as the light-cone spectrum, having different masses and an extra state at the first level, but does not have any obvious inconsistency. The difficulty is that there is no known way to give such a theory consistent interactions. Case *(ii)* is the one of interest in string theory, and the one we will recover from the more general approach of the next section.

The result at the next level is quite interesting: it depends on the constant A and also on the spacetime dimension D. Restricting to the value $A = -1$ found at the first excited level, the spectrum only agrees with the light-cone spectrum if in addition $D = 26$. If $D > 26$, there are negative-norm states; if $D < 26$ the OCQ spectrum has positive norm but more states than in light-cone quantization. The derivation is left to the reader. The OCQ at $A = -1$ and $D = 26$ is in fact the same as light-cone quantization at all mass levels,

$$\mathcal{H}_{\text{OCQ}} = \mathcal{H}_{\text{light-cone}} \ , \qquad (4.1.20)$$

as will be shown in section 4.4. It is only in this case that consistent interactions are known.

The extension to the closed string is straightforward. There are two sets of oscillators and two sets of Virasoro generators, so at each level the spectrum is the product of two copies of the open string spectrum (except for the normalization of p^2 in L_0^m). The first two levels are then

$$|0;k\rangle \,, \quad m^2 = -\frac{4}{\alpha'} \,; \tag{4.1.21a}$$

$$e_{\mu\nu}\alpha^\mu_{-1}\tilde{\alpha}^\nu_{-1}|0;k\rangle \,, \quad m^2 = 0 \,, \quad k^\mu e_{\mu\nu} = k^\nu e_{\mu\nu} = 0 \,, \tag{4.1.21b}$$

$$e_{\mu\nu} \cong e_{\mu\nu} + a_\mu k_\nu + k_\mu b_\nu \,, \quad a \cdot k = b \cdot k = 0 \,. \tag{4.1.21c}$$

The relevant values are again $A = -1$, $D = 26$. As in light-cone quantization, there are $(D-2)^2$ massless states forming a traceless symmetric tensor, an antisymmetric tensor, and a scalar.

Mnemonic

For more general string theories, it is useful to have a quick mnemonic for obtaining the zero-point constant. As will be derived in the following sections, the L_0^m condition can be understood as

$$(L_0^m + L_0^g)|\psi,\downarrow\rangle = 0 \,. \tag{4.1.22}$$

That is, one includes the ghost contribution with the ghosts in their ground state $|\downarrow\rangle$, which has $L_0^g = -1$. The L_0 generators differ from the Hamiltonian by a shift (2.6.10) proportional to the central charge, but the total central charge is zero in string theory so we can just as well write this condition as

$$(H^m + H^g)|\psi,\downarrow\rangle = 0 \,. \tag{4.1.23}$$

Now apply the mnemonic for zero-point energies given at the end of chapter 2. In particular, the ghosts always cancel the $\mu = 0, 1$ oscillators because they have the same periodicities but opposite statistics. So the rule is that A is given as the sum of the zero-point energies of the transverse oscillators. This is the same rule as in the light-cone, giving here $A = 24(-\frac{1}{24}) = -1$.

Incidentally, for the purposes of counting the physical states (but not for their precise form) one can simply ignore the ghost and $\mu = 0, 1$ oscillators and count transverse excitations as in the light-cone.

The condition (4.1.22) requires the weight of the matter state to be 1. Since the physical state condition requires the matter state to be a highest weight state, the vertex operator must then be a weight 1 or (1,1) tensor. This agrees with the condition from section 3.6 that the integrated vertex operator be conformally invariant, which gives another understanding of the condition $A = -1$.

4.2   BRST quantization

We now return to a more systematic study of the spectrum. The condition (4.1.2) is not sufficient to guarantee gauge invariance. It implies invariance for arbitrary fixed choices of g_{ab}, but this is not the most general gauge. In light-cone gauge, for example, we placed some conditions on X^μ and some conditions on g_{ab}. To consider the most general possible variation of the gauge condition, we must allow δg_{ab} to be an *operator*, that is, to depend on the fields in the path integral.

In order to derive the full invariance condition, it is useful to take a more general and abstract point of view. Consider a path integral with a local symmetry. The path integral fields are denoted ϕ_i, which in the present case would be $X^\mu(\sigma)$ and $g_{ab}(\sigma)$. Here we use a very condensed notation, where i labels the field and *also* represents the coordinate σ. The gauge invariance is $\epsilon^\alpha \delta_\alpha$, where again α includes the coordinate. By assumption the gauge parameters ϵ^α are real, since we can always separate a complex parameter into its real and imaginary parts. The gauge transformations satisfy an algebra

$$[\delta_\alpha, \delta_\beta] = f^\gamma_{\alpha\beta} \delta_\gamma \; . \tag{4.2.1}$$

Now fix the gauge by conditions

$$F^A(\phi) = 0 \; , \tag{4.2.2}$$

where once again A includes the coordinate. Following the same Faddeev–Popov procedure as in section 3.3, the path integral becomes

$$\int \frac{[d\phi_i]}{V_{\text{gauge}}} \exp(-S_1) \to \int [d\phi_i \, dB_A \, db_A \, dc^\alpha] \exp(-S_1 - S_2 - S_3) \; , \tag{4.2.3}$$

where S_1 is the original gauge-invariant action, S_2 is the gauge-fixing action

$$S_2 = -iB_A F^A(\phi) \; , \tag{4.2.4}$$

and S_3 is the Faddeev–Popov action

$$S_3 = b_A c^\alpha \delta_\alpha F^A(\phi) \; . \tag{4.2.5}$$

We have introduced the field B_A to produce an integral representation of the gauge-fixing $\delta(F^A)$.

There are two things to notice about this action. The first is that it is invariant under the *Becchi–Rouet–Stora–Tyutin (BRST) transformation*

$$\delta_{\text{B}} \phi_i = -i\epsilon c^\alpha \delta_\alpha \phi_i \; , \tag{4.2.6a}$$

$$\delta_{\text{B}} B_A = 0 \; , \tag{4.2.6b}$$

$$\delta_{\text{B}} b_A = \epsilon B_A \; , \tag{4.2.6c}$$

$$\delta_{\text{B}} c^\alpha = \frac{i}{2} \epsilon f^\alpha_{\beta\gamma} c^\beta c^\gamma \; . \tag{4.2.6d}$$

This transformation mixes commuting and anticommuting objects, so that ϵ must be taken to be anticommuting. There is a conserved ghost number, which is $+1$ for c^α, -1 for b_A and ϵ, and 0 for all other fields. The original action S_1 is invariant by itself, because the action of δ_B on ϕ_i is just a gauge transformation with parameter $i\epsilon c^\alpha$. The variation of S_2 cancels the variation of b_A in S_3, while the variations of $\delta_\alpha F^A$ and c^α in S_3 cancel.

The second key property is that

$$\delta_B(b_A F^A) = i\epsilon(S_2 + S_3) . \tag{4.2.7}$$

Now consider a small change δF in the gauge-fixing condition. The change in the gauge-fixing and ghost actions gives

$$\epsilon\delta\langle f|i\rangle = i\langle f|\delta_B(b_A\delta F^A)|i\rangle$$
$$= -\epsilon\langle f|\{Q_B, b_A\delta F^A\}|i\rangle , \tag{4.2.8}$$

where we have written the BRST variation as an anticommutator with the corresponding conserved charge Q_B.

Therefore, physical states must satisfy

$$\langle\psi|\{Q_B, b_A\delta F^A\}|\psi'\rangle = 0 . \tag{4.2.9}$$

In order for this to hold for arbitrary δF, it must be that

$$Q_B|\psi\rangle = Q_B|\psi'\rangle = 0 . \tag{4.2.10}$$

This is the essential condition: *physical states must be BRST-invariant*. We have assumed that $Q_B^\dagger = Q_B$. There are several ways to see that this must indeed be the case. One is that if $Q_B^\dagger$ were different, it would have to be some other symmetry, and there is no candidate. A better argument is that the fields c^α and b_A are like anticommuting versions of the gauge parameter ϵ^α and Lagrange multiplier B_A, and so inherit their reality properties.

As an aside, we may also add to the action a term proportional to

$$\epsilon^{-1}\delta_B(b_A B_B M^{AB}) = -B_A B_B M^{AB} \tag{4.2.11}$$

for any constant matrix M^{AB}. By the above argument, amplitudes between physical states are unaffected. The integral over B_A now produces a Gaussian rather than a delta function: these are the *Gaussian-averaged gauges*, which include for example the covariant α-gauges of gauge theory.

There is one more key idea. In order to move around in the space of gauge choices, the BRST charge must remain conserved. Thus it must commute with the change in the Hamiltonian,

$$0 = [Q_B, \{Q_B, b_A\delta F^A\}]$$
$$= Q_B^2 b_A\delta F^A - Q_B b_A\delta F^A Q_B + Q_B b_A\delta F^A Q_B - b_A\delta F^A Q_B^2$$
$$= [Q_B^2, b_A\delta F^A] . \tag{4.2.12}$$

In order for this to vanish for general changes of gauge, we need

$$Q_B^2 = 0 . \tag{4.2.13}$$

That is, the BRST charge is *nilpotent*; the possibility $Q_B^2 = $ constant is excluded because Q_B^2 has ghost number 2. The reader can check that acting twice with the BRST transformation (4.2.6) leaves all fields invariant. In particular,

$$\delta_B(\delta'_B c^\alpha) = -\frac{1}{2} \epsilon \epsilon' f^\alpha{}_{\beta\gamma} f^\gamma{}_{\delta\epsilon} c^\beta c^\delta c^\epsilon = 0 . \tag{4.2.14}$$

The product of ghosts is antisymmetric on the indices β, δ, ϵ, and the product of structure functions then vanishes by the Jacobi identity.

We should mention that we have made two simplifying assumptions about the gauge algebra (4.2.1). The first is that the structure constants $f^\alpha{}_{\beta\gamma}$ *are* constants, independent of the fields, and the second is that the algebra does not have, on the right-hand side, additional terms proportional to the equations of motion. More generally, both of these assumptions break down. In these cases, the BRST formalism as we have described it does not give a nilpotent transformation, and a generalization, the *Batalin–Vilkovisky (BV) formalism*, is needed. The BV formalism has had various applications in string theory, but we will not have a need for it.

The nilpotence of Q_B has an important consequence. A state of the form

$$Q_B |\chi\rangle \tag{4.2.15}$$

will be annihilated by Q_B for any χ and so is physical. However, it is orthogonal to all physical states including itself:

$$\langle \psi | (Q_B |\chi\rangle) = (\langle \psi | Q_B) |\chi\rangle = 0 \tag{4.2.16}$$

if $Q_B |\psi\rangle = 0$. All physical amplitudes involving such a *null state* thus vanish. Two physical states that differ by a null state,

$$|\psi'\rangle = |\psi\rangle + Q_B |\chi\rangle \tag{4.2.17}$$

will have the same inner products with all physical states and are therefore physically equivalent. So, just as for the OCQ, we identify the true physical space with a set of *equivalence classes*, states differing by a null state being equivalent. This is a natural construction for a nilpotent operator, and is known as the *cohomology* of Q_B. Other examples of nilpotent operators are the exterior derivative in differential geometry and the boundary operator in topology. In cohomology, the term *closed* is often used for states annihilated by Q_B and the term *exact* for states of the form (4.2.15). Thus, our prescription is

$$\mathscr{H}_{\text{BRST}} = \frac{\mathscr{H}_{\text{closed}}}{\mathscr{H}_{\text{exact}}} . \tag{4.2.18}$$

We will see explicitly in the remainder of the chapter that this space has the expected form in string theory. Essentially, the invariance condition removes one set of unphysical X^μ oscillators and one set of ghost oscillators, and the equivalence relation removes the other set of unphysical X^μ oscillators and the other set of ghost oscillators.

Point-particle example

Let us consider an example, the point particle. Expanding out the condensed notation above, the local symmetry is coordinate reparameterization $\delta\tau(\tau)$, so the index α just becomes τ and a basis of infinitesimal transformations is $\delta_{\tau_1}\tau(\tau) = \delta(\tau - \tau_1)$. These act on the fields as

$$\delta_{\tau_1}X^\mu(\tau) = -\delta(\tau - \tau_1)\partial_\tau X^\mu(\tau) \,, \quad \delta_{\tau_1}e(\tau) = -\partial_\tau[\delta(\tau - \tau_1)e(\tau)] \,. \quad (4.2.19)$$

Acting with a second transformation and forming the commutator, we have

$$[\delta_{\tau_1},\delta_{\tau_2}]X^\mu(\tau)$$
$$= -\Big[\delta(\tau - \tau_1)\partial_\tau\delta(\tau - \tau_2) - \delta(\tau - \tau_2)\partial_\tau\delta(\tau - \tau_1)\Big]\partial_\tau X^\mu(\tau)$$
$$\equiv \int d\tau_3 \, f^{\tau_3}{}_{\tau_1\tau_2}\delta_{\tau_3}X^\mu(\tau) \,. \quad (4.2.20)$$

From the commutator we have determined the structure function

$$f^{\tau_3}{}_{\tau_1\tau_2} = \delta(\tau_3 - \tau_1)\partial_{\tau_3}\delta(\tau_3 - \tau_2) - \delta(\tau_3 - \tau_2)\partial_{\tau_3}\delta(\tau_3 - \tau_1) \,. \quad (4.2.21)$$

The BRST transformation is then

$$\delta_B X^\mu = i\epsilon c\dot{X}^\mu \,, \quad (4.2.22a)$$
$$\delta_B e = i\epsilon(\dot{ce}) \,, \quad (4.2.22b)$$
$$\delta_B B = 0 \,, \quad (4.2.22c)$$
$$\delta_B b = \epsilon B \,, \quad (4.2.22d)$$
$$\delta_B c = i\epsilon c\dot{c} \,. \quad (4.2.22e)$$

The gauge $e(\tau) = 1$ is analogous to the unit gauge for the string, using the single coordinate freedom to fix the single component of the tetrad, so $F(\tau) = 1 - e(\tau)$. The gauge-fixed action is then

$$S = \int d\tau \left(\frac{1}{2}e^{-1}\dot{X}^\mu\dot{X}_\mu + \frac{1}{2}em^2 + iB(e - 1) - e\dot{b}c \right) \,. \quad (4.2.23)$$

We will find it convenient to integrate B out, thus fixing $e = 1$. This leaves only the fields X^μ, b, and c, with action

$$S = \int d\tau \left(\frac{1}{2}\dot{X}^\mu\dot{X}_\mu + \frac{1}{2}m^2 - \dot{b}c \right) \quad (4.2.24)$$

and BRST transformation

$$\delta_B X^\mu = i\epsilon c \dot{X}^\mu \, , \tag{4.2.25a}$$

$$\delta_B b = i\epsilon \left(-\frac{1}{2} \dot{X}^\mu \dot{X}_\mu + \frac{1}{2} m^2 - bc \right) \, , \tag{4.2.25b}$$

$$\delta_B c = i\epsilon c \dot{c} \, . \tag{4.2.25c}$$

Since B is no longer present, we have used the equation of motion from e to replace B in the transformation law for b. The reader can check that the new transformation (4.2.25) is a symmetry of the action and is nilpotent. This always works when the fields B^A are integrated out, though $\delta_B \delta'_B b$ is no longer identically zero but is proportional to the equations of motion. This is perfectly satisfactory: $Q_B^2 = 0$ holds as an operator equation, which is what we need.

The canonical commutators are

$$[p^\mu, X^\nu] = -i\eta^{\mu\nu} \, , \quad \{b, c\} = 1 \, , \tag{4.2.26}$$

with $p^\mu = i\dot{X}^\mu$, the i being from the Euclidean signature. The Hamiltonian is $H = \frac{1}{2}(p^2 + m^2)$, and the Noether procedure gives the BRST operator

$$Q_B = cH \, . \tag{4.2.27}$$

The structure here is analogous to what we will find for the string. The constraint (the missing equation of motion) is $H = 0$, and the BRST operator is c times this.

The ghosts generate a two-state system, so a complete set of states is $|k, \downarrow\rangle, |k, \uparrow\rangle$, where

$$p^\mu |k, \downarrow\rangle = k^\mu |k, \downarrow\rangle \, , \quad p^\mu |k, \uparrow\rangle = k^\mu |k, \uparrow\rangle \, , \tag{4.2.28a}$$

$$b|k, \downarrow\rangle = 0 \, , \quad b|k, \uparrow\rangle = |k, \downarrow\rangle \, , \tag{4.2.28b}$$

$$c|k, \downarrow\rangle = |k, \uparrow\rangle \, , \quad c|k, \uparrow\rangle = 0 \, . \tag{4.2.28c}$$

The action of the BRST operator on these is

$$Q_B|k, \downarrow\rangle = \frac{1}{2}(k^2 + m^2)|k, \uparrow\rangle \, , \quad Q_B|k, \uparrow\rangle = 0 \, . \tag{4.2.29}$$

From this it follows that the closed states are

$$|k, \downarrow\rangle \, , \quad k^2 + m^2 = 0 \, , \tag{4.2.30a}$$

$$|k, \uparrow\rangle \, , \quad \text{all } k^\mu \, , \tag{4.2.30b}$$

and that the exact states are

$$|k, \uparrow\rangle \, , \quad k^2 + m^2 \neq 0 \, . \tag{4.2.31}$$

The closed states that are not exact are

$$|k, \downarrow\rangle \, , \quad k^2 + m^2 = 0 \, ; \quad |k, \uparrow\rangle \, , \quad k^2 + m^2 = 0 \, . \tag{4.2.32}$$

That is, physical states must satisfy the mass-shell condition, but we have *two copies* of the expected spectrum. In fact, only states $|k, \downarrow\rangle$ satisfying the additional condition

$$b|\psi\rangle = 0 \tag{4.2.33}$$

appear in physical amplitudes. The origin of this additional condition is kinematic. For $k^2 + m^2 \neq 0$, the states $|k, \uparrow\rangle$ are exact — they are orthogonal to all physical states and their amplitudes must vanish identically. So the amplitudes can only be proportional to $\delta(k^2 + m^2)$. But amplitudes in field theory and in string theory, while they may have poles and cuts, never have delta functions (except in $D = 2$ where the kinematics is special). So they must vanish identically.

4.3 BRST quantization of the string

In string theory, the BRST transformation is

$$\delta_B X^\mu = i\epsilon(c\partial + \tilde{c}\bar{\partial})X^\mu , \tag{4.3.1a}$$

$$\delta_B b = i\epsilon(T^X + T^g) , \quad \delta_B \tilde{b} = i\epsilon(\tilde{T}^X + \tilde{T}^g) , \tag{4.3.1b}$$

$$\delta_B c = i\epsilon c\partial c , \quad \delta_B \tilde{c} = i\epsilon \tilde{c}\bar{\partial}\tilde{c} . \tag{4.3.1c}$$

The reader can derive this following the point-particle example. To the sum of the Polyakov and ghost actions, add the gauge-fixing term

$$\frac{i}{4\pi} \int d^2\sigma \, g^{1/2} B^{ab}(\delta_{ab} - g_{ab}) . \tag{4.3.2}$$

The transformation $\delta_B b_{ab} = \epsilon B_{ab}$ becomes (4.3.1) after integrating over B_{ab} and using the g_{ab} equation of motion to replace B_{ab} in the transformation. The Weyl ghost is just a Lagrange multiplier, making b_{ab} traceless. Again, the reader can check nilpotence up to the equations of motion. The similarity between the string BRST transformation (4.3.1) and the particle case (4.2.25) is evident. Replacing the X^μ with a general matter CFT, the BRST transformation of the matter fields is a conformal transformation with $v(z) = c(z)$, while T^m replaces T^X in the transformation of b.

Noether's theorem gives the BRST current

$$j_B = cT^m + \frac{1}{2} :cT^g: + \frac{3}{2}\partial^2 c ,$$

$$= cT^m + :bc\partial c: + \frac{3}{2}\partial^2 c , \tag{4.3.3}$$

and correspondingly for $\tilde{\jmath}_B$. The final term in the current is a total derivative and does not contribute to the BRST charge; it has been added by hand to make the BRST current a tensor. The OPEs of the BRST current with the ghost fields and with a general matter tensor field are

$$j_B(z)b(0) \sim \frac{3}{z^3} + \frac{1}{z^2}j^g(0) + \frac{1}{z}T^{m+g}(0) , \qquad (4.3.4a)$$

$$j_B(z)c(0) \sim \frac{1}{z}c\partial c(0) , \qquad (4.3.4b)$$

$$j_B(z)\mathcal{O}^m(0,0) \sim \frac{h}{z^2}c\mathcal{O}^m(0,0) + \frac{1}{z}\left[h(\partial c)\mathcal{O}^m(0,0) + c\partial\mathcal{O}^m(0,0)\right] . \quad (4.3.4c)$$

The simple poles reflect the BRST transformations (4.3.1) of these fields.
   The BRST operator is

$$Q_B = \frac{1}{2\pi i}\oint (dz\, j_B - d\bar{z}\, \tilde{j}_B) . \qquad (4.3.5)$$

By the usual contour argument, the OPE implies

$$\{Q_B, b_m\} = L_m^m + L_m^g . \qquad (4.3.6)$$

In terms of the ghost modes,

$$Q_B = \sum_{n=-\infty}^{\infty} (c_n L_{-n}^m + \tilde{c}_n \tilde{L}_{-n}^m)$$

$$+ \sum_{m,n=-\infty}^{\infty} \frac{(m-n)}{2} {}_\circ^\circ(c_m c_n b_{-m-n} + \tilde{c}_m \tilde{c}_n \tilde{b}_{-m-n}){}_\circ^\circ + a^B(c_0 + \tilde{c}_0) .$$

$$(4.3.7)$$

The ordering constant is $a^B = a^g = -1$, as follows from the anticommutator

$$\{Q_B, b_0\} = L_0^m + L_0^g . \qquad (4.3.8)$$

   There is an anomaly in the gauge symmetry when $c^m \neq 26$, so we expect some breakdown in the BRST formalism. The BRST current is still conserved: all of the terms in the current (4.3.3) are analytic for any value of the central charge. However, it is no longer *nilpotent*,

$$\{Q_B, Q_B\} = 0 \quad \text{only if } c^m = 26 . \qquad (4.3.9)$$

The shortest derivation of this uses the Jacobi identity, exercise 4.3. More directly it follows from the OPE

$$j_B(z)j_B(0) \sim -\frac{c^m - 18}{2z^3}c\partial c(0) - \frac{c^m - 18}{4z^2}c\partial^2 c(0) - \frac{c^m - 26}{12z}c\partial^3 c(0) . \quad (4.3.10)$$

This requires a bit of calculation, being careful about minus signs from anticommutation. The single pole implies that $\{Q_B, Q_B\} = 0$ when $c^m = 26$.

Note also the OPE

$$T(z)j_{\rm B}(0) \sim \frac{c^{\rm m} - 26}{2z^4}c(0) + \frac{1}{z^2}j_{\rm B}(0) + \frac{1}{z}\partial j_{\rm B}(0) \,, \tag{4.3.11}$$

which implies that $j_{\rm B}$ is a tensor only when $c^{\rm m} = 26$.

Let us point out some important features. The missing equation of motion is just the vanishing of the generator T_{ab} of the conformal transformations, the part of the local symmetry group not fixed by the gauge choice. This is a general result. When the gauge conditions fix the symmetry completely, one can show that the missing equations of motion are trivial due to the gauge invariance of the action. When they do not, so that there is a residual symmetry group, the missing equations of motion require its generators to vanish. They must vanish in the sense (4.1.2) of matrix elements, or more generally in the BRST sense. Denote the generators of the residual symmetry group by G_I. The G_I are called *constraints;* for the string these are L_m and $\tilde{L}_m$. They form an algebra

$$[G_I, G_J] = ig^K{}_{IJ}G_K \,. \tag{4.3.12}$$

Associated with each generator is a pair of ghosts, b_I and c^I, where

$$\{c^I, b_J\} = \delta^I{}_J \,, \quad \{c^I, c^J\} = \{b_I, b_J\} = 0 \,. \tag{4.3.13}$$

The general form of the BRST operator, as illustrated by the string case (4.3.7), is

$$Q_{\rm B} = c^I G_I^{\rm m} - \frac{i}{2}g^K{}_{IJ}c^I c^J b_K$$

$$= c^I \left(G_I^{\rm m} + \frac{1}{2}G_I^{\rm g} \right) \,, \tag{4.3.14}$$

where $G_I^{\rm m}$ is the matter part of G_I and

$$G_I^{\rm g} = -ig^K{}_{IJ}c^J b_K \tag{4.3.15}$$

is the ghost part. The $G_I^{\rm m}$ and $G_I^{\rm g}$ satisfy the same algebra (4.3.12) as the G_I. Using the commutators (4.3.12) and (4.3.13), one finds

$$Q_{\rm B}^2 = \frac{1}{2}\{Q_{\rm B}, Q_{\rm B}\} = -\frac{1}{2}g^K{}_{IJ}g^M{}_{KL}c^I c^J c^L b_M = 0 \,. \tag{4.3.16}$$

The last equality follows from the *GGG* Jacobi identity, which requires $g^K{}_{IJ}g^M{}_{KL}$ to vanish when antisymmetrized on IJL. We have neglected central charge terms; these need to be checked by hand.

To reiterate, a *constraint* algebra is a world-sheet symmetry algebra that is required to vanish in physical matrix elements. When we go on to generalize the bosonic string theory in volume two, it will be easiest to do so directly in terms of the constraint algebra, and we will write the BRST charge directly in the form (4.3.14).

BRST cohomology of the string

Let us now look at the BRST cohomology at the lowest levels of the string. The inner product is defined by specifying

$$(\alpha_m^\mu)^\dagger = \alpha_{-m}^\mu \,, \quad (\tilde\alpha_m^\mu)^\dagger = \tilde\alpha_{-m}^\mu \,, \tag{4.3.17a}$$

$$(b_m)^\dagger = b_{-m} \,, \quad (\tilde b_m)^\dagger = \tilde b_{-m} \,, \tag{4.3.17b}$$

$$(c_m)^\dagger = c_{-m} \,, \quad (\tilde c_m)^\dagger = \tilde c_{-m} \,. \tag{4.3.17c}$$

In particular, the Hermiticity of the BRST charge requires that the ghost fields be Hermitean as well. The Hermiticity of the ghost zero modes forces the inner products of the ground state to take the form

$$\text{open string:} \quad \langle 0;k|c_0|0;k'\rangle = (2\pi)^{26}\delta^{26}(k-k') \,, \tag{4.3.18a}$$

$$\text{closed string:} \quad \langle 0;k|\tilde c_0 c_0|0;k'\rangle = i(2\pi)^{26}\delta^{26}(k-k') \,. \tag{4.3.18b}$$

Here $|0;k\rangle$ denotes the matter ground state times the ghost ground state $|\downarrow\rangle$, with momentum k. The c_0 and $\tilde c_0$ insertions are necessary for a nonzero result. For example, $\langle 0;k|0;k'\rangle = \langle 0;k|(c_0 b_0 + b_0 c_0)|0;k'\rangle = 0$; the last equality follows because b_0 annihilates both the bra and the ket. The factor of i is needed in the ghost zero-mode inner product for Hermiticity. Inner products of general states are then obtained, as in the earlier calculation (4.1.13), by use of the commutation relations and the adjoints (4.3.17).

We will focus on the open string; the closed string discussion is entirely parallel but requires twice as much writing. As in the point-particle case (4.2.33), we will assert (and later show) that physical states must satisfy the additional condition

$$b_0|\psi\rangle = 0 \,. \tag{4.3.19}$$

This also implies

$$L_0|\psi\rangle = \{Q_B, b_0\}|\psi\rangle = 0 \,, \tag{4.3.20}$$

since Q_B and b_0 both annihilate $|\psi\rangle$. The operator L_0 is

$$L_0 = \alpha'(p^\mu p_\mu + m^2) \,, \tag{4.3.21}$$

where

$$\alpha' m^2 = \sum_{n=1}^\infty n\left(N_{bn} + N_{cn} + \sum_{\mu=0}^{25} N_{\mu n}\right) - 1 \,. \tag{4.3.22}$$

Thus the L_0 condition (4.3.20) determines the mass of the string. BRST invariance, with the extra condition (4.3.19), implies that every string state is on the mass shell. We will denote by $\hat{\mathscr{H}}$ the space of states satisfying the conditions (4.3.19) and (4.3.20). From the commutators $\{Q_B, b_0\} = L_0$ and $[Q_B, L_0] = 0$, it follows that Q_B takes $\hat{\mathscr{H}}$ into itself.

The inner product (4.3.18a) is not quite the correct one to use in $\hat{\mathscr{H}}$. It is not well defined: the ghost zero modes give zero, while the $\delta^{26}(k - k')$ contains a factor $\delta(0)$ because the momentum support is restricted to the mass shell. Therefore we use in $\hat{\mathscr{H}}$ a reduced inner product $\langle\ \|\ \rangle$ in which we simply ignore the X^0 and ghost zero modes. It is this inner product that will be relevant for the probabilistic interpretation. One can check that Q_B is still Hermitian with the reduced inner product, in the space $\hat{\mathscr{H}}$. Note that the mass-shell condition determines k^0 in terms of the spatial momentum $\mathbf{k}$ (we focus on the incoming case, $k^0 > 0$), and that we have used covariant normalization for states.

Let us now work out the first levels of the $D = 26$ flat spacetime string. At the lowest level, $N = 0$, we have

$$|0; \mathbf{k}\rangle\ , \quad -k^2 = -\frac{1}{\alpha'}\ . \tag{4.3.23}$$

This state is invariant,

$$Q_B|0; \mathbf{k}\rangle = 0\ , \tag{4.3.24}$$

because every term in Q_B contains either lowering operators or L_0. This also shows that there are no exact states at this level, so each invariant state corresponds to a cohomology class. These are just the states of the tachyon. The mass-shell condition is the same as found in the light-cone quantization of section 1.3 and from the open string vertex operators in section 3.6. The tachyon mass-squared, determined in a heuristic way in chapter 1, is here fixed by the normal ordering constant in Q_B, which is determined by the requirement of nilpotence.

At the next level, $N = 1$, there are $26 + 2$ states,

$$|\psi_1\rangle = (e \cdot \alpha_{-1} + \beta b_{-1} + \gamma c_{-1})|0; \mathbf{k}\rangle\ , \quad -k^2 = 0\ , \tag{4.3.25}$$

depending on a 26-vector e_μ and two constants, β and γ. The norm of this state is

$$
\begin{aligned}
\langle\psi_1\|\psi_1\rangle &= \langle 0; \mathbf{k}\|(e^* \cdot \alpha_1 + \beta^* b_1 + \gamma^* c_1)(e \cdot \alpha_{-1} + \beta b_{-1} + \gamma c_{-1})|0; \mathbf{k}'\rangle \\
&= (e^* \cdot e + \beta^* \gamma + \gamma^* \beta)\langle 0; \mathbf{k}\|0; \mathbf{k}'\rangle\ . \tag{4.3.26}
\end{aligned}
$$

Going to an orthogonal basis, there are 26 positive-norm states and 2 negative-norm states. The BRST condition is

$$
\begin{aligned}
0 = Q_B|\psi_1\rangle &= (2\alpha')^{1/2}(c_{-1}k \cdot \alpha_1 + c_1 k \cdot \alpha_{-1})|\psi_1\rangle \\
&= (2\alpha')^{1/2}(k \cdot e c_{-1} + \beta k \cdot \alpha_{-1})|0; \mathbf{k}\rangle\ . \tag{4.3.27}
\end{aligned}
$$

The terms proportional to c_0 sum to zero by the mass-shell condition, and are omitted. An invariant state therefore satisfies $k \cdot e = \beta = 0$. There are

26 linearly independent states remaining, of which 24 have positive norm
and 2 have zero norm, being orthogonal to all physical states including
themselves. The zero-norm invariant states are created by c_{-1} and $k \cdot \alpha_{-1}$.
A general $|\chi\rangle$ is of the same form (4.3.25) with constants e'_μ, β', γ', so the
general BRST-exact state at this level is

$$Q_B|\chi\rangle = (2\alpha')^{1/2}(k \cdot e'c_{-1} + \beta'k \cdot \alpha_{-1})|0;\mathbf{k}\rangle . \tag{4.3.28}$$

Thus the ghost state $c_{-1}|0;\mathbf{k}\rangle$ is BRST-exact, while the polarization is
transverse with the equivalence relation $e_\mu \cong e_\mu + (2\alpha')^{1/2}\beta'k_\mu$. This leaves
the 24 positive-norm states expected for a massless vector particle, in
agreement with the light-cone quantization and with the OCQ at $A = -1$.

This pattern is general: there are two extra positive-norm and two
extra negative-norm families of oscillators, as compared to the light-
cone quantization. The physical state condition eliminates two of these
and leaves two combinations with vanishing inner products. These null
oscillators are BRST-exact and are removed by the equivalence
relation.

An aside: the states $b_{-1}|0;\mathbf{k}\rangle$ and $c_{-1}|0;\mathbf{k}\rangle$ can be identified with the two
Faddeev–Popov ghosts that appear in the BRST-invariant quantization
of a massless vector field in field theory. The world-sheet BRST operator
acts on these states in the same way as the corresponding *spacetime*
BRST operator in gauge field theory. Acting on the full string Hilbert
space, the open string BRST operator is then some infinite-dimensional
generalization of the spacetime gauge theory BRST invariance, while the
closed string BRST operator is a generalization of the spacetime general
coordinate BRST invariance, in the free limit.

The generalization to the closed string is straightforward. We restrict
attention to the space $\hat{\mathscr{H}}$ of states satisfying[1]

$$b_0|\psi\rangle = \tilde{b}_0|\psi\rangle = 0 , \tag{4.3.29}$$

implying also

$$L_0|\psi\rangle = \tilde{L}_0|\psi\rangle = 0 . \tag{4.3.30}$$

In the closed string,

$$L_0 = \frac{\alpha'}{4}(p^2 + m^2) , \quad \tilde{L}_0 = \frac{\alpha'}{4}(p^2 + \tilde{m}^2) , \tag{4.3.31}$$

---

[1] S. Weinberg has pointed out that the simple argument given at the end of section 4.2 would
only lead to the weaker condition $(b_0 + \tilde{b}_0)|\psi\rangle = 0$. Nevertheless, we shall see from the detailed
form of string amplitudes in chapter 9 that there is a projection onto states satisfying conditions
(4.3.29).

where

$$\frac{\alpha'}{4}m^2 = \sum_{n=1}^{\infty} n\left(N_{bn} + N_{cn} + \sum_{\mu=0}^{25} N_{\mu n} \right) - 1 \,, \qquad (4.3.32a)$$

$$\frac{\alpha'}{4}\tilde{m}^2 = \sum_{n=1}^{\infty} n\left(\tilde{N}_{bn} + \tilde{N}_{cn} + \sum_{\mu=0}^{25} \tilde{N}_{\mu n} \right) - 1 \,. \qquad (4.3.32b)$$

Repeating the earlier exercise, we find at $m^2 = -4/\alpha'$ the tachyon, and at $m^2 = 0$ the 24×24 states of the graviton, dilaton, and antisymmetric tensor.

4.4 The no-ghost theorem

In this section we prove that the BRST cohomology of the string has a positive inner product and is isomorphic to the light-cone and OCQ spectra. We are identifying the BRST cohomology as the physical Hilbert space. We will also need to verify, in our study of string amplitudes, that the amplitudes are well defined on the cohomology (that is, that equivalent states have equal amplitudes) and that the S-matrix is unitary within the space of physical states.

We will work in the general framework described at the end of chapter 3. That is, the world-sheet theory consists of d free X^μ fields including $\mu = 0$, plus some compact unitary CFT K of central charge $26 - d$, plus the ghosts. The Virasoro generators are a sum

$$L_m = L_m^X + L_m^K + L_m^g \,. \qquad (4.4.1)$$

By compact we mean that L_0^K has a discrete spectrum. For example, in the case that K corresponds to strings on a compact manifold, the term $\alpha' p^2$ in L_0 is replaced by the Laplacian $-\alpha' \nabla^2$, which has a discrete spectrum on a compact space.

The general state is labeled

$$|N, I; k\rangle \,, \quad |N, \tilde{N}, I; k\rangle \,, \qquad (4.4.2)$$

in the open and closed strings respectively, where N (and $\tilde{N}$) refer to both the d-dimensional and ghost oscillators, k is the d-momentum, and I labels the states of the compact CFT with given boundary conditions. The b_0 condition (4.3.19) or (4.3.29) is imposed as above, implying the mass-shell condition

$$-\sum_{\mu=0}^{d-1} k_\mu k^\mu = m^2 \,, \tag{4.4.3a}$$

$$\alpha' m^2 = \sum_{n=1}^{\infty} n\left(N_{bn} + N_{cn} + \sum_{\mu=0}^{d-1} N_{\mu n} \right) + L_0^K - 1 \,, \tag{4.4.3b}$$

for the open string and

$$-\sum_{\mu=0}^{d-1} k_\mu k^\mu = m^2 = \tilde{m}^2 \,, \tag{4.4.4a}$$

$$\frac{\alpha'}{4} m^2 = \sum_{n=1}^{\infty} n\left(N_{bn} + N_{cn} + \sum_{\mu=0}^{d-1} N_{\mu n} \right) + L_0^K - 1 \,, \tag{4.4.4b}$$

$$\frac{\alpha'}{4} \tilde{m}^2 = \sum_{n=1}^{\infty} n\left(\tilde{N}_{bn} + \tilde{N}_{cn} + \sum_{\mu=0}^{d-1} \tilde{N}_{\mu n} \right) + \tilde{L}_0^K - 1 \,, \tag{4.4.4c}$$

for the closed string. That is, the contributions of the $d \le \mu \le 25$ oscillators are replaced by the eigenvalue of L_0^K or $\tilde{L}_0^K$ from the compact CFT. The only information we use about the compact CFT is that it is conformally invariant with the appropriate central charge, so there is a nilpotent BRST operator, and that it has a positive inner product. The basis I can be taken to be orthonormal, so the reduced inner products are

$$\langle 0, I\,; \mathbf{k} \| 0, I'\,; \mathbf{k}' \rangle = \langle 0, 0, I\,; \mathbf{k} \| 0, 0, I'\,; \mathbf{k}' \rangle = 2k^0 (2\pi)^{d-1} \delta^{d-1}(\mathbf{k} - \mathbf{k}') \delta_{I,I'} \,. \tag{4.4.5}$$

Now let us see what we expect for the physical Hilbert space. Define the transverse Hilbert space $\mathcal{H}^\perp$ to consist of those states in $\hat{\mathcal{H}}$ that have no longitudinal (X^0, X^1, b, or c) excitations. Since these oscillators are the source of the indefinite inner product, $\mathcal{H}^\perp$ has a positive inner product. Light-cone gauge-fixing eliminates the longitudinal oscillators directly — the light-cone Hilbert space is isomorphic to $\mathcal{H}^\perp$, as one sees explicitly in the flat-spacetime case from chapter 1. We will show, in the general case, that the BRST cohomology is isomorphic to $\mathcal{H}^\perp$. That is, it has the same number of states at each mass level, and has a positive inner product. This is the *no-ghost theorem*.

Proof

The proof has two parts. The first is to find the cohomology of a simplified BRST operator Q_1, which is quadratic in the oscillators, and the second is to show that the cohomology of the full Q_B is identical to that of Q_1.

Define the light-cone oscillators

$$\alpha_m^{\pm} = 2^{-1/2}(\alpha_m^0 \pm \alpha_m^1) \,, \qquad (4.4.6)$$

which satisfy

$$[\alpha_m^+, \alpha_n^-] = -m\delta_{m,-n} \,, \qquad [\alpha_m^+, \alpha_n^+] = [\alpha_m^-, \alpha_n^-] = 0 \,. \qquad (4.4.7)$$

We will use extensively the quantum number

$$N^{\text{lc}} = \sum_{\substack{m=-\infty \\ m \neq 0}}^{\infty} \frac{1}{m} \,{}^{\circ}_{\circ} \alpha_{-m}^+ \alpha_m^- \,{}^{\circ}_{\circ} \,. \qquad (4.4.8)$$

The number N^{lc} counts the number of $-$ excitations minus the number of $+$ excitations; it is not a Lorentz generator because the center-of-mass piece has been omitted. We choose a Lorentz frame in which the momentum component k^1 is nonzero.

Now decompose the BRST generator using the quantum number N^{lc}:

$$Q_{\text{B}} = Q_1 + Q_0 + Q_{-1} \,, \qquad (4.4.9)$$

where Q_j changes N^{lc} by j units,

$$[N^{\text{lc}}, Q_j] = jQ_j \,. \qquad (4.4.10)$$

Also, each of the Q_j increases the ghost number N^{g} by one unit:

$$[N^{\text{g}}, Q_j] = Q_j \,. \qquad (4.4.11)$$

Expanding out $Q_{\text{B}}^2 = 0$ gives

$$\left(Q_1^2\right) + \left(\{Q_1, Q_0\}\right) + \left(\{Q_1, Q_{-1}\} + Q_0^2\right) + \left(\{Q_0, Q_{-1}\}\right) + \left(Q_{-1}^2\right) = 0 \,. \qquad (4.4.12)$$

Each group in parentheses has a different N^{lc} and so must vanish separately. In particular, Q_1 itself is nilpotent, and so has a cohomology.

Explicitly,

$$Q_1 = -(2\alpha')^{1/2}k^+ \sum_{\substack{m=-\infty \\ m \neq 0}}^{\infty} \alpha_{-m}^- c_m \,. \qquad (4.4.13)$$

For $m < 0$ this destroys a $+$ mode and creates a c, and for $m > 0$ it creates a $-$ mode and destroys a b. One can find the cohomology directly by considering the action of Q_1 in the occupation basis. We will leave this as an exercise, and instead use a standard strategy which will also be useful in generalizing to Q_{B}. Define

$$R = \frac{1}{(2\alpha')^{1/2}k^+} \sum_{\substack{m=-\infty \\ m \neq 0}}^{\infty} \alpha_{-m}^+ b_m \qquad (4.4.14)$$

and

$$S \equiv \{Q_1, R\} = \sum_{m=1}^{\infty} \left(mb_{-m}c_m + mc_{-m}b_m - \alpha^+_{-m}\alpha^-_m - \alpha^-_{-m}\alpha^+_m \right)$$

$$= \sum_{m=1}^{\infty} m(N_{bm} + N_{cm} + N^+_m + N^-_m) . \tag{4.4.15}$$

The normal ordering constant is determined by noting that Q_1 and R both annihilate the ground state. Note that Q_1 commutes with S. We can then calculate the cohomology within each eigenspace of S, and the full cohomology is the union of the results. If $|\psi\rangle$ is Q_1-invariant with $S|\psi\rangle = s|\psi\rangle$, then for nonzero s

$$|\psi\rangle = \frac{1}{s}\{Q_1, R\}|\psi\rangle = \frac{1}{s}Q_1 R|\psi\rangle , \tag{4.4.16}$$

and so $|\psi\rangle$ is actually Q_1-exact. Therefore the Q_1 cohomology can be nonzero only at $s = 0$. By the definition (4.4.15) of S, the $s = 0$ states have no longitudinal excitations — the $s = 0$ space is just $\mathcal{H}^\perp$. The operator Q_1 annihilates all states in $\mathcal{H}^\perp$, so they are all Q_1-closed and there are no Q_1-exact states in this space. Therefore the cohomology is $\mathcal{H}^\perp$ itself. We have proven the no-ghost theorem, but for the operator Q_1, not Q_B.

The proof had two steps, the first being to show that the cohomology could only come from $s = 0$ states (the kernel of S), and the second to show that $s = 0$ states were Q_1-invariant. It is useful to prove the second step in a more abstract way, using the property that all $s = 0$ states have the same ghost number, in this case $-\frac{1}{2}$. Suppose $S|\psi\rangle = 0$. Since S and Q_1 commute we have

$$0 = Q_1 S|\psi\rangle = SQ_1|\psi\rangle . \tag{4.4.17}$$

The state $|\psi\rangle$ has ghost number $-\frac{1}{2}$, so $Q_1|\psi\rangle$ has ghost number $+\frac{1}{2}$. Since S is invertible at this ghost number, it must be that $Q_1|\psi\rangle = 0$ as we wished to show.

It remains to show that the cohomology of Q_B is the same as the cohomology of Q_1. The idea here is to use in place of S the operator

$$S + U \equiv \{Q_B, R\} . \tag{4.4.18}$$

Now, $U = \{Q_0 + Q_{-1}, R\}$ lowers N^{lc} by one or two units. In terms of $N^{lc}y$, S is diagonal and U is lower triangular. By general properties of lower triangular matrices, the kernel of $S + U$ can be no larger than the kernel of its diagonal part S. In fact they are isomorphic: if $|\psi_0\rangle$ is annihilated by S, then

$$|\psi\rangle = (1 - S^{-1}U + S^{-1}US^{-1}U - \ldots)|\psi_0\rangle \tag{4.4.19}$$

is annihilated by $S + U$. The factors of S^{-1} make sense because they always act on states of $N^{lc} < 0$, where S is invertible. For the same reason, $S + U$ is invertible except at ghost number $-\frac{1}{2}$. Eqs. (4.4.16) and (4.4.17) can now be applied to Q_B, with $S + U$ replacing S. They imply that the Q_B cohomology is isomorphic to the kernel of $S + U$, which is isomorphic to the kernel of S, which is isomorphic to the cohomology of Q_1, as we wished to show.

We must still check that the inner product is positive. All terms after the first on the right-hand side of eq. (4.4.19) have strictly negative N^{lc}. By the commutation relations, the inner product is nonzero only between states whose N^{lc} adds to zero. Then for two states (4.4.19) in the kernel of $S + U$,

$$\langle \psi \| \psi' \rangle = \langle \psi_0 \| \psi'_0 \rangle . \tag{4.4.20}$$

The positivity of the inner product on the kernel of $S + U$ then follows from that for the kernel of S, and we are done.

After adding the tilded operators to Q_B, N^{lc}, R, etc., the closed string proof is identical.

BRST–OCQ equivalence

We now prove the equivalence

$$\mathcal{H}_{OCQ} = \mathcal{H}_{BRST} = \mathcal{H}_{\text{light-cone}} . \tag{4.4.21}$$

To a state $|\psi\rangle$ in the matter Hilbert space, associate the state

$$|\psi, \downarrow\rangle \tag{4.4.22}$$

from the full matter plus ghost theory. Again the ghost vacuum $|\downarrow\rangle$ is annihilated by all of the ghost lowering operators, b_n for $n \geq 0$ and c_n for $n > 0$. Acting with Q_B,

$$Q_B|\psi, \downarrow\rangle = \sum_{n=0}^{\infty} c_{-n}(L_n^m - \delta_{n,0})|\psi, \downarrow\rangle = 0 . \tag{4.4.23}$$

All terms in Q_B that contain ghost lowering operators drop out, and the constant in the $n = 0$ term is known from the mode expansion of Q_B, eq. (4.3.7). Each OCQ physical state thus maps to a BRST-closed state. The ordering constant $A = -1$ arises here from the L_0 eigenvalue of the ghost vacuum.

To establish the equivalence (4.4.21), we need to show more. First, we need to show that we have a well-defined map of equivalence classes: that if ψ and ψ' are equivalent OCQ physical states, they map into the same BRST class:

$$|\psi, \downarrow\rangle - |\psi', \downarrow\rangle \tag{4.4.24}$$

must be BRST-exact. By (4.4.23), this state is BRST-closed. Further, since $|\psi\rangle' - |\psi\rangle$ is OCQ null, the state (4.4.24) has zero norm. From the BRST no-ghost theorem, the inner product on the cohomology is positive, so a zero-norm closed state is BRST-exact, as we needed to show.

To conclude that we have an isomorphism, we need to show that the map is one-to-one and onto. One-to-one means that OCQ physical states ψ and ψ' that map into the same BRST class must be in the same OCQ class — if

$$|\psi, \downarrow\rangle - |\psi', \downarrow\rangle = Q_B|\chi\rangle \,, \tag{4.4.25}$$

then $|\psi\rangle - |\psi\rangle'$ must be OCQ null. To see this, expand

$$|\chi\rangle = \sum_{n=1}^{\infty} b_{-n}|\chi_n, \downarrow\rangle + \dots \; ; \tag{4.4.26}$$

$|\chi\rangle$ has ghost number $-\frac{3}{2}$, so the ellipsis stands for terms with at least one c and two b excitations. Insert this into the form (4.4.25) and keep on both sides only terms with the ghost ground state. This gives

$$|\psi, \downarrow\rangle - |\psi', \downarrow\rangle = \sum_{m,n=1}^{\infty} c_m L^{\mathrm{m}}_{-m} b_{-n}|\chi_n, \downarrow\rangle$$

$$= \sum_{n=1}^{\infty} L^{\mathrm{m}}_{-n}|\chi_n, \downarrow\rangle \,. \tag{4.4.27}$$

Terms with ghost excitations must vanish separately, and so have been omitted. Thus, $|\psi\rangle - |\psi'\rangle = \sum_{n=1}^{\infty} L^{\mathrm{m}}_{-n}|\chi_n\rangle$ is OCQ null, and the map is one-to-one.

Finally, we must show that the map is onto, that every Q_B class contains at least one state of the form (4.4.22). In fact, the specific representatives (4.4.19), the states annihilated by $S + U$, are of this form. To see this, consider the quantum number $N' = 2N^- + N_b + N_c$, involving the total numbers of $-$, b, and c excitations. The operator R has $N' = -1$: terms in R with $m > 0$ reduce N_c by one unit and terms with $m < 0$ reduce N^- by one unit and increase N_b by one unit. Examining $Q_0 + Q_{-1}$, one finds various terms with $N' = 1$, but no greater. So $U = \{R, Q_0 + Q_{-1}\}$ cannot increase N'. Examining the state (4.4.19), and noting that S and $|\psi_0\rangle$ have $N' = 0$, we see that all terms on the right-hand side have $N' \leq 0$. By definition N' is nonnegative, so it must be that $N'|\psi\rangle = 0$. This implies no $-$, b, or c excitations, and so this state is of the form (4.4.22). Thus the equivalence (4.4.21) is shown.

The full power of the BRST method is needed for understanding the general structure of string amplitudes. However, for many practical purposes it is a great simplification to work with the special states of the form $|\psi,\downarrow\rangle$, and so it is useful to know that every BRST class contains at least one state in which the ghost modes are unexcited. We will call these OCQ-type states.

The OCQ physical state condition (4.1.5) requires that the matter state be a highest weight state with $L_0 = 1$. The corresponding vertex operator is a weight-1 tensor field $\mathcal{V}^m$ constructed from the matter fields. Including the ghost state $|\downarrow\rangle$, the full vertex operator is $c\mathcal{V}^m$. For the closed string the full vertex operator is $\tilde{c}c\mathcal{V}^m$ with $\mathcal{V}^m$ a $(1,1)$ tensor. The matter part of the vertex operator is the same as found in the Polyakov formalism in section 3.6, while the ghost part has a simple interpretation that we will encounter in the next chapter.

Eq. (4.4.19) defines an OCQ-physical state in each cohomology class. For the special case of flat spacetime this state can be constructed more explicitly by using the *Del Giudice–Di Vecchia–Fubini (DDF)* operators. To explain these we will need to develop some more vertex operator technology, so this is deferred to chapter 8.

Exercises

4.1 Extend the OCQ, for $A = -1$ and general D, to the second excited level of the open string. Verify the assertions made in the text about extra positive- and negative-norm states.

4.2 (a) In the OCQ, show that the state $L^m_{-1}|\chi\rangle$ is physical (and therefore null) if $|\chi\rangle$ is a highest weight state with $L^m_0 = 0$.
(b) Show that the state $(L^m_{-2} + \frac{3}{2}L^m_{-1}L^m_{-1})|\chi\rangle$ is physical and null if $|\chi\rangle$ is a highest weight state with $L^m_0 = -1$.

4.3 By writing out terms, demonstrate the Jacobi identity

$$\{ [Q_B, L_m], b_n\} - \{ [L_m, b_n], Q_B\} - [\{b_n, Q_B\}, L_m] = 0 \ .$$

This is the sum of cyclic permutations, with commutators or anticommutators as appropriate, and minus signs from anticommutation. Use the Jacobi identity and the (anti)commutators (2.6.24) and (4.3.6) of b_n with L_n and Q_B to show that $\{ [Q_B, L_m], b_n\}$ vanishes when the total central charge is zero. This implies that $[Q_B, L_m]$ contains no c modes. However, this commutator has $N^g = 1$, and so it must vanish: the charge Q_B is conformally invariant. Now use the $Q_B Q_B b_n$ Jacobi identity in the same way to show that Q_B is nilpotent when the total central charge is zero.

4.4 In more general situations, one encounters *graded Lie algebras*, with bosonic generators (fermion number F_I even) and fermionic generators (F_I odd). The algebra of constraints is

$$G_I G_J - (-1)^{F_I F_J} G_J G_I = i g^K{}_{IJ} G_K \ .$$

Construct a nilpotent BRST operator. You will need bosonic ghosts for the fermionic constraints.

4.5 (a) Carry out the BRST quantization for the first two levels of the closed string explicitly.

(b) Carry out the BRST quantization for the third level of the open string, $m^2 = 1/\alpha'$.

4.6 (a) Consider the operator $Q' = \alpha_{-1}^- c_1 + \alpha_1^- c_{-1}$, obtained by truncating Q_1 to the $m = \pm 1$ oscillators only. Calculate the cohomology of Q' directly, by considering its action on a general linear combination of the states

$$(b_{-1})^{N_b} (c_{-1})^{N_c} (\alpha_{-1}^+)^{N^+} (\alpha_{-1}^-)^{N^-} |0\rangle \; .$$

(b) Generalize this to the full Q_1.

5

The string S-matrix

In chapter 3 we wrote the string S-matrix as a path integral over compact two-dimensional surfaces with vertex operators. In this chapter we will reduce the path integral to gauge-fixed form.

5.1 The circle and the torus

We would like to identify a gauge slice, a choice of one configuration from each (diff×Weyl)-equivalent set as in figure 3.7. Locally we did this by fixing the metric, but globally there is a small mismatch between the space of metrics and the world-sheet gauge group.

Once again the point particle is a good illustration. We want to evaluate the Euclidean path integral

$$\int [de\, dX] \exp\left[-\frac{1}{2}\int d\tau\, (e^{-1}\dot{X}^\mu \dot{X}_\mu + em^2)\right] . \tag{5.1.1}$$

Consider a path forming a closed loop in spacetime, so the topology is a circle. The parameter τ can be taken to run from 0 to 1 with the endpoints identified. That is, $X^\mu(\tau)$ and $e(\tau)$ are periodic on $0 \le \tau \le 1$. The tetrad $e(\tau)$ has one component and there is one local symmetry, the choice of parameter, so as for the string there is just enough local symmetry to fix the tetrad completely. The tetrad transforms as $e'd\tau' = ed\tau$. The gauge choice $e' = 1$ thus gives a differential equation for $\tau'(\tau)$,

$$\frac{\partial \tau'}{\partial \tau} = e(\tau) . \tag{5.1.2}$$

Integrating this with the boundary condition $\tau'(0) = 0$ determines

$$\tau'(\tau) = \int_0^\tau d\tau''\, e(\tau'') . \tag{5.1.3}$$

145

The complication is that in general $\tau'(1) \neq 1$, so the periodicity is not preserved. In fact,

$$\tau'(1) = \int_0^1 d\tau \, e(\tau) = l \tag{5.1.4}$$

is the invariant length of the circle. So we cannot simultaneously set $e' = 1$ and keep the coordinate region fixed. We can hold the coordinate region fixed and set e' to the constant value $e' = l$, or set $e' = 1$ and let the coordinate region vary:

$$e' = l, \quad 0 \leq \tau \leq 1, \tag{5.1.5a}$$

or

$$e' = 1, \quad 0 \leq \tau \leq l. \tag{5.1.5b}$$

In either case, after fixing the gauge invariance we are left with an ordinary integral over l. In other words, not all tetrads on the circle are diff-equivalent. There is a one-parameter family of inequivalent tetrads, parameterized by l.

Both descriptions (5.1.5) will have analogs in the string. In practice, we will define the path integral using a description like (5.1.5a), where the fields are functions on a fixed coordinate region, and then convert to a description like (5.1.5b) where the metric is fixed and the moduli are encoded in the coordinate range.

There is a second complication with the gauge-fixing. The condition $e = \text{constant}$ is preserved by the rigid translation

$$\tau \to \tau + v \mod 1. \tag{5.1.6}$$

That is, we have not said where on the circle to put the origin of the coordinate system. Fixing the metric thus leaves a small part of the local symmetry unfixed. So there is a small mismatch in *both* directions, metrics that are not gauge-equivalent to one another and gauge transformations that are not fixed by the choice of metric.

Moving on to the string we take as an example the torus, where the same complications arise. Start with the coordinate region

$$0 \leq \sigma^1 \leq 2\pi, \quad 0 \leq \sigma^2 \leq 2\pi, \tag{5.1.7}$$

with $X^\mu(\sigma^1, \sigma^2)$ and $g_{ab}(\sigma_1, \sigma_2)$ periodic in both directions. Equivalently we can think of this as the σ plane with the identification of points,

$$(\sigma^1, \sigma^2) \cong (\sigma^1, \sigma^2) + 2\pi(m, n) \tag{5.1.8}$$

for integer m and n.

To what extent is the field space diff$\times$Weyl redundant? The theorem is that it is *not* possible to bring a general metric to unit form by a

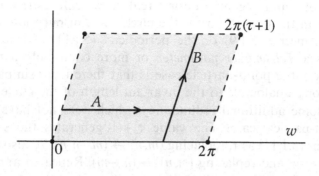

Fig. 5.1. Torus with modulus τ, in gauge $ds^2 = dwd\bar{w}$. The upper and lower edges are identified, as are the right- and left-hand edges. Closed curves A and B are marked for later reference.

diff×Weyl transformation that leaves invariant the periodicity (5.1.7), but it *is* possible to bring it to the form

$$ds^2 = |d\sigma^1 + \tau d\sigma^2|^2 \qquad (5.1.9)$$

where τ is a complex constant. For $\tau = i$ this would be the unit metric δ_{ab}.

To see this, repeat the steps used in the local discussion in section 3.3. We can first make the metric flat by a Weyl transformation satisfying $2\nabla^2\omega = R$. By expanding in eigenmodes of ∇^2, one sees that with the periodic boundary conditions this has a solution unique up to addition of a constant to ω. It is important here that $\int d^2\sigma \, g^{1/2}R$, which is 4π times the Euler number, vanishes for the torus. A transformation to new coordinates $\tilde{\sigma}^a$ then brings the metric to unit form. However, as in the point-particle case, there is no guarantee that this respects the original periodicity. Rather, we may now have

$$\tilde{\sigma}^a \cong \tilde{\sigma}^a + 2\pi(mu^a + nv^a) \qquad (5.1.10)$$

with general translations u^a and v^a. By rotating and rescaling the coordinate system, accompanied by a shift in ω to keep the metric normalized, we can always set $u = (1,0)$. This leaves two parameters, the components of v. Defining $w = \tilde{\sigma}^1 + i\tilde{\sigma}^2$, the metric is $dwd\bar{w}$ and the periodicity is

$$w \cong w + 2\pi(m + n\tau) , \qquad (5.1.11)$$

where $\tau = v^1 + iv^2$. The torus is a parallelogram in the w-plane with periodic boundary conditions, as shown in figure 5.1.

Alternatively, define σ^a by $w = \sigma^1 + \tau\sigma^2$. The original periodicity (5.1.8) is preserved but the metric now takes the more general form (5.1.9). The integration over metrics reduces to two ordinary integrals, over the real and imaginary parts of τ. The metric (5.1.9) is invariant under complex conjugation of τ and degenerate for τ real, so we can restrict attention to $\mathrm{Im}\,\tau > 0$. As in the case (5.1.5) of the circle, we can put these parameters either in the metric (5.1.9) or the periodicity (5.1.11). The parameter τ is known as a *Teichmüller parameter* or more commonly a *modulus*. A difference from the point-particle case is that there is no simple invariant expression for τ analogous to the invariant length of the circle.

There is some additional redundancy, which does not have an analog in the point-particle case. The value $\tau + 1$ generates the same set of identifications (5.1.11) as τ, replacing $(m, n) \to (m-n, n)$. So also does $-1/\tau$, defining $w' = \tau w$ and replacing $(m, n) \to (n, -m)$. Repeated application of these two transformations,

$$T : \quad \tau' = \tau + 1\,, \quad S : \quad \tau' = -1/\tau\,, \tag{5.1.12}$$

generates

$$\tau' = \frac{a\tau + b}{c\tau + d} \tag{5.1.13}$$

for all integer a, b, c, d such that $ad - bc = 1$.

We can also think about this as follows. The transformation

$$\begin{bmatrix} \sigma^1 \\ \sigma^2 \end{bmatrix} = \begin{bmatrix} d & b \\ c & a \end{bmatrix} \begin{bmatrix} \sigma'^1 \\ \sigma'^2 \end{bmatrix} \tag{5.1.14}$$

takes the metric (5.1.9) for σ into a metric of the same form in σ' but with the modulus τ'. This is a diffeomorphism of the torus. It is one-to-one as a consequence of $ad - bc = 1$ and it preserves the periodicity (5.1.8). However, it cannot be obtained from the identity by successive infinitesimal transformations — it is a so-called *large coordinate transformation*. The curve A in the coordinate σ maps to a curve in the σ' coordinate that runs a times in the A' direction and $-c$ times in the B' direction. These large coordinate transformations form the group $SL(2, \mathbf{Z})$, integer-valued 2×2 matrices with unit determinant. The group on the τ-plane is $SL(2, \mathbf{Z})/\mathbf{Z}_2 = PSL(2, \mathbf{Z})$, because τ' is unchanged if all the signs of a, b, c, d are reversed. The group of transformations (5.1.14) is known as the *modular group*.

Using the modular transformations (5.1.13), it can be shown that every τ is equivalent to exactly one point in the region F_0 shown in figure 5.2,

$$-\frac{1}{2} \le \mathrm{Re}\,\tau \le \frac{1}{2}\,, \quad |\tau| \ge 1\,, \tag{5.1.15}$$

except on the boundaries which are identified as shown in the figure. This is called a *fundamental region* for the upper half-plane mod $PSL(2, \mathbf{Z})$. Because of the identifications one should think of F_0 as being rolled up, and open only at $\mathrm{Im}\,\tau \to \infty$. The fundamental region F_0 is one representation of the moduli space of (diff×Weyl)-inequivalent metrics.

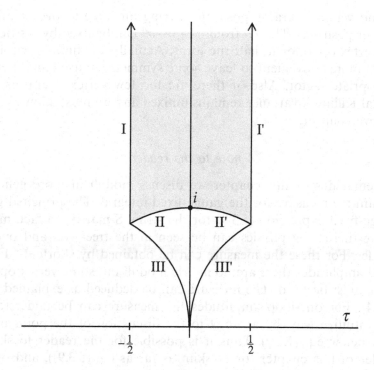

Fig. 5.2. The standard fundamental region F_0 for the moduli space of the torus, shaded. The lines I and I' are identified, as are the arcs II and II'. A different fundamental region, mapped into F_0 by the modular transformation S, is bounded by II, II', III, and III'.

There is a further complication as there was for the point particle. Requiring the metric to take the form (5.1.9) with τ in a given fundamental region does not fix all of the diff×Weyl invariance. The metric and periodicity are left invariant by *rigid* translations,

$$\sigma^a \to \sigma^a + v^a , \qquad (5.1.16)$$

so this two-parameter subgroup of diff×Weyl is not fixed. In addition, the discrete transformation $\sigma^a \to -\sigma^a$ leaves the metric invariant, and in the unoriented case also $(\sigma^1, \sigma^2) \to (-\sigma^1, \sigma^2)$ with $\tau \to -\bar\tau$. Thus there are again two kinds of mismatch between the metric degrees of freedom and the local symmetries: parameters in the metric that cannot be removed by the symmetries, and symmetries that are not fixed by the choice of metric. The unfixed symmetries are known as the *conformal Killing group (CKG)*.

When there are enough vertex operators in the amplitude, we can fully fix the gauge invariance by fixing some of the vertex operator positions. On the torus with n vertex operators, the invariance (5.1.16) can be used to fix one vertex operator position, leaving an integral over τ and the $n-1$ other positions. The $\mathbf{Z}_2$ from $\sigma^a \to -\sigma^a$ can be fixed by restricting a second vertex operator to half the torus. Actually, for finite overcounting it is often more convenient to leave some symmetry unfixed and divide by an appropriate factor. Also, if there are too few vertex operators, some conformal Killing invariance remains unfixed and we must allow explicitly for the overcounting.

A note to the reader

In the remainder of this chapter we discuss moduli in more generality, and obtain the measure for the gauge-fixed integral. The principal goal is the gauge-fixed expression (5.3.9) for the string S-matrix. In fact, most of the interesting string physics can be seen in the tree-level and one-loop amplitudes. For these the measure can be obtained by shortcuts. For the tree-level amplitudes there are no metric moduli but some vertex operator positions must be fixed. The measure can be deduced as explained below eq. (6.4.4). For one-loop amplitudes the measure can be understood in a fairly intuitive way, by analogy to the discussion of the point particle measure below eq. (7.3.9). Thus it is possible for the reader to skip the remainder of this chapter (or to skim as far as eq. (5.3.9)), and over all sections dealing with the Faddeev–Popov ghosts.

5.2   Moduli and Riemann surfaces

Now let us repeat the preceding discussion in a more general and abstract way. We start with the integral over all metrics on some given topology r. Call the space of metrics $\mathscr{G}_r$. For closed oriented surfaces we can label r simply by the number g of handles, also known as the *genus*. After taking into account the diff$\times$Weyl redundancy, we are left with the moduli space

$$\mathscr{M}_r = \frac{\mathscr{G}_r}{(\text{diff} \times \text{Weyl})_r} \, . \tag{5.2.1}$$

As in the case of the torus, this space is parameterized by a finite number of *moduli*. For the torus, $\mathscr{M}_1$ is the upper half-plane mod $PSL(2, \mathbf{Z})$, or equivalently any fundamental region, say F_0. There may also be a subgroup of diff$\times$Weyl, the CKG, that leaves the metric invariant.

When there are vertex operators in the path integral it is useful to treat their positions on the same footing as the moduli from the metric, referring to all as moduli. We specify the *metric moduli* when we need to make the distinction. One way to deal with the CKG, which is applicable if there are vertex operators in the path integral, is to specify the gauge further by fixing the coordinates of some vertex operators. The Polyakov path integral includes an integral over $\mathscr{G}_r$ and integrals of the n vertex operator positions over the world-sheet $\mathscr{M}$. The moduli space at topology r with n vertex operators is then

$$\mathscr{M}_{r,n} = \frac{\mathscr{G}_r \times \mathscr{M}^n}{(\text{diff} \times \text{Weyl})_r} \, . \tag{5.2.2}$$

As we have seen for the torus, diff_r is not in general connected. Picking out the connected component diff_{r0} that contains the identity, the quotient

$$\frac{\text{diff}_r}{\text{diff}_{r0}} \tag{5.2.3}$$

is the modular group.

It is interesting to study the diff×Weyl redundancy of the metric at the infinitesimal level. That is, we are looking for small variations of the metric that are not equivalent to diff×Weyl transformations and so correspond to changes in the moduli. We are also looking for small diff×Weyl transformations that do not change the metric; these would be infinitesimal elements of the CKG, called *conformal Killing vectors (CKVs)*. This will give further insight into the origin of the moduli and the CKG.

An infinitesimal diff×Weyl transformation changes the metric by

$$\delta g_{ab} = -2(P_1 \delta \sigma)_{ab} + (2\delta\omega - \nabla \cdot \delta\sigma)g_{ab} \, , \tag{5.2.4}$$

where P_1 is the traceless symmetric linear combination of derivatives (3.3.17). Moduli correspond to variations $\delta' g_{ab}$ of the metric that are orthogonal to all variations (5.2.4):

$$
\begin{aligned}
0 &= \int d^2\sigma \, g^{1/2} \, \delta' g_{ab} \left[-2(P_1 \delta\sigma)^{ab} + (2\delta\omega - \nabla \cdot \delta\sigma)g^{ab} \right] \\
&= \int d^2\sigma \, g^{1/2} \left[-2(P_1^T \delta' g)_a \delta\sigma^a + \delta' g_{ab} g^{ab} (2\delta\omega - \nabla \cdot \delta\sigma) \right] . \tag{5.2.5}
\end{aligned}
$$

The transpose $(P_1^T u)_a = -\nabla^b u_{ab}$ is defined as in exercise 3.2. In order for the overlap (5.2.5) to vanish for general $\delta\omega$ and $\delta\sigma$ we need

$$g^{ab} \delta' g_{ab} = 0 \, , \tag{5.2.6a}$$
$$(P_1^T \delta' g)_a = 0 \, . \tag{5.2.6b}$$

The first condition requires that $\delta' g_{ab}$ be traceless, and it is on traceless symmetric tensors that P_1^T acts. For each solution of these equations there will be a modulus.

CKVs are transformations (5.2.4) such that $\delta g_{ab} = 0$. The trace of this equation fixes $\delta\omega$ uniquely, leaving the *conformal Killing equation*

$$(P_1 \delta\sigma)_{ab} = 0 \, . \tag{5.2.7}$$

Eqs. (5.2.6) and (5.2.7) become simple for variations around conformal gauge,

$$\partial_{\bar{z}}\delta'g_{zz} = \partial_z\delta'g_{\bar{z}\bar{z}} = 0, \tag{5.2.8a}$$

$$\partial_{\bar{z}}\delta z = \partial_z\delta\bar{z} = 0, \tag{5.2.8b}$$

so variations of the moduli correspond to *holomorphic quadratic differentials* and CKVs to *holomorphic vector fields*. On the torus, the only holomorphic doubly periodic functions are the constants, so there are two real moduli and two real CKVs, in agreement with the discussion in the previous section.

The metric moduli correspond to the kernel (5.2.6) of P_1^T and the CKVs to the kernel (5.2.7) of P_1. The Riemann–Roch theorem relates the number of metric moduli $\mu = \dim\ker P_1^T$ and the number of CKVs $\kappa = \dim\ker P_1$ to the Euler number χ,

$$\mu - \kappa = -3\chi. \tag{5.2.9}$$

We will be able to derive this later in the chapter. For a closed oriented surface $-3\chi = 6g - 6$. This counting refers to real moduli, separating complex moduli such as τ into their real and imaginary parts.

Further, κ vanishes for $\chi < 0$ and μ vanishes for $\chi > 0$. To show this, we first cite the fact that it is always possible by a Weyl transformation to find a metric for which the scalar curvature R is constant. The sign of R is then the same as that of χ. Now, $P_1^T P_1 = -\frac{1}{2}\nabla^2 - \frac{1}{4}R$, so

$$\int d^2\sigma\, g^{1/2}(P_1\delta\sigma)_{ab}(P_1\delta\sigma)^{ab} = \int d^2\sigma\, g^{1/2}\delta\sigma_a(P_1^T P_1\delta\sigma)^a$$

$$= \int d^2\sigma\, g^{1/2}\left(\frac{1}{2}\nabla_a\delta\sigma_b\nabla^a\delta\sigma^b - \frac{R}{4}\delta\sigma_a\delta\sigma^a\right). \tag{5.2.10}$$

For negative χ the right-hand side is strictly positive, so $P_1\delta\sigma$ cannot vanish. A similar argument shows that $P_1^T\delta'g$ cannot vanish for positive χ. Combining these results we have

$$\chi > 0: \quad \kappa = 3\chi, \quad \mu = 0, \tag{5.2.11a}$$

$$\chi < 0: \quad \kappa = 0, \quad \mu = -3\chi. \tag{5.2.11b}$$

Riemann surfaces

The familiar way to describe moduli space would be to choose one metric from each equivalence class. This would be a family of metrics $\hat{g}_{ab}(t;\sigma)$, depending on the moduli t^k. For the torus, eq. (5.1.9) gives such a slice. It is often convenient to use an alternative description of the form (5.1.11), where the metric $dw d\bar{w}$ is fixed and the moduli are encoded in the coordinate region. We can formalize this idea as follows.

Let us first recall how *differentiable* manifolds are defined. One covers the manifold with a set of overlapping patches, with coordinates σ_m^a in the mth patch, with a running from 1 to the dimension of the manifold. When patches m and n overlap, the coordinates in the two patches are related by

$$\sigma_m^a = f_{mn}^a(\sigma_n) , \qquad (5.2.12)$$

where the transition functions f_{mn} are required to be differentiable a given number of times. For a *Riemannian* manifold a metric $g_{m,ab}(\sigma_m)$ is also given in each patch, with the usual tensor transformation law relating the values in the overlaps.

For a *complex* manifold, there are complex coordinates z_m^a in each patch, where now a runs from 1 to half the dimension of the manifold. The transition functions are required to be holomorphic,

$$z_m^a = f_{mn}^a(z_n) . \qquad (5.2.13)$$

One can now define holomorphic functions on the manifold, since the holomorphicity will not depend on which coordinates z_m^a are used. Just as two differentiable manifolds are equivalent if there is a one-to-one differentiable map between them, two complex manifolds are equivalent if there is a one-to-one holomorphic map between them. In particular, making a holomorphic change of coordinates within each patch gives an equivalent surface.

In the two dimensional case (one complex coordinate), a complex manifold is known as a *Riemann surface*. In this case there is a one-to-one correspondence

$$\text{Riemann surfaces} \leftrightarrow \text{Riemannian manifolds mod Weyl} . \qquad (5.2.14)$$

On the right we put only 'mod Weyl' because 'mod diff' is already implicit in the definition of a Riemannian manifold. To see this isomorphism, start with a Riemannian manifold. We know from our discussion of conformal gauge that we can find in each coordinate patch a coordinate z_m such that

$$ds^2 \propto dz_m d\bar{z}_m . \qquad (5.2.15)$$

In neighboring patches the coordinate need not be the same, but because $ds^2 \propto dz_m d\bar{z}_m \propto dz_n d\bar{z}_n$, the transition functions will be holomorphic.[1] This is the map from Riemannian manifolds to complex manifolds in two real dimensions. For the inverse map, one can take the metric $dz_m d\bar{z}_m$ in

---

[1] To be precise, an antiholomorphic transition function is also possible. For the oriented string we forbid this by choosing a standard orientation; for the unoriented string it is allowed. The term Riemann surface in the strict sense refers to an oriented manifold without boundary.

the mth patch and smooth it in the overlaps (this is a standard construction, the partition of unity) to produce a Riemannian manifold. The two characterizations of a Riemann surface are thus equivalent.

The description of the torus in terms of the complex coordinate w of the parallelogram (figure 5.1) illustrates the idea of a complex manifold. Imagine taking a single coordinate patch which is a little larger than the fundamental parallelogram in figure 5.1. The periodicity conditions

$$w \cong w + 2\pi \cong w + 2\pi\tau \tag{5.2.16}$$

are then the transition functions on the overlap between the opposite edges of the patch. The surface is defined by these transition functions. To define the original path integral over metrics, it is simplest to start by regarding the metric as a function of fixed coordinates such as the square (5.1.7), or at least on a fixed set of coordinate patches with fixed transition functions. To study the quantum field theory on a given surface, it then is easiest to work with the unit metric with moduli-dependent transition functions, as in the parallelogram (5.1.11).

One can also think of a Riemann surface as follows. Defining the world-sheet as a union of patches, we can use the diff×Weyl freedom to reach the metric $dz\,d\bar{z}$ in each. The gauge choices on overlapping patches can then differ only by diff×Weyl invariances of this metric, which as we have discussed are the conformal transformations. A Riemann surface is therefore the natural place for a CFT to live, because the fields in a CFT have definite transformation properties under conformal transformations.

5.3   The measure for moduli

We now revisit the gauge-fixing of the Polyakov path integral. We have learned that the gauge redundancy does not completely eliminate the path integral over metrics but leaves behind a finite-dimensional integral over moduli space. It is necessary to refine the discussion from section 3.3 in order to take this into account.

The Polyakov path integral for the S-matrix is

$$S_{j_1 \dots j_n}(k_1, \dots, k_n)$$
$$= \sum_{\substack{\text{compact} \\ \text{topologies}}} \int \frac{[d\phi \, dg]}{V_{\text{diff} \times \text{Weyl}}} \exp(-S_{\text{m}} - \lambda\chi) \prod_{i=1}^{n} \int d^2\sigma_i \, g(\sigma_i)^{1/2} \mathscr{V}_{j_i}(k_i, \sigma_i) \,, \tag{5.3.1}$$

which is the earlier expression (3.5.5) but now written with a general $c = \tilde{c} = 26$ matter theory, the matter fields being denoted ϕ. In gauge-fixing, the integral over metrics and positions is converted to an integral over the gauge group, the moduli, and the unfixed positions,

$$[dg]\, d^{2n}\sigma \to [d\zeta]\, d^\mu t\, d^{2n-\kappa}\sigma . \tag{5.3.2}$$

After factoring out the gauge volume, the Jacobian for this transformation becomes the measure on moduli space.

The steps are the same as in section 3.3, now taking account of the moduli and the CKVs. In particular, the gauge choice will now fix κ of the vertex operator coordinates, $\sigma_i^a \to \hat{\sigma}_i^a$. Label the set of fixed coordinates (a, i) by f. Define the Faddeev–Popov measure on moduli space by

$$1 = \Delta_{\text{FP}}(g, \sigma) \int_F d^\mu t \int_{\text{diff}\times\text{Weyl}} [d\zeta]\, \delta(g - \hat{g}(t)^\zeta) \prod_{(a,i)\in f} \delta(\sigma_i^a - \hat{\sigma}_i^{\zeta a}) . \tag{5.3.3}$$

By definition, every metric is (diff$\times$Weyl)-equivalent to $\hat{g}(t)$ for at least one value of t and ζ. The delta function for the fixed coordinates then picks out a unique value of ζ. Actually, as discussed at the end of section 5.1, there may be a residual discrete group of symmetries of finite order n_R, so the delta functions are nonzero at n_R points.

Inserting the expression (5.3.3) into the path integral (5.3.1) and following the same steps as before leads to

$$S_{j_1\ldots j_n}(k_1, \ldots, k_n) = \sum_{\substack{\text{compact}\\\text{topologies}}} \int_F d^\mu t\, \Delta_{\text{FP}}(\hat{g}(t), \hat{\sigma}) \int [d\phi] \int \prod_{(a,i)\notin f} d\sigma_i^a$$

$$\times \exp\left(-S_{\text{m}}[\phi, \hat{g}(t)] - \lambda\chi\right) \prod_{i=1}^n \left[\hat{g}(\sigma_i)^{1/2} \mathscr{V}_{j_i}(k_i; \sigma_i)\right] . \tag{5.3.4}$$

The integral over metrics and vertex operator coordinates is now reduced to an integral over the moduli space F for the metric and over the unfixed coordinates, with the measure given by Δ_{FP}. In the vertex operators, κ of the positions are fixed.

Now we evaluate the Faddeev–Popov measure. The delta functions are nonzero at n_R points that are related by symmetry, so we consider one such point and divide by n_R. Expand the definition (5.3.3) of Δ_{FP} near the point. The general metric variation is equal to a local symmetry variation plus a change in the moduli t^k,

$$\delta g_{ab} = \sum_{k=1}^\mu \delta t^k \partial_{t^k} \hat{g}_{ab} - 2(\hat{P}_1 \delta\sigma)_{ab} + (2\delta\omega - \hat{\nabla}\cdot\delta\sigma)\hat{g}_{ab} . \tag{5.3.5}$$

The inverse Faddeev–Popov determinant is then

$$\Delta_{\text{FP}}(g, \sigma)^{-1}$$

$$= n_R \int d^\mu \delta t\, [d\delta\omega\, d\delta\sigma]\, \delta(\delta g_{ab}) \prod_{(a,i)\in f} \delta(\delta\sigma^a(\hat{\sigma}_i))$$

$$= n_R \int d^\mu \delta t \, d^\kappa x \, [d\beta' \, d\delta\sigma]$$

$$\times \exp\left[2\pi i(\beta', 2\hat{P}_1 \delta\sigma - \delta t^k \partial_k \hat{g}) + 2\pi i \sum_{(a,i)\in f} x_{ai} \delta\sigma^a(\hat\sigma_i) \right] , \qquad (5.3.6)$$

the inner product in the second line having been defined in exercise 3.2. We have followed the same steps as in the local discussion in section 3.3, writing the delta functions and functionals as integrals over x and β_{ab} and then integrating out $\delta\omega$ to obtain the constraint that β'_{ab} be traceless.

As before, invert the integral (5.3.6) by replacing all bosonic variables with Grassmann variables:

$$\delta\sigma^a \to c^a , \qquad (5.3.7a)$$

$$\beta'_{ab} \to b_{ab} , \qquad (5.3.7b)$$

$$x_{ai} \to \eta_{ai} , \qquad (5.3.7c)$$

$$\delta t^k \to \xi^k . \qquad (5.3.7d)$$

Then with a convenient normalization of the fields,

$$\Delta_{\rm FP}(\hat{g}, \hat\sigma) = \frac{1}{n_R} \int [db \, dc] \, d^\mu\xi \, d^\kappa\eta$$

$$\times \exp\left[-\frac{1}{4\pi}(b, 2\hat{P}_1 c - \xi^k \partial_k \hat{g}) + \sum_{(a,i)\in f} \eta_{ai} c^a(\hat\sigma_i) \right]$$

$$= \frac{1}{n_R} \int [db \, dc] \, \exp(-S_{\rm g}) \prod_{k=1}^\mu \frac{1}{4\pi}(b, \partial_k \hat{g}) \prod_{(a,i)\in f} c^a(\hat\sigma_i) . \qquad (5.3.8)$$

In the final line we have integrated over the Grassmann parameters η_{ai} and ξ^k.

The appropriate measure for integration on moduli space is generated by this ghost path integral with insertions. We will not try to keep track of the overall sign at intermediate steps; since this is supposed to be a Jacobian, we implicitly choose the overall sign to give a positive result. The full expression for the S-matrix is then

$$S_{j_1 \dots j_n}(k_1, \dots, k_n) = \sum_{\substack{\text{compact} \\ \text{topologies}}} \int_F \frac{d^\mu t}{n_R} \int [d\phi \, db \, dc] \, \exp(-S_{\rm m} - S_{\rm g} - \lambda\chi)$$

$$\times \prod_{(a,i)\notin f} \int d\sigma_i^a \prod_{k=1}^\mu \frac{1}{4\pi}(b, \partial_k \hat{g}) \prod_{(a,i)\in f} c^a(\hat\sigma_i) \prod_{i=1}^n \hat{g}(\sigma_i)^{1/2} \mathscr{V}_{j_i}(k_i, \sigma_i) .$$

$$(5.3.9)$$

This result extends readily to all bosonic string theories — closed and open, oriented and unoriented — the only difference being which topologies are included and which vertex operators allowed. Eq. (5.3.9) is a useful and elegant result. The complications caused by the moduli and the CKVs are taken into account by the c and b insertions in the path integral. For each fixed coordinate, $d\sigma_i^a$ is replaced by c_i^a, while each metric modulus gives rise to a b insertion.

Expression in terms of determinants

In eq. (5.3.8) we have represented the Faddeev–Popov determinant as an integral over Grassmann parameters and fields. We now reduce it to a product of finite-dimensional and functional determinants. When we go on to study the string S-matrix in coming chapters, this direct path integral approach is just one of the methods we will use.

Expand the ghost fields in suitable complete sets,

$$c^a(\sigma) = \sum_J c_J C_J^a(\sigma) , \quad b_{ab}(\sigma) = \sum_K b_K B_{Kab}(\sigma) . \tag{5.3.10}$$

The complete sets C_J^a and B_{Kab} are defined as follows. The derivative P_1 appearing in the ghost action takes vectors into traceless symmetric tensors. Its transpose does the reverse. The ghost action can be written with either,

$$S_g = \frac{1}{2\pi}(b, P_1 c) = \frac{1}{2\pi}(P_1^T b, c) . \tag{5.3.11}$$

We cannot diagonalize P_1 in a diff-invariant way because it turns one kind of field into another, but we can diagonalize $P_1^T P_1$ and $P_1 P_1^T$:

$$P_1^T P_1 C_J^a = v_J'^2 C_J^a , \quad P_1 P_1^T B_{Kab} = v_K^2 B_{Kab} . \tag{5.3.12}$$

The eigenfunctions can be chosen real, and are normalized in the respective inner products

$$(C_J, C_{J'}) = \int d^2\sigma \, g^{1/2} C_J^a C_{J'a} = \delta_{JJ'} , \tag{5.3.13a}$$

$$(B_K, B_{K'}) = \int d^2\sigma \, g^{1/2} B_{Kab} B_{K'}^{ab} = \delta_{KK'} . \tag{5.3.13b}$$

Now note that

$$(P_1 P_1^T) P_1 C_J = P_1 (P_1^T P_1) C_J = v_J'^2 P_1 C_J , \tag{5.3.14}$$

so $P_1 C_J$ is an eigenfunction of $P_1 P_1^T$. In the same way, $P_1^T B_K$ is an eigenfunction of $P_1^T P_1$. Thus there is a one-to-one correspondence between the eigenfunctions, *except* when $P_1 C_J = 0$ or $P_1^T B_K = 0$. The latter correspond to zero eigenvalues of $P_1^T P_1$ or $P_1 P_1^T$. These are just the CKVs and holomorphic quadratic differentials, and so their numbers are respectively κ and μ. Let us label the eigenfunctions with zero eigenvalue as C_{0j} or B_{0k}, while the nonzero eigenvalues are labeled with $J, K = 1,\dots.$

For the latter, the normalized eigenfunctions are related

$$B_{Jab} = \frac{1}{v_J}(P_1 C_J)_{ab} , \quad v_J = v'_J \neq 0 .$$ (5.3.15)

In terms of modes, the ghost path integral Δ_{FP} becomes

$$\int \prod_{k=1}^{\mu} db_{0k} \prod_{j=1}^{\kappa} dc_{0j} \prod_J db_J \, dc_J \, \exp\left(-\frac{v_J b_J c_J}{2\pi}\right) \prod_{k'=1}^{\mu} \frac{1}{4\pi}(b, \partial_{k'}\hat{g}) \prod_{(a,i)\in f} c^a(\sigma_i) .$$ (5.3.16)

From section A.2, a Grassmann integral vanishes unless the variable appears in the integrand. The c_{0j} and b_{0k} do not appear in the action but only in the insertions. In fact, the number of insertions of each type in the integral (5.3.16), κ and μ respectively, just matches the number of ghost zero modes. We have just enough insertions to give a nonzero answer, and only the zero-mode part of the ghost fields contributes in the insertions. Thus,

$$\Delta_{\text{FP}} = \int \prod_{k=1}^{\mu} db_{0k} \prod_{k'=1}^{\mu} \left[\sum_{k''=1}^{\mu} \frac{b_{0k''}}{4\pi}(B_{0k''}, \partial_{k'}\hat{g})\right]$$

$$\times \int \prod_{j=1}^{\kappa} dc_{0j} \prod_{(a,i)\in f} \left[\sum_{j'=1}^{\kappa} c_{0j'} C_{0j'}^a(\sigma_i)\right]$$

$$\times \int \prod_J db_J \, dc_J \, \exp\left(-\frac{v_J b_J c_J}{2\pi}\right) .$$ (5.3.17)

Summing over all ways of saturating the zero-mode integrals with the Grassmann variables generates a finite-dimensional determinant in each case, while the nonzero modes produce an infinite product, a functional determinant. In all,

$$\Delta_{\text{FP}} = \det \frac{(B_{0k}, \partial_{k'}\hat{g})}{4\pi} \, \det C_{0j}^a(\sigma_i) \, \det'\left(\frac{P_1^T P_1}{4\pi^2}\right)^{1/2} .$$ (5.3.18)

Note that $C_{0j}^a(\sigma_i)$ is a square matrix, in that $(a,i) \in f$ runs over κ values as does j. The prime on the functional determinant denotes omission of zero eigenvalues.

The Riemann–Roch theorem

We can now give a path integral derivation of the Riemann–Roch theorem. For the ghost current of a holomorphic bc system (without $\tilde{b}, \tilde{c}$), the current conservation anomaly derived in exercise 3.6 is

$$\nabla_a j^a = \frac{1-2\lambda}{4} R .$$ (5.3.19)

Noether's theorem relates conserved currents to invariances. For a current that is not conserved, one finds by the same argument that

$$\frac{\delta\left([d\phi]\exp(-S)\right)}{[d\phi]\exp(-S)} = \frac{i\epsilon}{2\pi}\int d^2\sigma\, g^{1/2}\nabla_a j^a \rightarrow -i\epsilon\frac{2\lambda-1}{2}\chi\,. \qquad (5.3.20)$$

The ghost number symmetry acts as $\delta b = -i\epsilon b$, $\delta c = i\epsilon c$. When the path integral is nonvanishing, the transformations of the measure, action, and insertions must cancel. Thus we learn that from the anomaly the number of c minus the number of b insertions is $3\chi/2$. The path integral calculation relates the number of insertions to the number of zero modes and so gives for the same difference

$$\frac{1}{2}(\kappa - \mu) = \frac{1}{2}(\dim\ker P_1 - \dim\ker P_1^T)\,. \qquad (5.3.21)$$

The factor of $\frac{1}{2}$ appears because we have taken the holomorphic theory only; the antiholomorphic theory gives an equal contribution. Equating the result from the anomaly with the result from the zero modes gives the Riemann–Roch theorem $\kappa - \mu = 3\chi$. For the general bc system one finds in the same way that

$$\dim\ker P_n - \dim\ker P_n^T = (2n+1)\chi\,, \qquad (5.3.22)$$

the operators being defined in exercise 3.2.

5.4 More about the measure

Here we collect some general properties of the gauge-fixed string amplitude. These are primarily useful for the more formal considerations in chapter 9.

Gauge invariance

The Faddeev–Popov procedure guarantees that the gauge-fixed amplitude is BRST-invariant and independent of gauge choice, but it is useful to check explicitly that it has all the expected invariance properties. First, it is independent of the choice of coordinates t on moduli space. For new coordinates $t'^k(t)$,

$$d^\mu t' = \left|\det\frac{\partial t'}{\partial t}\right|d^\mu t\,,$$

$$\prod_{k=1}^\mu \frac{1}{4\pi}(b, \partial_k'\hat{g}) = \det\left(\frac{\partial t}{\partial t'}\right)\prod_{k=1}^\mu \frac{1}{4\pi}(b, \partial_k\hat{g})\,, \qquad (5.4.1)$$

and the two Jacobians cancel (up to a sign, which as we have noted must be fixed by hand to give a positive measure). In other words, due to the b-ghost insertions the integrand transforms as a density on moduli space.

Second, let us see that the measure is invariant under a Weyl transformation of the gauge slice. Under a Weyl transformation,

$$\delta \hat{g}'_{ab}(t;\sigma) = 2\delta\omega(t;\sigma)\hat{g}_{ab}(t;\sigma) , \tag{5.4.2}$$

the variation of the action and measure gives an insertion of the local operator $T^a{}_a$, which vanishes in $D = 26$ by the equations of motion. We need therefore worry only about the effect on the various insertions. The vertex operator insertions are Weyl-invariant by construction. The insertions $c^a(\sigma_i)$ are Weyl-invariant as discussed below eq. (3.3.24). For the b insertions,

$$
\begin{aligned}
(b, \partial_k \hat{g}') &= \int d^2\sigma \, \hat{g}'^{1/2} \, b_{ab} \hat{g}'^{ac} \hat{g}'^{bd} \partial_k \hat{g}'_{cd} \\
&= \int d^2\sigma \, \hat{g}^{1/2} \, b_{ab} (\hat{g}^{ac} \hat{g}^{bd} \partial_k \hat{g}_{cd} + 2\hat{g}^{ab} \partial_k \omega) \\
&= (b, \partial_k \hat{g}) ,
\end{aligned}
\tag{5.4.3}
$$

the last equality following from tracelessness of b.

Third, let us check invariance under an infinitesimal diff transformation, $\delta\sigma = \xi(t;\sigma)$. The extension to a general diff transformation is straightforward. The only terms in the amplitude (5.3.9) that are not immediately invariant are the b insertions and the vertex operators whose coordinates have been fixed. The former transform as

$$
\begin{aligned}
\delta(b, \partial_k \hat{g}) &= -2(b, P_1 \partial_k \xi) \\
&= -2(P_1^T b, \partial_k \xi) \\
&= 0
\end{aligned}
\tag{5.4.4}
$$

using the b equation of motion. The b equation of motion comes from $\delta S/\delta c = 0$, so there will be source terms at the c insertions; these are precisely what is needed to account for the effect of the coordinate transformation on the fixed vertex operators.

BRST invariance

We now verify the full BRST invariance of the amplitude (5.3.9). The path integral measure and action are known to be invariant from the local analysis, so we have to consider the effect of the BRST transformation on the insertions. In the gauge-fixed path integral (5.3.9), the unintegrated vertex operators are accompanied by a factor of c or $\tilde{c}c$. This is precisely what we described at the end of section 4.4 for the BRST-invariant vertex operator corresponding to an OCQ state. The integrated vertex operators,

on the other hand, are not BRST-invariant, but we leave it as an exercise
to show that they have the BRST variation

$$\delta_B \mathscr{V}_m = i\epsilon \partial_a (c^a \mathscr{V}_m) \,, \tag{5.4.5}$$

which vanishes upon integration.

Finally there is the variation of the b insertion,

$$\delta_B(b, \partial_k \hat{g}) = i\epsilon(T, \partial_k \hat{g}) \,. \tag{5.4.6}$$

This insertion of the energy-momentum tensor just produces a derivative
with respect to the modulus t^k, so the BRST variation vanishes up to a
possible term from the boundary of moduli space. We will examine the
boundaries of moduli space in the coming chapters. In most cases there is
no surface contribution, due to what is known as the *canceled propagator
argument*. In certain circumstances this argument does not apply, and we
will see how to deal with this.

It is important to note that the gauge-fixed result (5.3.9) could be written
down directly from the requirement of BRST invariance, without reference
to the gauge-invariant form. As we saw when evaluating the ghost path
integral, there must be at least μ b insertions and κ c insertions, or else
the path integral vanishes. In the amplitude (5.3.9) there are precisely
the right number of ghost insertions to give a nonzero result. Once we
include the necessary b factors, the BRST variation brings in the energy-
momentum tensor convoluted with the t^k-derivative of the metric. This
is proportional to a derivative with respect to the moduli, so we get an
invariant result only by integrating over moduli space as we have done.
The result (5.3.9) can be generalized in various ways, for example to the
case that part of the conformal Killing invariance is left unfixed. We will
illustrate this explicitly in chapter 7 for the example of the torus with no
vertex operators.

BRST invariance implies that amplitudes for BRST-equivalent states
are equal. If we add to any state a null piece $Q_B|\chi\rangle$, the effect is to
insert in the path integral the variation $\delta_B \mathscr{V}_\chi$; this integrates to zero.
This is important because the physical Hilbert space is identified with the
cohomology, so equivalent states should have equal amplitudes.

Measure for Riemann surfaces

We derived the Faddeev–Popov measure in the framework where the
gauge-fixed world-sheet is described by a moduli-dependent metric. We
now recast the result in the Riemann surface framework, where the structure is encoded in moduli-dependent transition functions.

To express the measure at a given point t_0 in moduli space, let us take
a set of coordinate patches, with complex coordinate z_m in the mth patch

and holomorphic transition functions, where the metric $\hat{g}(t_0)$ is Weyl-equivalent to $dz_m d\bar{z}_m$ in each patch. Consider now a change in the moduli. We will first describe this as a change in the metric with fixed transition functions, and then convert this to a change in the transition functions with fixed metric. In the first description, define the *Beltrami differential*

$$\mu_{ka}{}^b = \frac{1}{2}\hat{g}^{bc}\partial_k\hat{g}_{ac} \ . \tag{5.4.7}$$

The b insertion for δt^k becomes

$$\frac{1}{2\pi}(b,\mu_k) = \frac{1}{2\pi}\int d^2z \left(b_{zz}\mu_{k\bar{z}}{}^z + b_{\bar{z}\bar{z}}\mu_{kz}{}^{\bar{z}}\right) \ . \tag{5.4.8}$$

In the second description, after a change δt^k in the moduli there will be new coordinates in each patch

$$z'_m = z_m + \delta t^k v_{km}^{z_m}(z_m, \bar{z}_m) \ . \tag{5.4.9}$$

The superscript on v_{km}^a is a vector index; note that v_{km}^a is defined only in the mth patch. By the definition of a Riemann surface, $dz'_m d\bar{z}'_m$ is Weyl-equivalent to the metric at $t_0 + \delta t$,

$$dz'_m d\bar{z}'_m \propto dz_m d\bar{z}_m + \delta t^k \left(\mu_{kz_m}{}^{\bar{z}_m} dz_m dz_m + \mu_{k\bar{z}_m}{}^{z_m} d\bar{z}_m d\bar{z}_m\right) \ . \tag{5.4.10}$$

The coordinate change v_{km}^a is thus related to the Beltrami differential by

$$\mu_{kz_m}{}^{\bar{z}_m} = \partial_{z_m} v_{km}^{\bar{z}_m}, \qquad \mu_{k\bar{z}_m}{}^{z_m} = \partial_{\bar{z}_m} v_{km}^{z_m} \ . \tag{5.4.11}$$

This is the infinitesimal version of *Beltrami's equation*. It implies that $v_{km}^{z_m}$ is not holomorphic — otherwise, it would not correspond to a change in the moduli. Also, $v_{km}^{z_m}$ has a holomorphic part not determined by Beltrami's equation, and $v_{km}^{\bar{z}_m}$ an antiholomorphic part; these just correspond to the freedom to make holomorphic reparameterizations within each patch.

Integrating by parts, the b insertion (5.4.8) becomes

$$\frac{1}{2\pi}(b,\mu_k) = \frac{1}{2\pi i}\sum_m \oint_{C_m} \left(dz_m v_{km}^{z_m} b_{z_m z_m} - d\bar{z}_m v_{km}^{\bar{z}_m} b_{\bar{z}_m \bar{z}_m}\right) \ , \tag{5.4.12}$$

the contour C_m circling the mth patch counterclockwise. By eq. (5.4.9), the derivative of the coordinate with respect to the moduli at a given point is

$$\frac{dz_m}{dt^k} = v_{km}^{z_m} \ . \tag{5.4.13}$$

The change in the transition functions under a change in moduli is therefore

$$\frac{\partial z_m}{\partial t^k}\bigg|_{z_n} = v_{km}^{z_m} - \frac{\partial z_m}{\partial z_n}\bigg|_t v_{kn}^{z_n} = v_{km}^{z_m} - v_{kn}^{z_m} \ . \tag{5.4.14}$$

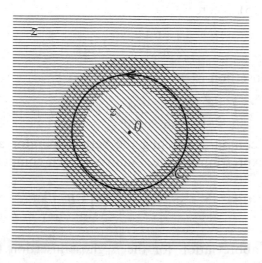

Fig. 5.3. Coordinate patches z' (diagonal hatching) and z (horizontal hatching) with vertex operator at $z' = 0$. In the annular region, $z = z' + z_v$.

The contour integrals (5.4.12) around adjacent patches then combine to give

$$\frac{1}{2\pi}(b, \mu_k) = \frac{1}{2\pi i} \sum_{(mn)} \int_{C_{mn}} \left(dz_m \frac{\partial z_m}{\partial t^k}\bigg|_{z_n} b_{z_m z_m} - d\bar{z}_m \frac{\partial \bar{z}_m}{\partial t^k}\bigg|_{z_n} b_{\bar{z}_m \bar{z}_m} \right) , \quad (5.4.15)$$

which is proportional to the derivatives of the transition functions. The sum runs over all pairs of overlapping patches. The contour C_{mn} runs between patches m and n, counterclockwise from the point of view of m. The (mn) term is symmetric in m and n, though this is not manifest. Each contour either closes or ends in a triple overlap of patches m, n, and p, where the contours C_{mn}, C_{np}, and C_{pm} meet at a point. The b insertion is now expressed entirely in terms of the data defining the Riemann surface, the transition functions modulo holomorphic equivalence.

As an illustration, we can use this to put the moduli for the metric and for the vertex operator positions on a more equal footing. Consider a vertex operator at position z_v in a coordinate system z. Let us make a small new coordinate patch z' centered on the vertex operator, as shown in figure 5.3. We keep the vertex operator at coordinate point $z' = 0$, and encode its position in the transition functions

$$z = z' + z_v \quad (5.4.16)$$

for the annular overlap region. The measure for the moduli z_v, $\bar{z}_v$ is given by the form (5.4.15), where

$$\left.\frac{\partial z'}{\partial z_v}\right|_z = -1 \ . \tag{5.4.17}$$

Thus the ghost insertion is

$$\int_C \frac{dz'}{2\pi i} b_{z'z'} \int_C \frac{d\bar{z}'_m}{-2\pi i} b_{\bar{z}'\bar{z}'} = b_{-1}\tilde{b}_{-1}\cdot \ , \tag{5.4.18}$$

where C is any contour encircling the vertex operator as shown in the figure. The full expression for the S-matrix is then compactly written as

$$S(1;\dots;n) = \sum_{\substack{\text{compact} \\ \text{topologies}}} e^{-\lambda\chi} \int_F \frac{d^m t}{n_R} \left\langle \prod_{k=1}^m B_k \prod_{i=1}^n \hat{\mathcal{V}}_i \right\rangle . \tag{5.4.19}$$

Here B_k is the abbreviation for the b ghost insertion (5.4.15), and $\hat{\mathcal{V}}$ denotes $\tilde{c}c\mathcal{V}_m$ for a closed string or $t_a c^a \mathcal{V}_m$ for an open string. That is, we are now treating *all* vertex operators as fixed, having traded their coordinates for extra parameters in the transition functions. The number m of moduli is

$$m = \mu + 2n_c + n_o - \kappa = -3\chi + 2n_c + n_o \ , \tag{5.4.20}$$

where n_c and n_o are the number of closed and open string vertex operators respectively.

The expression (5.4.19) shows that the hatted vertex operator, with $\tilde{c}c$ or $t_a c^a$, is the basic one. This is what comes out of the state–operator correspondence, and it is BRST-invariant. If the vertex operator is integrated, the ghost insertions (5.4.18) simply remove the $\tilde{c}c$ or $t_a c^a$. For example,

$$b_{-1}\tilde{b}_{-1} \cdot \tilde{c}c\mathcal{V}_m = \mathcal{V}_m \ , \tag{5.4.21}$$

leaving the integrated form of the vertex operator. We derived the result (5.4.19) from the Polyakov path integral, so the vertex operators come out in OCQ form, but we may now use arbitrary BRST-invariant vertex operators.

Exercises

5.1 Consider the sum (5.1.1), but now over all particle paths beginning and ending at given points in spacetime. In this case fixing the tetrad leaves no coordinate freedom.
(a) Derive the analog of the gauge-fixed path integral (5.3.9).
(b) Reduce the ghost path integral to determinants as in eq. (5.3.18).
(c) Reduce the X^μ path integral to determinants (as will be done in section 6.2 for the string).

(d) Evaluate the finite and functional determinants and show that the result is the scalar propagator for a particle of mass m.

5.2 Repeat the steps in the previous problem for paths forming a closed loop. In order to fix the residual gauge symmetry, introduce a fake vertex operator by differentiating with respect to m^2. The answer is given as eq. (7.3.9). If you do both exercises 5.1 and 5.2, discuss the relation between the results for the moduli measure.

5.3 (a) Show from the OPE with the BRST current that the vertex operator $\tilde{c}c\mathcal{V}_m$ is BRST-invariant for $\mathcal{V}_m$ a (1,1) matter tensor.
(b) Show that the vertex operator satisfies eq. (5.4.5).

6

Tree-level amplitudes

We are now ready to study string interactions. In this chapter we consider the lowest order amplitudes, coming from surfaces with positive Euler number. We first describe the relevant Riemann surfaces and calculate the CFT expectation values that will be needed. We next study scattering amplitudes, first for open strings and then for closed. Along the way we introduce an important generalization in the open string theory, the Chan–Paton factors. At the end of the section we return to CFT, and discuss some general properties of expectation values.

6.1 Riemann surfaces

There are three Riemann surfaces of positive Euler number: the sphere, the disk, and the projective plane.

The sphere

The sphere S_2 can be covered by two coordinate patches as shown in figure 6.1. Take the disks $|z| < \rho$ and $|u| < \rho$ for $\rho > 1$ and join them by identifying points such that

$$u = 1/z \ . \tag{6.1.1}$$

In fact, we may as well take $\rho \to \infty$. The coordinate z is then good everywhere except at the 'north pole,' $u = 0$. We can work mainly in the z patch except to check that things are well behaved at the north pole.

We can think of the sphere as a Riemann surface, taking the flat metric in both patches and connecting them by a conformal (coordinate plus Weyl) transformation. Or, we can think of it as a Riemannian manifold, with a globally defined metric. A general conformal gauge metric is

$$ds^2 = \exp(2\omega(z, \bar{z}))dz\,d\bar{z} \ . \tag{6.1.2}$$

166

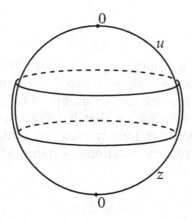

Fig. 6.1. The sphere built from z and u coordinate patches.

Since $dz d\bar{z} = |z|^4 du d\bar{u}$, the condition for the metric to be nonsingular at $u = 0$ is that $\exp(2\omega(z,\bar{z}))$ fall as $|z|^{-4}$ for $z \to \infty$. For example,

$$ds^2 = \frac{4r^2 dz d\bar{z}}{(1 + z\bar{z})^2} = \frac{4r^2 du d\bar{u}}{(1 + u\bar{u})^2} \tag{6.1.3}$$

describes a sphere of radius r and curvature $R = 2/r^2$.

According to the general discussion in section 5.2, the sphere has no moduli and six CKVs, so that in particular every metric is (diff×Weyl)-equivalent to the round metric (6.1.3). Let us see this at the infinitesimal level. As in eq. (5.2.8) one is looking for holomorphic tensor fields $\delta g_{zz}(z)$ and holomorphic vector fields $\delta z(z)$. These must be defined on the whole sphere, so we need to consider the transformation to the u-patch,

$$\delta u = \frac{\partial u}{\partial z}\delta z = -z^{-2}\delta z \,, \tag{6.1.4a}$$

$$\delta g_{uu} = \left(\frac{\partial u}{\partial z}\right)^{-2}\delta g_{zz} = z^4 \delta g_{zz} \,. \tag{6.1.4b}$$

Any holomorphic quadratic differential δg_{zz} would have to be holomorphic in z but vanish as z^{-4} at infinity, and so must vanish identically. A CKV δz, on the other hand, is holomorphic at $u = 0$ provided it grows no more rapidly than z^2 as $z \to \infty$. The general CKV is then

$$\delta z = a_0 + a_1 z + a_2 z^2 \,, \tag{6.1.5a}$$

$$\delta \bar{z} = a_0^* + a_1^* \bar{z} + a_2^* \bar{z}^2 \,, \tag{6.1.5b}$$

with three complex or six real parameters as expected from the Riemann–Roch theorem.

These infinitesimal transformations exponentiate to give the *Möbius group*

$$z' = \frac{\alpha z + \beta}{\gamma z + \delta} \tag{6.1.6}$$

for complex $\alpha, \beta, \gamma, \delta$. Rescaling $\alpha, \beta, \gamma, \delta$ leaves the transformation unchanged, so we can fix $\alpha\delta - \beta\gamma = 1$ and identify under an overall sign reversal of $\alpha, \beta, \gamma, \delta$. This defines the group $PSL(2, \mathbf{C})$. This is the most general coordinate transformation that is holomorphic on all of S_2. It is one-to-one with the point at infinity included. Three of the six parameters correspond to ordinary rotations, forming an $SO(3)$ subgroup of $PSL(2, \mathbf{C})$.

The disk

It is useful to construct the disk D_2 from the sphere by identifying points under a reflection. For example, identify points z and z' such that

$$z' = 1/\bar{z} \ . \tag{6.1.7}$$

In polar coordinates $z = re^{i\phi}$, this inverts the radius and leaves the angle fixed, so the unit disk $|z| \leq 1$ is a fundamental region for the identification. The points on the unit circle are fixed by the reflection, so this becomes a boundary. It is often more convenient to use the conformally equivalent reflection

$$z' = \bar{z} \ . \tag{6.1.8}$$

The upper half-plane is now a fundamental region, and the real axis is the boundary.

The CKG of the disk is the subgroup of $PSL(2, \mathbf{C})$ that leaves the boundary of the disk fixed. For the reflection (6.1.8) this is just the subgroup of (6.1.6) with $\alpha, \beta, \gamma, \delta$ all real, which is $PSL(2, \mathbf{R})$, the Möbius group with real parameters. One CKV is the ordinary rotational symmetry of the disk. Again, all metrics are equivalent — there are no moduli.

The projective plane

The projective plane RP_2 can also be obtained as a $\mathbf{Z}_2$ identification of the sphere. Identify points z and z' with

$$z' = -1/\bar{z} \ . \tag{6.1.9}$$

These points are diametrically opposite in the round metric (6.1.3). There are no fixed points and so no boundary in the resulting space, but the space is not oriented. One fundamental region for the identification is the unit disk $|z| \leq 1$, with points $e^{i\phi}$ and $-e^{i\phi}$ identified. Another choice is the upper half-z-plane. There are no moduli. The CKG is the subgroup of

$PSL(2,\mathbf{C})$ that respects the identification (6.1.9); this is just the ordinary rotation group $SO(3)$.

Both the disk and projective plane have been represented as the sphere with points identified under a $\mathbf{Z}_2$ transformation, or *involution*. In fact, every world-sheet can be obtained from a closed oriented world-sheet by identifying under one or two $\mathbf{Z}_2$s. The method of images can then be used to obtain the Green's functions.

6.2 Scalar expectation values

The basic quantities that we need are the expectation values of products of vertex operators. In order to develop a number of useful techniques and points of view we will calculate these in three different ways: by direct path integral evaluation, by using holomorphicity properties, and later in the chapter by operator methods.

The path integral method has already been used in section 5.3 for the Faddeev–Popov determinant and in the appendix for the harmonic oscillator. Start with a generating functional

$$Z[J] = \left\langle \exp\left(i \int d^2\sigma\, J(\sigma)\cdot X(\sigma)\right) \right\rangle \qquad (6.2.1)$$

for arbitrary $J_\mu(\sigma)$. For now we work on an arbitrary compact two-dimensional surface M, and in an arbitrary spacetime dimension d. Expand $X^\mu(\sigma)$ in terms of a complete set $X_I(\sigma)$,

$$X^\mu(\sigma) = \sum_I x_I^\mu X_I(\sigma)\,, \qquad (6.2.2a)$$

$$\nabla^2 X_I = -\omega_I^2 X_I\,, \qquad (6.2.2b)$$

$$\int_M d^2\sigma g^{1/2}\, X_I X_{I'} = \delta_{II'}\,. \qquad (6.2.2c)$$

Then

$$Z[J] = \prod_{I,\mu} \int dx_I^\mu \exp\left(-\frac{\omega_I^2 x_I^\mu x_{I\mu}}{4\pi\alpha'} + ix_I^\mu J_{I\mu}\right), \qquad (6.2.3)$$

where

$$J_I^\mu = \int d^2\sigma J^\mu(\sigma) X_I(\sigma)\,. \qquad (6.2.4)$$

The integrals are Gaussian except for the constant mode

$$X_0 = \left(\int d^2\sigma g^{1/2}\right)^{-1/2}, \qquad (6.2.5)$$

which has vanishing action and so gives a delta function. Carrying out the integrations leaves

$$
Z[J] = i(2\pi)^d \delta^d(J_0) \prod_{I \neq 0} \left(\frac{4\pi^2 \alpha'}{\omega_I^2} \right)^{d/2} \exp\left(-\frac{\pi \alpha' J_I \cdot J_I}{\omega_I^2} \right)
$$

$$
= i(2\pi)^d \delta^d(J_0) \left(\det' \frac{-\nabla^2}{4\pi^2 \alpha'} \right)^{-d/2}
$$

$$
\times \exp\left(-\frac{1}{2} \int d^2\sigma \, d^2\sigma' \, J(\sigma) \cdot J(\sigma') G'(\sigma, \sigma') \right) . \quad (6.2.6)
$$

As discussed in section 2.1 and further in section 3.2, the timelike modes x_I^0 give rise to wrong-sign Gaussians and are defined by the contour rotation[1] $x_I^0 \to -i x_I^d$, $I \neq 0$. The primed Green's function excludes the zero mode contribution,

$$
G'(\sigma_1, \sigma_2) = \sum_{I \neq 0} \frac{2\pi \alpha'}{\omega_I^2} X_I(\sigma_1) X_I(\sigma_2) . \quad (6.2.7)
$$

It satisfies the differential equation

$$
-\frac{1}{2\pi\alpha'} \nabla^2 G'(\sigma_1, \sigma_2) = \sum_{I \neq 0} X_I(\sigma_1) X_I(\sigma_2)
$$

$$
= g^{-1/2} \delta^2(\sigma_1 - \sigma_2) - X_0^2 , \quad (6.2.8)
$$

where the completeness of the X_I has been used. The ordinary Green's function with a delta function source does not exist. It would correspond to the electrostatic potential of a single charge, but on a compact surface the field lines from the source have no place to go. The X_0^2 term can be thought of as a neutralizing background charge distribution.

The sphere

Specializing to the sphere, the solution to the differential equation (6.2.8) is

$$
G'(\sigma_1, \sigma_2) = -\frac{\alpha'}{2} \ln|z_{12}|^2 + f(z_1, \bar{z}_1) + f(z_2, \bar{z}_2) , \quad (6.2.9)
$$

---

[1] A curious factor of i has been inserted into eq. (6.2.6) by hand because it is needed in the S-matrix, but it can be understood formally as arising from the same rotation. If we rotated the entire field $X^0 \to -iX^d$, the Jacobian should be 1, by the usual argument about rescaling fields (footnote 1 of the appendix). However, we do not want to rotate x_0^0, because this mode produces the energy delta function. So we have to rotate it back, giving a factor of i.

where

$$f(z, \bar{z}) = \frac{\alpha' X_0^2}{4} \int d^2 z' \, \exp[2\omega(z', \bar{z}')] \ln |z - z'|^2 + k \; . \qquad (6.2.10)$$

The constant k is determined by the property that G' is orthogonal to X_0, but in any case we will see that the function f drops out of all expectation values. It comes from the background charge, but the delta function from the zero-mode integration forces overall neutrality, $J_0^\mu = 0$, and the background makes no net contribution.

Now consider the path integral with a product of tachyon vertex operators,

$$A_{S_2}^n(k, \sigma) = \left\langle \left[e^{ik_1 \cdot X(\sigma_1)} \right]_{\mathrm{r}} \left[e^{ik_2 \cdot X(\sigma_2)} \right]_{\mathrm{r}} \cdots \left[e^{ik_n \cdot X(\sigma_n)} \right]_{\mathrm{r}} \right\rangle_{S_2} . \qquad (6.2.11)$$

This corresponds to

$$J(\sigma) = \sum_{i=1}^n k_i \delta^2(\sigma - \sigma_i) \; . \qquad (6.2.12)$$

The amplitude (6.2.6) then becomes

$$A_{S_2}^n(k, \sigma) = iC_{S_2}^X (2\pi)^d \delta^d(\textstyle\sum_i k_i)$$

$$\times \exp\left(-\sum_{\substack{i,j=1 \\ i<j}}^n k_i \cdot k_j G'(\sigma_i, \sigma_j) - \frac{1}{2} \sum_{i=1}^n k_i^2 G_{\mathrm{r}}'(\sigma_i, \sigma_i) \right) . \qquad (6.2.13)$$

The constant here is

$$C_{S_2}^X = X_0^{-d} \left(\det' \frac{-\nabla^2}{4\pi^2 \alpha'} \right)_{S_2}^{-d/2} . \qquad (6.2.14)$$

The determinant can be regularized and computed, but we will not need to do this explicitly. We are employing here the simple renormalization used in section 3.6, so that the self-contractions involve

$$G_{\mathrm{r}}'(\sigma, \sigma') = G'(\sigma, \sigma') + \frac{\alpha'}{2} \ln d^2(\sigma, \sigma') \; . \qquad (6.2.15)$$

Note that

$$G_{\mathrm{r}}'(\sigma, \sigma) = 2f(z, \bar{z}) + \alpha' \omega(z, \bar{z}) \qquad (6.2.16)$$

is finite by design. The path integral on the sphere is then

$$A_{S_2}^n(k, \sigma) = iC_{S_2}^X (2\pi)^d \delta^d(\textstyle\sum_i k_i) \exp\left(-\frac{\alpha'}{2} \sum_i k_i^2 \omega(\sigma_i) \right) \prod_{\substack{i,j=1 \\ i<j}}^n |z_{ij}|^{\alpha' k_i \cdot k_j} . \qquad (6.2.17)$$

The function f has dropped out as promised. The dependence on the conformal factor $\omega(\sigma)$ is precisely that found in section 3.6 from the Weyl anomaly in the vertex operator. It will cancel the variation of $g^{1/2}$ for an on-shell operator.[2] We can take a metric that is flat in a large region containing all the vertex operators ('pushing the curvature to infinity') and the term in $\omega(\sigma_i)$ drops out.

Higher vertex operators are exponentials times derivatives of X^μ, so we also need

$$\left\langle \prod_{i=1}^{n} \left[e^{ik_i \cdot X(z_i, \bar{z}_i)} \right]_{\mathrm{r}} \prod_{j=1}^{p} \partial X^{\mu_j}(z'_j) \prod_{k=1}^{q} \bar{\partial} X^{\nu_k}(\bar{z}''_k) \right\rangle_{S_2} . \qquad (6.2.18)$$

This is given by summing over all contractions, where every ∂X or $\bar{\partial} X$ must be contracted either with an exponential or with another ∂X or $\bar{\partial} X$. The XX contraction is simply $-\frac{1}{2}\alpha' \ln |z|^2$, the fs again dropping out in the final expression. The result can be summarized as

$$iC_{S_2}^X (2\pi)^d \delta^d(\textstyle\sum_i k_i) \prod_{\substack{i,j=1 \\ i<j}}^{n} |z_{ij}|^{\alpha' k_i \cdot k_j}$$

$$\times \left\langle \prod_{j=1}^{p} \left[v^{\mu_j}(z'_j) + q^{\mu_j}(z'_j) \right] \prod_{k=1}^{q} \left[\tilde{v}^{\nu_k}(\bar{z}''_k) + \tilde{q}^{\nu_k}(\bar{z}''_k) \right] \right\rangle \qquad (6.2.19)$$

Here

$$v^\mu(z) = -i\frac{\alpha'}{2} \sum_{i=1}^{n} \frac{k_i^\mu}{z - z_i} , \qquad \tilde{v}^\mu(\bar{z}) = -i\frac{\alpha'}{2} \sum_{i=1}^{n} \frac{k_i^\mu}{\bar{z} - \bar{z}_i} \qquad (6.2.20)$$

come from the contractions with the exponentials. The expectation values of $q^\mu = \partial X^\mu - v^\mu$ are given by the sum over all contractions using $-\eta^{\mu\nu}(z - z')^{-2}\alpha'/2$, and those of $\tilde{q}^\nu$ by the conjugate.

Now we repeat the calculation in a different way, using holomorphicity. As an example consider

$$\langle \partial X^\mu(z_1) \partial X^\nu(z_2) \rangle_{S_2} . \qquad (6.2.21)$$

The OPE determines this to be

$$-\frac{\alpha' \eta^{\mu\nu}}{2z_{12}^2} \langle 1 \rangle_{S_2} + g(z_1, z_2) , \qquad (6.2.22)$$

where $g(z_1, z_2)$ is holomorphic in both variables. In the u-patch,

$$\partial_u X^\mu = -z^2 \partial_z X^\mu , \qquad (6.2.23)$$

so the condition of holomorphicity at $u = 0$ is that expectation values of $\partial_z X^\mu$ fall as z^{-2} at infinity. More generally, a tensor of weight $(h, 0)$ must

---

[2] The constant $C_{S_2}^X$ also depends on the conformal factor, because the X^μ CFT by itself has nonzero central charge, but in the full string theory this will cancel.

behave as z^{-2h} at infinity. Focusing on the z_1-dependence in eq. (6.2.22) at fixed z_2, this implies that $g(z_1, z_2)$ falls as z_1^{-2} at infinity and so vanishes by holomorphicity. This determines the expectation value up to normalization. Comparing with the path integral result (6.2.19), there is agreement, though the normalization $\langle 1 \rangle_{S_2}$ is seen to diverge as $\delta(0)$. This is just the zero-mode divergence from the infinite volume of spacetime.

To obtain the expectation value of the product of vertex operators

$$A_{S_2}^n(k, \sigma) = \left\langle \prod_{i=1}^n :e^{ik_i \cdot X(z_i, \bar{z}_i)}: \right\rangle_{S_2} \tag{6.2.24}$$

by this method is less direct, because the exponentials are not holomorphic. In fact they can be factored into holomorphic and antiholomorphic parts, but this is subtle and is best introduced in operator language as we will do in chapter 8. We use the holomorphicity of the translation current. Consider the expectation value with one current $\partial X^\mu(z)$ added. The OPE of the current with the vertex operators determines the singularities in z,

$$\left\langle \partial X^\mu(z) \prod_{i=1}^n :e^{ik_i \cdot X(z_i, \bar{z}_i)}: \right\rangle_{S_2} = -\frac{i\alpha'}{2} A_{S_2}^n(k, \sigma) \sum_{i=1}^n \frac{k_i^\mu}{z - z_i}$$
$$+ \text{ terms holomorphic in } z. \tag{6.2.25}$$

Now look at $z \to \infty$. The condition that $\partial_u X^\mu$ be holomorphic at $u = 0$ again requires that the expectation value vanish as z^{-2} when $z \to \infty$. The holomorphic term in eq. (6.2.25) must then vanish, and from the vanishing of the order z^{-1} term we recover momentum conservation,

$$A_{S_2}^n(k, \sigma) \sum_{i=1}^n k_i^\mu = 0. \tag{6.2.26}$$

Let us also say this in a slightly different way. Consider the contour integral of the spacetime translation current

$$p^\mu = \frac{1}{2\pi i} \oint_C (dz\, j_z^\mu - d\bar{z}\, j_{\bar{z}}^\mu), \tag{6.2.27}$$

where the contour C encircles all the exponential operators. There are two ways to evaluate this. The first is to contract C until it is a small circle around each vertex operator, which picks out the $(z - z_i)^{-1}$ term from each OPE and gives $\sum_{i=1}^n k_i^\mu$. The second is to expand it until it is a small circle in the u-patch: by holomorphicity at $u = 0$ it must vanish. This sort of 'contour-pulling' argument is commonly used in CFT.

We have used the OPE to determine the singularities and then used holomorphicity to get the full z-dependence. Now we look at the second term of the OPE as $z \to z_1$. Expanding the result (6.2.25) gives

$$-\frac{i\alpha'}{2} A^n_{S_2}(k, \sigma) \left(\frac{k_1^\mu}{z - z_1} + \sum_{i=2}^{n} \frac{k_i^\mu}{z_1 - z_i} + O(z - z_1) \right). \qquad (6.2.28)$$

Contracting this with ik_1^μ, the term of order $(z - z_1)^0$ must agree with the OPE

$$ik_1 \cdot \partial X(z) \; :e^{ik_1 \cdot X(z_1, \bar{z}_1)}:$$

$$= \frac{\alpha' k_1^2}{2(z - z_1)} \; :e^{ik_1 \cdot X(z_1, \bar{z}_1)}: \; + \partial_{z_1} \; :e^{ik_1 \cdot X(z_1, \bar{z}_1)}: \; + O(z - z_1) \,. \qquad (6.2.29)$$

This implies

$$\partial_{z_1} A^n_{S_2}(k, \sigma) = \frac{\alpha'}{2} A^n_{S_2}(k, \sigma) \sum_{i=2}^{n} \frac{k_1 \cdot k_i}{z_{1i}} \,. \qquad (6.2.30)$$

Integrating, and using the conjugate equation and momentum conservation, determines the path integral up to normalization,

$$A^n_{S_2}(k, \sigma) \propto \delta^d(\textstyle\sum_i k_i) \prod_{\substack{i,j=1 \\ i<j}}^{n} |z_{ij}|^{\alpha' k_i \cdot k_j}, \qquad (6.2.31)$$

in agreement with the first method. Note that to compare we must push the curvature to infinity ($\omega(\sigma_i) = 0$) so that $[\;]_r = \; : \, :$.

It is worth noting that the intermediate steps in the path integral method depend in a detailed way on the particular choice of Riemannian metric. The metric is needed in order to preserve coordinate invariance in these steps. A different, Weyl-equivalent, metric gives a different Laplacian and different eigenfunctions, though at the very end this dependence must drop out in string theory. The second method uses only the basic data of a Riemann surface, its holomorphic structure.

The disk

The generalization to the disk is straightforward. We will represent the disk by taking the above representation of the sphere and restricting z to the upper half of the complex plane. The Neumann boundary term is accounted for by an image charge,

$$G'(\sigma_1, \sigma_2) = -\frac{\alpha'}{2} \ln |z_1 - z_2|^2 - \frac{\alpha'}{2} \ln |z_1 - \bar{z}_2|^2 \,, \qquad (6.2.32)$$

up to terms that drop out due to momentum conservation. Then

$$\left\langle \prod_{i=1}^{n} : e^{ik_i \cdot X(z_i, \bar{z}_i)} : \right\rangle_{D_2} = iC_{D_2}^X (2\pi)^d \delta^d(\textstyle\sum_i k_i) \prod_{i=1}^{n} |z_i - \bar{z}_i|^{\alpha' k_i^2/2}$$

$$\times \prod_{\substack{i,j=1 \\ i<j}}^{n} |z_i - z_j|^{\alpha' k_i \cdot k_j} |z_i - \bar{z}_j|^{\alpha' k_i \cdot k_j} . \tag{6.2.33}$$

For expectation values with $\partial_a X^\mu$s one again sums over contractions, now using the Green's function (6.2.32). Note that $\partial X^\mu(z) \bar{\partial} X^\nu(\bar{z}')$ (or $q^\mu(z) \tilde{q}^\nu(\bar{z}')$) has a nonzero contraction.

For two points on the boundary, the two terms in the Green's function (6.2.32) are equal, and the Green's function diverges at zero separation even after normal ordering subtracts the first term. For this reason, boundary operators must be defined with *boundary normal ordering*, where the subtraction is doubled:

$$\,_\star^\star X^\mu(y_1) X^\nu(y_2)\,_\star^\star = X^\mu(y_1) X^\nu(y_2) + 2\alpha' \eta^{\mu\nu} \ln|y_1 - y_2| , \tag{6.2.34}$$

with y denoting a coordinate on the real axis. The combinatorics are the same as for other forms of normal ordering. Boundary normal-ordered expectation values of boundary operators have the same good properties (nonsingularity) as conformal normal-ordered operators in the interior.

An expectation value with exponentials both in the interior and on the boundary, each having the appropriate normal ordering, is given by taking the appropriate limit of the interior result (6.2.33) and dropping the $|z_i - \bar{z}_i|^{\alpha' k_i^2/2}$ factor for the boundary operators. Explicitly, for exponentials all on the boundary,

$$\left\langle \prod_{i=1}^{n} \,_\star^\star e^{ik_i \cdot X(y_i)} \,_\star^\star \right\rangle_{D_2} = iC_{D_2}^X (2\pi)^d \delta^d(\textstyle\sum_i k_i) \prod_{\substack{i,j=1 \\ i<j}}^{n} |y_i - y_j|^{2\alpha' k_i \cdot k_j} . \tag{6.2.35}$$

More generally,

$$\left\langle \prod_{i=1}^{n} \,_\star^\star e^{ik_i \cdot X(y_i)} \,_\star^\star \prod_{j=1}^{p} \partial_y X^{\mu_j}(y'_j) \right\rangle_{D_2} = iC_{D_2}^X (2\pi)^d \delta^d(\textstyle\sum_i k_i)$$

$$\times \prod_{\substack{i,j=1 \\ i<j}}^{n} |y_{ij}|^{2\alpha' k_i \cdot k_j} \left\langle \prod_{j=1}^{p} \left[v^{\mu_j}(y'_j) + q^{\mu_j}(y'_j) \right] \right\rangle_{D_2} , \tag{6.2.36}$$

where now

$$v^\mu(y) \underset{}{=} -i\alpha' \sum_{i=1}^{n} \frac{k_i^\mu}{y - y_i} \tag{6.2.37}$$

and the qs are contracted using $-2\alpha'(y - y')^{-2}\eta^{\mu\nu}$.

The projective plane

The method of images gives the Green's function as

$$G'(\sigma_1, \sigma_2) = -\frac{\alpha'}{2} \ln |z_1 - z_2|^2 - \frac{\alpha'}{2} \ln |1 + z_1 \bar{z}_2|^2 . \qquad (6.2.38)$$

Now

$$\left\langle \prod_{i=1}^{n} :e^{ik_i \cdot X(z_i, \bar{z}_i)}: \right\rangle_{RP_2} = iC_{RP_2}^X (2\pi)^d \delta^d(\textstyle\sum_i k_i) \prod_{i=1}^{n} |1 + z_i \bar{z}_i|^{\alpha' k_i^2/2}$$

$$\times \prod_{\substack{i,j=1 \\ i<j}}^{n} |z_i - z_j|^{\alpha' k_i \cdot k_j} |1 + z_i \bar{z}_j|^{\alpha' k_i \cdot k_j} , \qquad (6.2.39)$$

and so on for more general expectation values. There is no boundary because the involution has no fixed points — there is no way for a point z to approach its image $-\bar{z}^{-1}$.

6.3   The bc CFT

The sphere

The path integral for the ghosts has already been set up in section 5.3. According to the Riemann–Roch theorem, the simplest nonvanishing expectation value will be

$$\langle c(z_1)c(z_2)c(z_3)\tilde{c}(\bar{z}_4)\tilde{c}(\bar{z}_5)\tilde{c}(\bar{z}_6) \rangle_{S_2} . \qquad (6.3.1)$$

Up to normalization (a functional determinant), the result (5.3.18) was just the zero-mode determinant,

$$\det C_{0j}^a(\sigma_i) . \qquad (6.3.2)$$

The six CKVs were found in section 6.1. In a complex basis they are

$$C^z = 1, z, z^2 , \quad C^{\bar{z}} = 0 , \qquad (6.3.3a)$$
$$C^z = 0 , \quad C^{\bar{z}} = 1, \bar{z}, \bar{z}^2 . \qquad (6.3.3b)$$

In this basis the determinant splits into two 3×3 blocks, and the expectation value becomes

$$C_{S_2}^g \det \begin{vmatrix} 1 & 1 & 1 \\ z_1 & z_2 & z_3 \\ z_1^2 & z_2^2 & z_3^2 \end{vmatrix} \det \begin{vmatrix} 1 & 1 & 1 \\ \bar{z}_4 & \bar{z}_5 & \bar{z}_6 \\ \bar{z}_4^2 & \bar{z}_5^2 & \bar{z}_6^2 \end{vmatrix} = C_{S_2}^g z_{12}z_{13}z_{23}\bar{z}_{45}\bar{z}_{46}\bar{z}_{56} . \qquad (6.3.4)$$

The constant $C_{S_2}^g$ includes the functional determinant and also a finite-dimensional Jacobian (independent of the positions) that arises because the basis (6.3.3) is not orthonormal.

This is the only bc path integral that we will actually need for the tree-level amplitudes, but for completeness we give

$$\left\langle \prod_{i=1}^{p+3} c(z_i) \prod_{j=1}^{p} b(z'_j) \cdot (\text{anti}) \right\rangle_{S_2} = C_{S_2}^g \frac{z_{p+1,p+2}z_{p+1,p+3}z_{p+2,p+3}}{(z_1 - z'_1)\dots(z_p - z'_p)} \cdot (\text{anti})$$

$$\pm \text{ permutations}, \qquad (6.3.5)$$

obtained by contracting the bs with cs. The antiholomorphic part 'anti' has the same form. We are being a little careless with the *overall* sign; again, for the Faddeev–Popov determinant we will in any case take an absolute value.

To give an alternative derivation using holomorphicity, we again consider first the conservation law, derived by inserting into the amplitude (6.3.5) the ghost number contour integral $\oint_C dz\, j(z)/2\pi i$, where C encircles all the vertex operators and $j = -:bc:$ is again the ghost number current. From the OPE with b and c, this just counts the number of c minus the number of b fields, giving $n_c - n_b$ times the original amplitude. Now pull the contour into the u-patch using the conformal transformation (2.5.17),

$$\oint_C \frac{dz}{2\pi i} j_z = -\oint_C \frac{du}{2\pi i} j_u + 3 \;\to\; 3 \,. \qquad (6.3.6)$$

The shift $+3$ comes about because j is not a tensor, having a z^{-3} term in its OPE with T. In the last step we use holomorphicity in the u-patch. The nonzero amplitudes therefore have $n_c - n_b = 3$ and similarly $n_{\tilde{c}} - n_{\tilde{b}} = 3$, in agreement with the Riemann–Roch theorem.

From the OPE, the ghost expectation value (6.3.1) is holomorphic in each variable, and it must have a zero when two identical anticommuting fields come together. Thus it must be of the form

$$z_{12}z_{13}z_{23}\bar{z}_{45}\bar{z}_{46}\bar{z}_{56} F(z_1, z_2, z_3)\tilde{F}(\bar{z}_4, \bar{z}_5, \bar{z}_6) \,, \qquad (6.3.7)$$

with F holomorphic and $\tilde{F}$ antiholomorphic functions of the positions. As $z_1 \to \infty$ this goes as $z_1^2 F$. However, c is a tensor of weight -1, so the amplitude can be no larger than z_1^2 at infinity. Thus $F(z_i)$ must be independent of z_1, and so also of z_2 and z_3; arguing similarly for $\tilde{F}$ we obtain again the result (6.3.4). The same argument gives for the general case (6.3.5) the result

$$C_{S_2}^g \prod_{\substack{i,i'=1 \\ i<i'}}^{p+3} z_{ii'} \prod_{\substack{i,i'=1 \\ j<j'}}^{p} z'_{jj'} \prod_{i=1}^{p+3}\prod_{i=1}^{p} (z_i - z'_j)^{-1} \cdot (\text{anti}) \,, \qquad (6.3.8)$$

which has the correct poles, zeros, and behavior at infinity. The permutations (6.3.5) evidently sum up to give this single term.

We have considered the $\lambda=2$ theory that is relevant to the ghosts of the bosonic string, but either method generalizes readily to any λ.

The disk

The simplest way to obtain the bc amplitudes on the disk is with the doubling trick. As in eq. (2.7.30), we can represent both the holomorphic and antiholomorphic fields in the upper half-plane by holomorphic fields in the whole plane, using

$$\tilde{b}(\bar{z}) = b(z') , \quad \tilde{c}(\bar{z}) = c(z') , \quad z' = \bar{z} , \text{ Im } z > 0 . \tag{6.3.9}$$

The expectation value of the holomorphic fields follows as on the sphere,

$$\langle c(z_1)c(z_2)c(z_3) \rangle_{D_2} = C^g_{D_2} z_{12} z_{13} z_{23} \tag{6.3.10}$$

for all z. Then for example

$$\langle c(z_1)c(z_2)\tilde{c}(\bar{z}_3) \rangle_{D_2} = \langle c(z_1)c(z_2)c(z'_3) \rangle_{D_2}$$
$$= C^g_{D_2} z_{12}(z_1 - \bar{z}_3)(z_2 - \bar{z}_3) . \tag{6.3.11}$$

More general correlators are obtained in the same way.

The projective plane

The doubling trick again can be used. The involution $z' = -\bar{z}^{-1}$ implies that

$$\tilde{b}(\bar{z}) = \left(\frac{\partial z'}{\partial \bar{z}}\right)^2 b(z') = z'^4 b(z') , \quad \tilde{c}(\bar{z}) = \left(\frac{\partial z'}{\partial \bar{z}}\right)^{-1} c(z') = z'^{-2} c(z') . \tag{6.3.12}$$

Again

$$\langle c(z_1)c(z_2)c(z_3) \rangle_{RP_2} = C^g_{RP_2} z_{12} z_{13} z_{23} \tag{6.3.13}$$

and so

$$\langle c(z_1)c(z_2)\tilde{c}(\bar{z}_3) \rangle_{RP_2} = z_3'^{-2} \langle c(z_1)c(z_2)c(z'_3) \rangle_{RP_2}$$
$$= C^g_{RP_2} z_{12}(1 + z_1\bar{z}_3)(1 + z_2\bar{z}_3) . \tag{6.3.14}$$

6.4   The Veneziano amplitude

Open string amplitudes are slightly simpler than closed string amplitudes, so we begin with these.

We represent the disk as the upper half-plane so the boundary coordinate y is real. There are no moduli. After fixing the metric, the CKG $PSL(2, \mathbf{R})$ can be used to fix the three vertex operators to arbitrary positions y_1, y_2, y_3 on the boundary, except that this group does not change the cyclic ordering of the vertex operators so we must sum over the two orderings. For three open string tachyons on the disk, the general expression (5.3.9) for the string S-matrix thus reduces to

$$S_{D_2}(k_1; k_2; k_3) = g_0^3 e^{-\lambda} \left\langle {}^*_* c^1 e^{ik_1 \cdot X}(y_1){}^*_* {}^*_* c^1 e^{ik_2 \cdot X}(y_2){}^*_* {}^*_* c^1 e^{ik_3 \cdot X}(y_3){}^*_* \right\rangle_{D_2}$$
$$+ (k_2 \leftrightarrow k_3), \qquad (6.4.1)$$

each fixed coordinate integration being replaced by the corresponding c-ghost. Each vertex operator includes a factor of g_0, the open string coupling. The factor $e^{-\lambda}$ is from the Euler number term in the action. Of course g_0 and e^λ are related, $g_0^2 \propto e^\lambda$, but we will determine the constant of proportionality as we go along.

The expectation values in (6.4.1) were found in the previous sections (again, we take the absolute value for the ghosts), giving

$$S_{D_2}(k_1; k_2; k_3) = ig_0^3 C_{D_2}(2\pi)^{26}\delta^{26}(\textstyle\sum_i k_i)$$
$$\times |y_{12}|^{1+2\alpha' k_1 \cdot k_2} |y_{13}|^{1+2\alpha' k_1 \cdot k_3} |y_{23}|^{1+2\alpha' k_2 \cdot k_3} + (k_2 \leftrightarrow k_3), \quad (6.4.2)$$

where $C_{D_2} = e^{-\lambda} C_{D_2}^X C_{D_2}^g$. Momentum conservation and the mass-shell condition $k_i^2 = 1/\alpha'$ imply that

$$2\alpha' k_1 \cdot k_2 = \alpha'(k_3^2 - k_1^2 - k_2^2) = -1 \qquad (6.4.3)$$

and the same for the other $k_i \cdot k_j$, so this reduces to

$$S_{D_2}(k_1; k_2; k_3) = 2ig_0^3 C_{D_2}(2\pi)^{26}\delta^{26}(\textstyle\sum_i k_i) . \qquad (6.4.4)$$

This is independent of the gauge choice y_i, which is of course a general property of the Faddeev–Popov procedure. The Weyl invariance is crucial here — if the vertex operators were taken off the mass shell, the amplitude would depend on the choice of y_i.

We could have used independence of the y_i to determine the Faddeev–Popov determinant without calculation. Using the mass-shell condition and momentum conservation, the X^μ expectation value is proportional to $|y_{12}y_{13}y_{23}|^{-1}$, and so the measure must be reciprocal to this. This same measure then applies for $n > 3$ vertex operators, because in all cases three positions are fixed.

The four-tachyon amplitude is obtained in the same way,

$$S_{D_2}(k_1; k_2; k_3; k_4)$$

$$= g_0^4 e^{-\lambda} \int_{-\infty}^{\infty} dy_4 \left\langle \prod_{i=1}^{3} {}^*_* c^1(y_i) e^{ik_i \cdot X(y_i)}{}^*_* {}^*_* e^{ik_4 \cdot X(y_4)}{}^*_* \right\rangle + (k_2 \leftrightarrow k_3)$$

$$= ig_0^4 C_{D_2}(2\pi)^{26}\delta^{26}(\textstyle\sum_i k_i)|y_{12}y_{13}y_{23}| \int_{-\infty}^{\infty} dy_4 \prod_{i<j} |y_{ij}|^{2\alpha' k_i \cdot k_j}$$

$$+ (k_2 \leftrightarrow k_3) . \qquad (6.4.5)$$

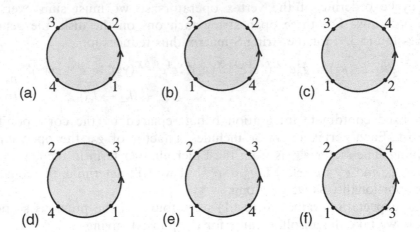

Fig. 6.2.   The six cyclically inequivalent orderings of four open string vertex operators on the disk. The coordinate y increases in the direction of the arrow, except that at point 3 it jumps from $+\infty$ to $-\infty$.

This is again independent of $y_{1,2,3}$ after a change of variables (a Möbius transformation) on y_4. It is convenient to take $y_1 = 0$, $y_2 = 1$, and $y_3 \rightarrow \infty$. The amplitude is conventionally written in terms of the *Mandelstam variables*

$$s = -(k_1 + k_2)^2 \ , \quad t = -(k_1 + k_3)^2 \ , \quad u = -(k_1 + k_4)^2 \ . \qquad (6.4.6)$$

These are not independent: momentum conservation and the mass-shell condition imply that

$$s + t + u = \sum_i m_i^2 = -\frac{4}{\alpha'} \ . \qquad (6.4.7)$$

Using $2\alpha' k_i \cdot k_j = -2 + \alpha'(k_i + k_j)^2$, the amplitude becomes

$$S_{D_2}(k_1 ; k_2 ; k_3 ; k_4) = i g_o^4 C_{D_2} (2\pi)^{26} \delta^{26} (\textstyle\sum_i k_i)$$
$$\times \left[\int_{-\infty}^{\infty} dy_4 \, |y_4|^{-\alpha' u - 2} |1 - y_4|^{-\alpha' t - 2} + (t \rightarrow s) \right] . \qquad (6.4.8)$$

The integral splits into three ranges, $-\infty < y_4 < 0$, $0 < y_4 < 1$, and $1 < y_4 < \infty$. For these three ranges the vertex operators are ordered as in figures 6.2(a), (b), and (c) respectively. Möbius invariance can be used to take each of these ranges into any other, so they give contributions that are equal up to interchange of vertex operators. The $(t \leftrightarrow s)$ term gives figures 6.2(d), (e), and (f). In all,

$$S_{D_2}(k_1 ; k_2 ; k_3 ; k_4) = 2 i g_o^4 C_{D_2} (2\pi)^{26} \delta^{26} (\textstyle\sum_i k_i) \Big[I(s,t) + I(t,u) + I(u,s) \Big] ,$$
$$(6.4.9)$$

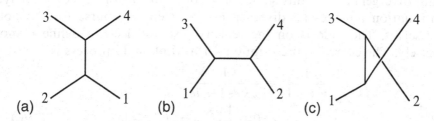

Fig. 6.3. Processes giving poles in the (a) *s*-, (b) *t*-, and (c) *u*-channels.

where

$$I(s,t) = \int_0^1 dy\, y^{-\alpha's-2}(1-y)^{-\alpha't-2}. \tag{6.4.10}$$

The three terms come from figures 6.2(c), (f), 6.2(b), (d), and 6.2(a), (e) respectively.

The integral $I(s,t)$ converges if $\alpha's < -1$ and $\alpha't < -1$. As $\alpha's \to -1$, the integral diverges at $y = 0$. To study the divergence, take a neighborhood of $y = 0$ and approximate the integrand:

$$I(s,t) = \int_0^r dy\, y^{-\alpha's-2} \;+\; \text{terms analytic at } \alpha's = -1$$

$$= -\frac{r^{-\alpha's-1}}{\alpha's+1} \;+\; \text{terms analytic at } \alpha's = -1$$

$$= -\frac{1}{\alpha's+1} \;+\; \text{terms analytic at } \alpha's = -1. \tag{6.4.11}$$

In (6.4.11), we have evaluated the integral in the convergent region. We see that the divergence is a pole at $s = -1/\alpha'$, the mass-squared of the open string tachyon. The variable s is just the square of the center-of-mass energy for scattering $1 + 2 \to 3 + 4$, so this pole is a resonance due to an intermediate tachyon state. Again, it is an artifact of the bosonic string that this lightest string state is tachyonic, and not relevant to the discussion. The pole is due to the process shown in figure 6.3(a), in which tachyons 1 and 2 join to become a single tachyon, which then splits into tachyons 3 and 4.

Because the singularity at $\alpha's = -1$ is just a pole, $I(s,t)$ can be analytically continued past this point into the region $\alpha's > -1$. The amplitude is defined via this analytic continuation. The divergence of the amplitude *at* the pole is an essential physical feature of the amplitude, a resonance corresponding to propagation of the intermediate string state over long spacetime distances. The divergence of the integral *past* the pole is not; it is just an artifact of this particular integral representation of the amplitude. The continuation poses no problem. In fact, we will see that every string divergence is of this same basic form, so this one kind of analytic continuation removes all divergences — except of course for the poles themselves. The pole is on the real axis, so we need to define it more precisely. The correct ϵ prescription for a Minkowski process is

$$\frac{1}{\alpha's+1} \equiv \frac{1}{\alpha's+1+i\epsilon}$$

$$\equiv \mathrm{P}\frac{1}{\alpha's+1} - i\pi\delta(\alpha's+1) ,\qquad (6.4.12)$$

where P denotes the principal value. Unitarity (which we will develop more systematically in chapter 9) requires this pole to be present and determines its coefficient in terms of the amplitude for two tachyons to scatter into one:

$$S_{D_2}(k_1;k_2;k_3;k_4) = i\int \frac{d^{26}k}{(2\pi)^{26}} \frac{S_{D_2}(k_1;k_2;k)S_{D_2}(-k;k_3;k_4)}{-k^2+\alpha'^{-1}+i\epsilon}$$

$$+ \text{ terms analytic at } k^2 = 1/\alpha' . \quad (6.4.13)$$

Gathering together the factors in the four-tachyon amplitude, including an equal contribution to the pole from $I(u,s)$, and using the three-tachyon result (6.4.4), the condition (6.4.13) gives

$$C_{D_2} = e^{-\lambda}C_{D_2}^X C_{D_2}^g = \frac{1}{\alpha'g_0^2} . \qquad (6.4.14)$$

The three-tachyon amplitude is then

$$S_{D_2}(k_1;k_2;k_3) = \frac{2ig_0}{\alpha'}(2\pi)^{26}\delta^{26}(\textstyle\sum_i k_i) . \qquad (6.4.15)$$

The various functional determinants have dropped out. Using unitarity, all normalizations can be expressed in terms of the constant g_0 appearing in the vertex operators. The determinants can in fact be computed by careful regularization and renormalization, and the relative normalizations of different topologies agree with those from unitarity.

Continuing past the pole, we encounter further singularities. Taylor expanding the integrand at $y = 0$ gives

$$I(s,t) = \int_0^r dy \left[y^{-\alpha's-2} + (\alpha't+2)y^{-\alpha's-1} + \dots \right] . \qquad (6.4.16)$$

The second term gives a pole at $\alpha's = 0$,

$$I(s,t) = \frac{u-t}{2s} + \text{ terms analytic at } \alpha's = 0 . \qquad (6.4.17)$$

From the further terms in the Taylor expansion, the amplitude has poles at

$$\alpha's = -1,0,1,2,\dots . \qquad (6.4.18)$$

These are precisely the positions of the open string states. The integral $I(s,t)$ also has poles in the variable t at the same positions (6.4.18), coming from the endpoint $y = 1$. These are due to the process of figure 6.3(b). The other two terms in the amplitude (6.4.9) give further contributions to the s- and t-channel poles, and also give poles in the u-channel, figure 6.3(c). Because the residue at $s = 0$ in (6.4.17) is odd in $u - t$, this pole actually cancels that from $I(s,u)$, as do all poles at even multiples of $1/\alpha'$. This will not be true for the more general open string theories to be introduced in the next section.

Define the Euler beta function

$$B(a,b) = \int_0^1 dy\, y^{a-1}(1-y)^{b-1} , \qquad (6.4.19)$$

so that

$$I(s,t) = B(-\alpha_o(s), -\alpha_o(t)) , \qquad \alpha_o(x) = 1 + \alpha'x . \qquad (6.4.20)$$

This can be expressed in terms of gamma functions. Defining $y = v/w$ for fixed w gives

$$w^{a+b-1}B(a,b) = \int_0^w dv\, v^{a-1}(w-v)^{b-1} . \qquad (6.4.21)$$

Multiplying both sides by e^{-w}, integrating $\int_0^\infty dw$, and regrouping gives

$$\Gamma(a+b)B(a,b) = \int_0^\infty dv\, v^{a-1}e^{-v} \int_0^\infty d(w-v)(w-v)^{b-1}e^{-(w-v)}$$
$$= \Gamma(a)\Gamma(b) . \qquad (6.4.22)$$

The four-tachyon amplitude is then

$$S_{D_2}(k_1;k_2;k_3;k_4) = \frac{2ig_o^2}{\alpha'}(2\pi)^{26}\delta^{26}(\textstyle\sum_i k_i)$$
$$\times \Big[B(-\alpha_o(s), -\alpha_o(t)) + B(-\alpha_o(s), -\alpha_o(u)) + B(-\alpha_o(t), -\alpha_o(u)) \Big], \qquad (6.4.23)$$

where

$$B(-\alpha_o(x), -\alpha_o(y)) = \frac{\Gamma(-\alpha'x - 1)\Gamma(-\alpha'y - 1)}{\Gamma(-\alpha'x - \alpha'y - 2)} . \qquad (6.4.24)$$

This is the *Veneziano amplitude*, originally written to model certain features of strong interaction phenomenology.

The high energy behavior of the Veneziano amplitude is important. There are two regions of interest, the *Regge limit*,

$$s \to \infty , \qquad t \text{ fixed} , \qquad (6.4.25)$$

and the *hard scattering limit*,

$$s \to \infty , \qquad t/s \text{ fixed} . \qquad (6.4.26)$$

If we consider the scattering process $1 + 2 \rightarrow 3 + 4$ (so that k_1^0 and k_2^0 are positive and k_3^0 and k_4^0 negative), then in the 1-2 center-of-mass frame,

$$s = E^2 , \quad t = (4m^2 - E^2) \sin^2 \frac{\theta}{2} , \quad u = (4m^2 - E^2) \cos^2 \frac{\theta}{2} , \quad (6.4.27)$$

where E is the center-of-mass energy and θ is the angle between particle 1 and particle 3. The Regge limit is high energy and small angle, while the hard scattering limit is high energy and fixed angle. Using Stirling's approximation, $\Gamma(x+1) \approx x^x e^{-x} (2\pi x)^{1/2}$, the behavior in the Regge region is

$$S_{D_2}(k_1; k_2; k_3; k_4) \propto s^{\alpha_0(t)} \Gamma(-\alpha_0(t)) , \quad (6.4.28)$$

where again $\alpha_0(t) = \alpha' t + 1$. That is, the amplitude varies as a power of s, the power being t-dependent. This is *Regge behavior*. At the poles of the gamma function, the amplitude is an integer power of s, corresponding to exchange of a string of integer spin $\alpha_0(t)$.

In the hard scattering limit,

$$S_{D_2}(k_1; k_2; k_3; k_4) \approx \exp\left[-\alpha'(s \ln s\alpha' + t \ln t\alpha' + u \ln u\alpha')\right] = \exp[-\alpha' s f(\theta)] , \quad (6.4.29)$$

where

$$f(\theta) \approx -\sin^2 \frac{\theta}{2} \ln \sin^2 \frac{\theta}{2} - \cos^2 \frac{\theta}{2} \ln \cos^2 \frac{\theta}{2} \quad (6.4.30)$$

is positive. The result (6.4.29) is notable. High energy, fixed angle scattering probes the internal structure of the objects being scattered. Rutherford discovered the atomic nucleus with hard alpha–atom scattering. Hard electron–nucleon scattering at SLAC revealed the quark constituents of the nucleon. In quantum field theory, hard scattering amplitudes fall as a power of s. Even a composite object like the nucleon, if its constituents are pointlike, has power law amplitudes. The exponential falloff (6.4.29) is very much softer. The result (6.4.29) suggests a smooth object of size $\alpha'^{1/2}$, as one would expect.

We started with the three-particle amplitude, skipping over the zero-, one-, and two-particle amplitudes. We will discuss these amplitudes, and their interpretation, in section 6.6.

6.5 Chan–Paton factors and gauge interactions

In this section we will consider the interactions of the massless vector state of the open string. To make the discussion a bit more interesting, we first introduce a generalization of the open string theory.

At the end of chapter 3 we introduced a very large class of bosonic string theories, but in this first look at the interactions we are focusing

Fig. 6.4. Open string with Chan–Paton degrees of freedom.

on the simplest case of 26 flat dimensions. One can think of this in terms of symmetry: this theory has the maximal 26-dimensional Poincaré invariance. In the closed bosonic string this is the unique theory with this symmetry. An outline of the proof is as follows. The world-sheet Noether currents for spacetime translations have components of weights $(1, 0)$ and $(0, 1)$. By an argument given in section 2.9, these currents are then holomorphic in z or $\bar{z}$. We have seen in the calculations in this chapter that this is enough to determine all the expectation values.

In open string theory, however, there is a generalization. The open string has boundaries, endpoints. In quantum systems with distinguished points it is natural to have degrees of freedom residing at those points in addition to the fields propagating in the bulk. At each end of the open string let us add a new degree of freedom, known as a *Chan–Paton* degree of freedom, which can be in one of n states. A basis of string states is then

$$|N; k; ij\rangle , \tag{6.5.1}$$

where i and j denote the states of the left- and right-hand endpoints, running from 1 to n. The energy-momentum tensor is defined to be the same as before, with no dependence on the new degrees of freedom. Conformal invariance is therefore automatic. Poincaré invariance is automatic as well, as the Chan–Paton degrees of freedom are invariant. Although these new degrees of freedom have trivial world-sheet dynamics, they will have a profound effect on the spacetime physics.

In string theories of the strong interaction, the motivation for this was to introduce $SU(3)$ flavor quantum numbers: the endpoints are like quarks and antiquarks, connected by a color-electric flux tube. Now we are motivating it in the general framework of considering all possibilities with given symmetries. We will give a new interpretation to the Chan–Paton degrees of freedom in chapter 8, and a possible further refinement in chapter 14.

There are now n^2 scalar tachyons, n^2 massless vector bosons, and so forth. The n^2 Hermitian matrices λ^a_{ij}, normalized to

$$\mathrm{Tr}(\lambda^a \lambda^b) = \delta^{ab} , \tag{6.5.2}$$

are a complete set of states for the two endpoints. These are the representation matrices of $U(n)$, so one might guess that the massless vectors are associated with a $U(n)$ gauge symmetry; we will soon see that this is the case.

Define the basis

$$|N;k;a\rangle = \sum_{i,j=1}^{n} |N;k;ij\rangle\lambda_{ij}^{a} . \tag{6.5.3}$$

Now consider the four-tachyon amplitude shown in figure 6.2(a), in which the vertex operators are arranged in the cyclic order 1234. Because the Chan–Paton degrees of freedom do not appear in the Hamiltonian, their state does not evolve between the vertex operators: the right-hand end-point of tachyon 1 must be in the same state as the left-hand endpoint of tachyon 2, and so forth. Thus, the amplitude 6.2(a) will now contain an additional factor of

$$\mathrm{Tr}\,(\lambda^{a_1}\lambda^{a_2}\lambda^{a_3}\lambda^{a_4}) \tag{6.5.4}$$

from the overlap of the Chan–Paton wavefunctions for each tachyon. This rule generalizes to an arbitrary amplitude: each vertex operator now contains a *Chan–Paton factor* λ_{ij}^{a} from the wavefunction of the endpoint degrees of freedom, and the amplitude for each world-sheet is multiplied by a trace of the Chan–Paton factors around each boundary.

The three-tachyon amplitude becomes

$$S_{D_2}(k_1,a_1;\,k_2,a_2;\,k_3,a_3) =$$
$$\frac{ig_o}{\alpha'}(2\pi)^{26}\delta^{26}(\textstyle\sum_i k_i)\mathrm{Tr}\,(\lambda^{a_1}\lambda^{a_2}\lambda^{a_3} + \lambda^{a_1}\lambda^{a_3}\lambda^{a_2}) , \tag{6.5.5}$$

the two cyclic orderings now having different Chan–Paton traces. The four-tachyon amplitude is

$$S_{D_2}(k_1,a_1;\,k_2,a_2;\,k_3,a_3;\,k_4,a_4) = \frac{ig_o^2}{\alpha'}(2\pi)^{26}\delta^{26}(\textstyle\sum_i k_i)$$
$$\times\Big[\mathrm{Tr}\,(\lambda^{a_1}\lambda^{a_2}\lambda^{a_4}\lambda^{a_3} + \lambda^{a_1}\lambda^{a_3}\lambda^{a_4}\lambda^{a_2})B(-\alpha_o(s),-\alpha_o(t))$$
$$+\,\mathrm{Tr}\,(\lambda^{a_1}\lambda^{a_3}\lambda^{a_2}\lambda^{a_4} + \lambda^{a_1}\lambda^{a_4}\lambda^{a_2}\lambda^{a_3})B(-\alpha_o(t),-\alpha_o(u))$$
$$+\,\mathrm{Tr}\,(\lambda^{a_1}\lambda^{a_2}\lambda^{a_3}\lambda^{a_4} + \lambda^{a_1}\lambda^{a_4}\lambda^{a_3}\lambda^{a_2})B(-\alpha_o(s),-\alpha_o(u))\Big] . \tag{6.5.6}$$

Considering again the unitarity relation (6.4.13), the pole at $s = -1/\alpha'$ acquires a factor of

$$\frac{1}{4}\mathrm{Tr}\,\Big(\{\lambda^{a_1},\lambda^{a_2}\}\{\lambda^{a_3},\lambda^{a_4}\}\Big) \tag{6.5.7}$$

on the left and a factor of

$$\frac{1}{4}\sum_a \mathrm{Tr}\,\Big(\{\lambda^{a_1},\lambda^{a_2}\}\lambda^a\Big)\,\mathrm{Tr}\,\Big(\{\lambda^{a_3},\lambda^{a_4}\}\lambda^a\Big) , \tag{6.5.8}$$

on the right, the sum being over the Chan–Paton wavefunction of the intermediate state. The completeness of the λ_{ij}^a and the normalization (6.5.2) imply that

$$\mathrm{Tr}\,(A\lambda^a)\,\mathrm{Tr}\,(B\lambda^a) = \mathrm{Tr}\,(AB) \tag{6.5.9}$$

for any matrices A and B, and so the amplitudes are still unitary.

Gauge interactions

The amplitude for a gauge boson and two tachyons is

$$S_{D_2}(k_1, a_1, e_1; k_2, a_2; k_3, a_3) = -ig_o' g_o^2 e^{-\lambda} e_{1\mu}$$
$$\times \left\langle {}^*_* c^1 \dot{X}^\mu e^{ik_1\cdot X}(y_1)^*_* \, {}^*_* c^1 e^{ik_2\cdot X}(y_2)^*_* \, {}^*_* c^1 e^{ik_3\cdot X}(y_3)^*_* \right\rangle_{D_2} \mathrm{Tr}\,(\lambda^{a_1}\lambda^{a_2}\lambda^{a_3})$$
$$+ (k_2, a_2) \leftrightarrow (k_3, a_3) \, . \tag{6.5.10}$$

We have used the gauge boson vertex operator (3.6.26), but for now are allowing an independent normalization constant g_o'. Using the results of section 6.2, the X path integral is

$$\left\langle {}^*_* \dot{X}^\mu e^{ik_1\cdot X}(y_1)^*_* \, {}^*_* e^{ik_2\cdot X}(y_2)^*_* \, {}^*_* e^{ik_3\cdot X}(y_3)^*_* \right\rangle_{D_2} = -2i\alpha' \left(\frac{k_2^\mu}{y_{12}} + \frac{k_3^\mu}{y_{13}} \right)$$
$$\times iC_{D_2}^X (2\pi)^{26} \delta^{26}(\textstyle\sum_i k_i) |y_{12}|^{2\alpha' k_1\cdot k_2} |y_{13}|^{2\alpha' k_1\cdot k_3} |y_{23}|^{2\alpha' k_2\cdot k_3} \, . \tag{6.5.11}$$

Using momentum conservation, the mass-shell conditions, and the physical state condition $k_1\cdot e_1 = 0$, the amplitude becomes

$$S_{D_2}(k_1, a_1, e_1; k_2, a_2; k_3, a_3)$$
$$= -ig_o' e_1\cdot k_{23}(2\pi)^{26}\delta^{26}(\textstyle\sum_i k_i)\mathrm{Tr}\left(\lambda^{a_1}[\lambda^{a_2}, \lambda^{a_3}]\right), \tag{6.5.12}$$

where $k_{ij} \equiv k_i - k_j$. This is again independent of the vertex operator positions.

The $s = 0$ pole in the four-tachyon amplitude no longer vanishes. The terms that canceled now have the Chan–Paton factors in different orders, so the pole is proportional to

$$\mathrm{Tr}\left([\lambda^{a_1}, \lambda^{a_2}][\lambda^{a_3}, \lambda^{a_4}]\right) \, . \tag{6.5.13}$$

Relating the coefficient of this pole to the amplitude (6.5.12) by unitarity, one obtains

$$g_o' = (2\alpha')^{1/2} g_o \, . \tag{6.5.14}$$

This is the same relative normalization as from the state–operator mapping: there is only one independent coupling constant.

For the three-gauge-boson coupling, a similar calculation gives

$$S_{D_2}(k_1, a_1, e_1; k_2, a_2, e_2; k_3, a_3, e_3)$$

$$= ig_0'(2\pi)^{26}\delta^{26}(\textstyle\sum_i k_i)\Big(e_1 \cdot k_{23} \, e_2 \cdot e_3 + e_2 \cdot k_{31} \, e_3 \cdot e_1 + e_3 \cdot k_{12} \, e_1 \cdot e_2$$

$$+ \frac{\alpha'}{2} e_1 \cdot k_{23} \, e_2 \cdot k_{31} \, e_3 \cdot k_{12}\Big) \mathrm{Tr}\left(\lambda^{a_1}[\lambda^{a_2}, \lambda^{a_3}]\right) . \qquad (6.5.15)$$

Up to first order in the momenta, the amplitudes we have found are reproduced by the spacetime action

$$S = \frac{1}{g_0'^2}\int d^{26}x \left[-\frac{1}{2}\mathrm{Tr}\left(D_\mu\varphi D^\mu\varphi\right) + \frac{1}{2\alpha'}\mathrm{Tr}\left(\varphi^2\right) + \frac{2^{1/2}}{3\alpha'^{1/2}}\mathrm{Tr}\left(\varphi^3\right)\right.$$

$$\left. - \frac{1}{4}\mathrm{Tr}\left(F_{\mu\nu}F^{\mu\nu}\right)\right], \qquad (6.5.16)$$

where the tachyon field φ and the Yang–Mills vector potential A_μ are written as $n \times n$ matrices, e. g. $A_\mu = A_\mu^a \lambda^a$. Also, $D_\mu\varphi = \partial_\mu\varphi - i[A_\mu, \varphi]$ and $F_{\mu\nu} = \partial_\mu A_\nu - \partial_\nu A_\mu - i[A_\mu, A_\nu]$.

This is the action for a $U(n)$ gauge field coupled to a scalar in the adjoint representation. Adding in the Chan–Paton factors has produced just the gauge-invariant expressions needed. The gauge invariance is automatic because the decoupling of unphysical states is guaranteed in string perturbation theory, as we will develop further.

At momenta k small compared to the string scale, the only open string states are the massless gauge bosons. As discussed in section 3.7, it is in this limit that the physics should reduce to an effective field theory of the massless states. We have therefore been somewhat illogical in including the tachyon in the action (6.5.16); we did so for illustration, but now let us focus on the gauge bosons. The four-gauge-boson amplitude has a form analogous to the Veneziano amplitude but with additional structure from the polarization tensors. Expanding it in powers of $\alpha'k^2$, the first term, which survives in the *zero-slope limit* $\alpha' \to 0$, is the sum of pole terms in s, t, and u plus a constant. Consistency guarantees that it is precisely the four-gauge-boson amplitude obtained in field theory from the Yang–Mills Lagrangian $F_{\mu\nu}F^{\mu\nu}$. Note, however, the term of order $\alpha'k^3$ in the three-gauge-boson amplitude (6.5.15). This implies a higher derivative term

$$-\frac{2i\alpha'}{3g_0'^2}\mathrm{Tr}\left(F_\mu{}^\nu F_\nu{}^\omega F_\omega{}^\mu\right) \qquad (6.5.17)$$

in the Lagrangian. Similarly, expanding the four-point amplitude reveals an infinite sum of higher order interactions (beyond (6.5.17), these do not contribute to the three-gauge-boson amplitude for kinematic reasons). String loop amplitudes also reduce, in the low energy region, to the loops obtained from the effective Lagrangian. By the usual logic of effective actions, the higher derivative terms are less important at low energy. The scale where they become important, $\alpha' k^2 \approx 1$, is just where new physics (massive string states) appears and the effective action is no longer applicable.

If we have a cutoff, why do we need renormalization theory? Renormalization theory still has content and in fact this is its real interpretation: it means that low energy physics is independent of the details of the high energy theory, except for the parameters in the effective Lagrangian. This is a mixed blessing: it means that we can use ordinary quantum field theory to make predictions at accelerator energies without knowing the form of the Planck scale theory, but it also means that we cannot probe the Planck scale theory with physics at particle accelerators.

The string spectrum and amplitudes have an obvious global $U(n)$ symmetry,

$$\lambda^a \to U \lambda^a U^\dagger , \tag{6.5.18}$$

which leaves the Chan–Paton traces and the norms of states invariant. From the detailed form of the amplitudes, we have learned that this is actually a *local* symmetry in spacetime. We will see that this promotion of a global world-sheet symmetry to a local spacetime symmetry is a rather general phenomenon in string theory.

All open string states transform as the $n \times n$ adjoint representation under the $U(n)$ symmetry. Incidentally, $U(n)$ is not a simple Lie algebra: $U(n) = SU(n) \times U(1)$. The $U(1)$ gauge bosons, $\lambda_{ij} = \delta_{ij}/n^{1/2}$, decouple from the amplitudes (6.5.12) and (6.5.15). The adjoint representation of $U(1)$ is trivial, so all string states are neutral under the $U(1)$ symmetry.

The unoriented string

It is interesting to generalize to the unoriented string. Consider first the theory without Chan–Paton factors. In addition to Möbius invariance, the X^μ CFT on the sphere or disk is invariant under the orientation-reversing symmetry $\sigma^1 \to \pi - \sigma^1$ in the open string or $\sigma^1 \to 2\pi - \sigma^1$ in the closed string. This *world-sheet parity* symmetry is generated by an operator Ω. From the mode expansions it follows that

$$\Omega \alpha_n^\mu \Omega^{-1} = (-1)^n \alpha_n^\mu \tag{6.5.19}$$

in the open string and

$$\Omega \alpha_n^\mu \Omega^{-1} = \tilde{\alpha}_n^\mu \tag{6.5.20}$$

in the closed string. The symmetry extends to the ghosts as well, but to avoid distraction we will not discuss them explicitly, since they only contribute a fixed factor to the tree-level amplitudes.

The tachyon vertex operators are even under world-sheet parity in either the closed or open string (this is obvious for the integrated vertex operators, without ghosts; the fixed operators must then transform in the same way), determining the sign of the operator. All states can then be classified by their parity eigenvalue $\omega = \pm 1$. The relation (6.5.19) in the open string implies that

$$\Omega|N;k\rangle = \omega_N|N;k\rangle, \quad \omega_N = (-1)^{1+\alpha'm^2}. \qquad (6.5.21)$$

World-sheet parity is multiplicatively conserved. For example, the three-tachyon amplitude is nonzero, consistent with $(+1)^3 = 1$. On the other hand, the massless vector has $\omega = -1$ and so we would expect the vector–tachyon–tachyon amplitude (6.5.10) and three-vector amplitude (6.5.15) to vanish in the absence of Chan–Paton degrees of freedom, as indeed they do (the λ^a are replaced by 1 and the commutators vanish). The different cyclic orderings, related to one another by world-sheet parity, cancel.

Given a consistent oriented string theory, we can make a new *unoriented* string theory by restricting the spectrum to the states of $\omega = +1$. States with odd $\alpha'm^2$ remain, while states with even $\alpha'm^2$, including the photon, are absent. The conservation of ω guarantees that if all external states have $\omega = +1$, then the intermediate states in tree-level amplitudes will also have $\omega = +1$. Unitarity of the unoriented theory thus follows from that of the oriented theory, at least at tree level (and in fact to all orders, as we will outline in chapter 9).

The main point of interest in the unoriented theory is the treatment of the Chan–Paton factors. Since we have identified these with the respective endpoints of the open string, world-sheet parity must reverse them,

$$\Omega|N;k;ij\rangle = \omega_N|N;k;ji\rangle . \qquad (6.5.22)$$

Again, this is a symmetry of all amplitudes in the oriented theory. To form the unoriented theory, we again restrict the spectrum to world-sheet parity eigenvalue $\omega = +1$. Take a basis for λ^a_{ij} in which each matrix is either symmetric, $s^a = +1$, or antisymmetric, $s^a = -1$. Then

$$\Omega|N;k;a\rangle = \omega_N s^a|N;k;a\rangle . \qquad (6.5.23)$$

The world-sheet parity eigenvalue is $\omega = \omega_N s^a$, and the unoriented spectrum is

$$\alpha'm^2 \text{ even:} \quad \lambda^a \text{ antisymmetric}, \qquad (6.5.24a)$$
$$\alpha'm^2 \text{ odd:} \quad \lambda^a \text{ symmetric} . \qquad (6.5.24b)$$

For the massless gauge bosons the Chan–Paton factors are the $n \times n$ antisymmetric matrices and so the gauge group is $SO(n)$. The states at even mass levels transform as the adjoint representation of the orthogonal group $SO(n)$, and the states at odd mass levels transform as the traceless symmetric tensor plus singlet representations.

The oriented theory has a larger set of orientation-reversing symmetries, obtained as a combination of Ω and a $U(n)$ rotation,

$$\Omega_\gamma |N;k;ij\rangle = \omega_N \gamma_{jj'} |N;k;j'i'\rangle \gamma^{-1}_{i'i} . \tag{6.5.25}$$

We can form more general unoriented theories by restricting the spectrum to $\omega_\gamma = +1$, which is again consistent with the interactions. Acting twice with Ω_γ gives

$$\Omega_\gamma^2 |N;k;ij\rangle = [(\gamma^T)^{-1}\gamma]_{ii'} |N;k;i'j'\rangle (\gamma^{-1}\gamma^T)_{j'j} . \tag{6.5.26}$$

We will insist that $\Omega_\gamma^2 = 1$ for reasons to be explained below. This then implies that

$$\gamma^T = \pm\gamma . \tag{6.5.27}$$

That is, γ is symmetric or antisymmetric.

A general change of Chan–Paton basis,

$$|N;k;ij\rangle' = U^{-1}_{ii'} |N;k;i'j'\rangle U_{j'j} , \tag{6.5.28}$$

transforms γ to

$$\gamma' = U^T \gamma U . \tag{6.5.29}$$

In the symmetric case, it is always possible to find a basis such that $\gamma = 1$, giving the theory already considered above. In the antisymmetric case there is a basis in which

$$\gamma = M \equiv i \begin{bmatrix} 0 & I \\ -I & 0 \end{bmatrix} . \tag{6.5.30}$$

Here I is the $k \times k$ identity matrix and $n = 2k$ must be even because γ is an invertible antisymmetric matrix. We take a basis for the Chan–Paton wavefunctions such that $M(\lambda^a)^T M = s^{a'}\lambda^a$ with $s^{a'} = \pm 1$. Then the world-sheet parity eigenvalue is $\omega_\gamma = \omega_N s^{a'}$, and the unoriented spectrum is

$$\alpha'm^2 \text{ even:} \quad M(\lambda^a)^T M = -\lambda^a , \tag{6.5.31a}$$
$$\alpha'm^2 \text{ odd:} \quad M(\lambda^a)^T M = +\lambda^a . \tag{6.5.31b}$$

At the even mass levels, including the gauge bosons, this defines the adjoint representation of the symplectic group $Sp(k)$.

The argument that we must have $\Omega_\gamma^2 = 1$ to construct the unoriented theory is as follows. Since $\Omega^2 = 1$, it must be that Ω_γ^2 acts only on the

Chan–Paton states, not the oscillators. In fact, from eq. (6.5.26), it acts on the Chan–Paton wavefunction as

$$\lambda \to (\gamma^T)^{-1}\gamma\lambda\gamma^{-1}\gamma^T = \lambda \,, \tag{6.5.32}$$

where the last equality must hold in the unoriented theory, since all states in this theory are invariant under Ω_γ and so also under Ω_γ^2. Now we assert that the allowed Chan–Paton wavefunctions must form a complete set. The point is that two open strings, by the splitting–joining interaction of figure 3.4(c), can exchange endpoints. In this way one can get to a complete set, that is, to any Chan–Paton state $|ij\rangle$ (this argument is slightly heuristic, but true). By Schur's lemma, if eq. (6.5.32) holds for a complete set, then $\gamma^{-1}\gamma^T = 1$ and so $\Omega_\gamma^2 = 1$.

We might try to obtain other gauge groups by taking different sets of λ^a. In fact, the oriented $U(n)$ and unoriented $SO(n)$ and $Sp(k)$ theories constructed above are the only possibilities. Generalization of the completeness argument in eqs. (6.5.7)–(6.5.9) shows these to be the most general solutions of the unitarity conditions. In particular, exceptional Lie algebras, which are of interest in grand unification and which will play a major role in volume two, cannot be obtained with Chan–Paton factors (in perturbation theory). In closed string theory there is another mechanism that gives rise to gauge bosons, and it allows other groups.

6.6   Closed string tree amplitudes

The discussion of closed string amplitudes is parallel to the above. The amplitude for three closed string tachyons is

$$S_{S_2}(k_1;k_2;k_3) = g_c^3 e^{-2\lambda}\left\langle \prod_{i=1}^{3} :\tilde{c}c e^{ik_i\cdot X}(z_i,\bar{z}_i): \right\rangle_{S_2} . \tag{6.6.1}$$

In this case the CKG $PSL(2,\mathbf{C})$ (the Möbius group) can be used to fix the three vertex operators to arbitrary positions $z_{1,2,3}$. Taking the expectation values from section 6.2, the result is again independent of the vertex operator positions,

$$S_{S_2}(k_1;k_2;k_3) = ig_c^3 C_{S_2}(2\pi)^{26}\delta^{26}(\textstyle\sum_i k_i) \,, \tag{6.6.2}$$

where $C_{S_2} = e^{-2\lambda}C_{S_2}^X C_{S_2}^g$.

For four closed string tachyons,

$$S_{S_2}(k_1;k_2;k_3;k_4) = g_c^4 e^{-2\lambda}\int_{\mathbf{C}} d^2 z_4 \left\langle \prod_{i=1}^{3} :\tilde{c}c e^{ik_i\cdot X}(z_i,\bar{z}_i): \, :e^{ik_4\cdot X}(z_4,\bar{z}_4): \right\rangle_{S_2} ,$$

$$\tag{6.6.3}$$

where the integral runs over the complex plane $\mathbf{C}$. Evaluating the expectation value and setting $z_1 = 0$, $z_2 = 1$, $z_3 = \infty$, this becomes

$$S_{S_2}(k_1; k_2; k_3; k_4) = ig_c^4 C_{S_2}(2\pi)^{26}\delta^{26}(\textstyle\sum_i k_i)J(s, t, u) ,\qquad (6.6.4)$$

where

$$J(s, t, u) = \int_{\mathbf{C}} d^2 z_4 \, |z_4|^{-\alpha' u/2 - 4}|1 - z_4|^{-\alpha' t/2 - 4} .\qquad (6.6.5)$$

Here $s + t + u = -16/\alpha'$, but we indicate the dependence of J on all three variables to emphasize its symmetry among them. This amplitude converges when $s, t, u < -4/\alpha'$. It has poles in the variable u from $z_4 \to 0$, in the variable t from $z_4 \to 1$, and in the variable s from $z_4 \to \infty$. The poles are at the values

$$\alpha's , \ \alpha't , \ \alpha'u = -4, 0, 4, 8, \ldots ,\qquad (6.6.6)$$

which are the masses-squared of the closed string states. The pole at $\alpha's = -4$ is

$$ig_c^4 C_{S_2} \int_{|z_4| > 1/\epsilon} d^2 z_4 \, |z_4|^{\alpha's/2} \sim -\frac{8\pi ig_c^4 C_{S_2}}{\alpha's + 4} .\qquad (6.6.7)$$

Unitarity gives

$$C_{S_2} = \frac{8\pi}{\alpha' g_c^2} ,\qquad (6.6.8)$$

and so

$$S_{S_2}(k_1; k_2; k_3) = \frac{8\pi ig_0}{\alpha'}(2\pi)^{26}\delta^{26}(\textstyle\sum_i k_i) .\qquad (6.6.9)$$

Like the Veneziano amplitude, the amplitude for four closed string tachyons can be expressed in terms of gamma functions (exercise 6.10):

$$S_{S_2}(k_1; k_2; k_3; k_4) = \frac{8\pi ig_c^2}{\alpha'}(2\pi)^{26}\delta^{26}(\textstyle\sum_i k_i)C(-\alpha_c(t), -\alpha_c(u)) ,\qquad (6.6.10)$$

where $\alpha_c(x) = 1 + \alpha'x/4$ and

$$C(a, b) = \int_{\mathbf{C}} d^2 z \, |z|^{2a-2}|1 - z|^{2b-2}$$
$$= 2\pi \frac{\Gamma(a)\Gamma(b)\Gamma(c)}{\Gamma(a + b)\Gamma(a + c)\Gamma(b + c)} , \quad a + b + c = 1 .\qquad (6.6.11)$$

This is the *Virasoro–Shapiro amplitude*. There is just a single term, with poles in the s-, t-, and u-channels coming from the gamma functions in the numerator. Like the Veneziano amplitude, the Virasoro–Shapiro amplitude has Regge behavior in the Regge limit,

$$S_{S_2}(k_1; k_2; k_3; k_4) \propto s^{2\alpha_c(t)}\frac{\Gamma(-\alpha_c(t))}{\Gamma(1 + \alpha_c(t))} ,\qquad (6.6.12)$$

and exponential behavior in the hard scattering limit,

$$S_{S_2}(k_1; k_2; k_3; k_4) \propto \exp\left[-\frac{\alpha'}{2}(s \ln s\alpha' + t \ln t\alpha' + u \ln u\alpha')\right] . \qquad (6.6.13)$$

On the sphere, the amplitude for a massless closed string and two closed string tachyons is

$$S_{S_2}(k_1, e_1; k_2; k_3) = g_c^2 g_c' e^{-2\lambda} e_{1\mu\nu} \left\langle :\tilde{c}c\partial X^\mu \bar{\partial} X^\nu e^{ik_1 \cdot X}(z_1, \bar{z}_1): \right.$$

$$\left. :\tilde{c}c e^{ik_2 \cdot X}(z_2, \bar{z}_2): :\tilde{c}c e^{ik_3 \cdot X}(z_3, \bar{z}_3): \right\rangle_{S_2}$$

$$= -\frac{\pi i \alpha'}{2} g_c' e_{1\mu\nu} k_{23}^\mu k_{23}^\nu (2\pi)^{26} \delta^{26}(\textstyle\sum_i k_i) , \qquad (6.6.14)$$

where $e_{1\mu\nu} e_1^{\mu\nu} = 1$. Expanding the Virasoro–Shapiro amplitude (6.6.10) on the $s = 0$ pole and using unitarity determines

$$g_c' = \frac{2}{\alpha'} g_c , \qquad (6.6.15)$$

again in agreement with the state–operator mapping with overall constant g_c. The amplitude (6.6.14) would be obtained in field theory from the spacetime action $S + S_T$, where S is the action (3.7.20) for the massless fields, and where

$$S_T = -\frac{1}{2} \int d^{26}x \, (-G)^{1/2} e^{-2\Phi} \left(G^{\mu\nu} \partial_\mu T \partial_\nu T - \frac{4}{\alpha'} T^2 \right) , \qquad (6.6.16)$$

is the action for the closed string tachyon T. For example, the amplitude for a graviton of polarization $e_{\mu\nu}$ is obtained from this action by expanding

$$\tilde{G}_{\mu\nu} = \eta_{\mu\nu} - 2\kappa e_{\mu\nu} e^{ik\cdot x} . \qquad (6.6.17)$$

Note that this is the Einstein metric, whose action (3.7.25) is independent of the dilaton. The normalization of the fluctuation is determined by that of the graviton kinetic term in the spacetime action. Specifically, if one takes $\tilde{G}_{\mu\nu} - \eta_{\mu\nu} = -2\kappa e_{\mu\nu} f(x)$ with $e_{\mu\nu} e^{\mu\nu} = 1$, the effective action for f has the canonical normalization $\frac{1}{2}$ for a real scalar. The field theory amplitude matches the string result (6.6.14) and relates the normalization of the vertex operators to the gravitational coupling,

$$\kappa = \pi \alpha' g_c' = 2\pi g_c . \qquad (6.6.18)$$

The amplitude for three massless closed strings is

$$S_{S_2}(k_1, e_1; k_2, e_2; k_3, e_3) = \frac{i\kappa}{2}(2\pi)^{26} \delta^{26}(\textstyle\sum_i k_i) e_{1\mu\nu} e_{2\alpha\beta} e_{3\gamma\delta} T^{\mu\alpha\gamma} T^{\nu\beta\delta} , \qquad (6.6.19)$$

where

$$T^{\mu\alpha\gamma} = k_{23}^\mu \eta^{\alpha\gamma} + k_{31}^\alpha \eta^{\gamma\mu} + k_{12}^\gamma \eta^{\mu\alpha} + \frac{\alpha'}{8} k_{23}^\mu k_{31}^\alpha k_{12}^\gamma . \qquad (6.6.20)$$

The order k^2 terms in this amplitude correspond to the spacetime action (3.7.25), while the k^4 and k^6 terms come from a variety of higher derivative interactions, including terms quadratic and cubic in the spacetime curvature. These higher corrections to the action can also be determined by calculating higher loop corrections to the world-sheet beta functions (3.7.14).

The tensor structure of the closed string amplitude (6.6.19) is just two copies of that in the open string amplitude (6.5.15), if one sets $\alpha' = 2$ in the closed string and $\alpha' = \frac{1}{2}$ in the open. The same holds for the amplitude (6.6.14). This is a consequence of the factorization of free-field expectation values on the sphere into holomorphic and antiholomorphic parts. A similar factorization holds for four or more closed strings *before* integration over the vertex operator positions. Further, by careful treatment of the contour of integration it is possible to find relations between the integrated amplitudes. For the four-tachyon amplitudes the integrals above are related

$$J(s, t, u, \alpha') = -2 \sin \pi\alpha_c(t) I(s, t, 4\alpha') I(t, u, 4\alpha') , \qquad (6.6.21)$$

after use of the gamma function identity $\Gamma(x)\Gamma(1-x)\sin(\pi x) = \pi$; we now indicate the explicit dependence of the integrals on α'. For the general integral appearing in four-point closed string amplitudes there is a relation

$$\int_{\mathbf{C}} d^2z \, z^{a-1+m_1} \bar{z}^{a-1+n_1} (1-z)^{b-1+m_2} (1-\bar{z})^{b-1+n_2}$$

$$= 2\sin[\pi(b+n_2)] B(a+m_1, b+m_2) B(b+n_2, 1-a-b-n_1-n_2) .$$

$$m_1 - n_1 \in \mathbf{Z} , \quad m_2 - n_2 \in \mathbf{Z} \qquad (6.6.22)$$

This implies a corresponding relation between four-point open and closed string amplitudes

$$A_c(s, t, u, \alpha', g_c) = \frac{\pi i g_c^2 \alpha'}{g_o^4} \sin[\pi\alpha_c(t)] A_o(s, t, \tfrac{1}{4}\alpha', g_o) A_o(t, u, \tfrac{1}{4}\alpha', g_o)^* ,$$

$$(6.6.23)$$

where the open string amplitudes include just one of the six cyclic permutations, with poles in the indicated channels.

Consistency

In chapter 9 we will discuss the convergence and gauge invariance of tree-level amplitudes in a general way, but as an introduction we will now use the OPE to see how these work for the lowest levels. Consider the operator product

$$:e^{ik_1 \cdot X(z_1, \bar{z}_1)}: \, :e^{ik_4 \cdot X(z_4, \bar{z}_4)}: \, = \, |z_{14}|^{\alpha' k_1 \cdot k_4} : \Big(1 + iz_{14} k_1 \cdot \partial X + i\bar{z}_{14} k_1 \cdot \bar{\partial} X$$

$$- z_{14}\bar{z}_{14} k_1 \cdot \partial X \, k_1 \cdot \bar{\partial} X + \dots \Big) e^{i(k_1+k_4) \cdot X}(z_4, \bar{z}_4): \, . \qquad (6.6.24)$$

This appears in the amplitude with z_{14} integrated. The integral converges as $z_{14} \to 0$ when

$$\alpha' k_1 \cdot k_4 = \frac{\alpha'}{2}(k_1 + k_4)^2 - 4 > -2 \qquad (6.6.25)$$

and has a pole at that point. The coefficient of the pole is just the tachyon vertex operator. Thus, if a pair of tachyons in any amplitude has total momentum $(k_1 + k_4)^2 = 4/\alpha'$, there will be a pole proportional to the amplitude with one fewer tachyon as required by unitarity. A pole in momentum space corresponds to long distance in spacetime, so this is a process in which two tachyons scatter into one, which then propagates and interacts with the remaining particles. Carrying the OPE further, the $O(z_{14})$ and $O(\bar{z}_{14})$ terms do not produce poles because the angular integration gives zero residue, the $O(z_{14}\bar{z}_{14})$ term gives a massless pole, and so on.

Now let us look a little more closely at the way the local spacetime symmetries are maintained in the string amplitudes. The various amplitudes we have calculated all vanish if any of the polarizations are of the form $e_\mu = k_\mu$, or $e_{\mu\nu} = \zeta_\mu k_\nu + k_\mu \tilde{\zeta}_\nu$ with $k \cdot \zeta = k \cdot \tilde{\zeta} = 0$. The corresponding spacetime actions are thus invariant under the Yang–Mills, coordinate, and antisymmetric tensor symmetries. As discussed in section 3.6, the vertex operator for a longitudinal polarization is the sum of a total derivative and a term that vanishes by the equations of motion. Upon integration, the total derivative vanishes but the equation of motion term might have a source at one of the other vertex operators in the path integral. The operator product (6.6.24) vanishes rapidly when $k_1 \cdot k_4$ is large. Using this property, there will be for any pair of vertex operators a kinematic region in which all possible contact terms are suppressed. The amplitude for any null polarization then vanishes *identically* in this region, and since all amplitudes are analytic except for poles (and branch points at higher order) the amplitudes for null polarizations must vanish everywhere. We see that the singularities required by unitarity, as well as any possible divergences or violations of spacetime gauge invariance, arise from the limits $z \to 0, 1, \infty$, where two vertex operators come together. For the sphere with four marked points (vertex operators) these are the boundaries of moduli space. The analytic continuation argument used here is known for historic reasons as the *canceled propagator argument*.

Closed strings on D_2 and RP_2

The lowest order closed–open interactions come from the disk with both closed and open vertex operators. The low energy effective action for these can be deduced from general considerations. With a trivial closed string background we found the usual gauge kinetic term,

$$-\frac{1}{4g_0'^2}\int d^{26}x\, \mathrm{Tr}\,(F_{\mu\nu}F^{\mu\nu})\,. \tag{6.6.26}$$

Obviously the metric must couple to this in a covariant way. In addition, the coupling of the dilaton can be deduced. Recall that $g_0'^2 \propto e^{\Phi_0}$, with Φ_0 the expectation value of the dilaton. So we should replace this (inside the integral) with $\Phi = \Phi_0 + \tilde{\Phi}$. The action

$$-\frac{1}{4g_0'^2}\int d^{26}x\,(-G)^{1/2}e^{-\tilde{\Phi}}\mathrm{Tr}\,(F_{\mu\nu}F^{\mu\nu}) \tag{6.6.27}$$

thus incorporates all interactions not involving derivatives of the closed string fields; indices are now raised and lowered with $G_{\mu\nu}$. This action reflects the general principle that the effective action from Euler number χ is weighted by $g_0^{-\chi} \propto e^{-\chi\Phi}$.

The disk and projective plane also make a contribution to purely closed string interactions. The amplitudes for n closed strings are of order $g_0^{-2}g_c^n \sim g_c^{n-1}$, one power of g_c higher than the sphere. Closed string loop amplitudes, to be considered in the next chapter, are of order g_c^2 times the sphere, because emitting and reabsorbing a closed string adds two factors of g_c. Thus the disk and projective plane are 'half-loop order.'

Of particular interest are the amplitudes on the disk and the projective plane with a single closed string vertex operator. Fixing the position of the vertex operator removes only two of the three CKVs, the residual gauge symmetry consisting of rotations about the vertex operator position. Thus we have to divide the amplitude by the volume of this residual CKG. We have not shown how to do this explicitly, but will work it out for the torus in the next chapter. For the disk with one closed string this is a finite factor and the result is nonzero. The amplitude is a numerical factor times $g_c g_0^{-2}$, which in turn is a pure number, times powers of α' as required by dimensional analysis. We will not work out these numerical factors here, but will obtain them in an indirect way in chapter 8.

Thus there is an amplitude for a single closed string to appear from the vacuum, either through the disk or the projective plane, necessarily with zero momentum. Such an amplitude is known as a *tadpole*. In other words, the background closed string fields are corrected from their original values at order g_c. Again we can write an effective action, which is simply

$$-\Lambda\int d^{26}x\,(-G)^{1/2}e^{-\tilde{\Phi}}\,, \tag{6.6.28}$$

the Φ dependence deduced as above. This is a potential for the dilaton. We will consider it further in the next chapter.

The amplitude for a single closed string on the sphere, on the other hand, is zero. The residual CKG is a noncompact subgroup of $PSL(2,\mathbf{C})$ and so one has to divide by an infinite volume. A nonzero result would have been a logical inconsistency, a zeroth order correction to the background fields. Similarly the amplitude for two closed strings on the sphere (a zeroth order correction to the mass) vanishes, as do the corresponding disk amplitudes, one or two open strings. The amplitudes with no vertex operators at all are also meaningful — they just calculate the term of order $\tilde{\Phi}^0$ in the Taylor expansion of the action (6.6.28). The disk amplitude with no vertex operators is thus nonvanishing; this requires a somewhat formal treatment of the conformal Killing volume.

6.7 General results

In this section we obtain some general results concerning CFT on the sphere and disk.

Möbius invariance

We have seen that the sphere has a group of globally defined conformal transformations, the Möbius group $PSL(2, \mathbf{C})$,

$$z' = \frac{\alpha z + \beta}{\gamma z + \delta} \tag{6.7.1}$$

for complex $\alpha, \beta, \gamma, \delta$ with $\alpha\delta - \beta\gamma = 1$. This is the most general conformal transformation that is one-to-one on all of S_2, the complex z-plane plus the point at infinity. Expectation values must be invariant under any Möbius transformation:

$$\langle \mathscr{A}_i(z_1, \bar{z}_1) \ldots \mathscr{A}_k(z_n, \bar{z}_n) \rangle_{S_2} = \langle \mathscr{A}'_i(z_1, \bar{z}_1) \ldots \mathscr{A}'_k(z_n, \bar{z}_n) \rangle_{S_2}. \tag{6.7.2}$$

We will consider the consequences of this symmetry for expectation values with one, two, three, or four local operators.

For a single operator of weight $(h_i, \tilde{h}_i)$, the rescaling plus rotation $z' = \gamma z$ gives

$$\langle \mathscr{A}_i(0, 0) \rangle_{S_2} = \langle \mathscr{A}'_i(0, 0) \rangle_{S_2} = \gamma^{-h_i} \bar{\gamma}^{-\tilde{h}_i} \langle \mathscr{A}_i(0, 0) \rangle_{S_2}. \tag{6.7.3}$$

The one-point function therefore vanishes unless $h_i = \tilde{h}_i = 0$. This is another way to see that the one-point string amplitude vanishes on the sphere, because the matter factor is the expectation value of a $(1, 1)$ operator.

For $n = 2$, we can use a translation plus $z' = \gamma z$ to bring any pair of operators to the points 0 and 1, giving

$$\langle \mathscr{A}_i(z_1, \bar{z}_1) \mathscr{A}_j(z_2, \bar{z}_2) \rangle_{S_2} = z_{12}^{-h_i - h_j} \bar{z}_{12}^{-\tilde{h}_i - \tilde{h}_j} \langle \mathscr{A}_i(1, 1) \mathscr{A}_j(0, 0) \rangle_{S_2}, \tag{6.7.4}$$

so the position dependence is completely determined. Single-valuedness implies that $J_i + J_j \in \mathbf{Z}$, where $J_i = h_i - \tilde{h}_i$. There is a further constraint on the two-point function from the conformal transformation $z' = z + \epsilon(z - z_1)(z - z_2) + O(\epsilon^2)$, which leaves z_1 and z_2 fixed. For general operators this is complicated, but for *tensor fields* $\mathcal{O}_p$ and $\mathcal{O}_q$ it simply implies that

$$\langle \mathcal{O}_p(z_1, \bar{z}_1) \mathcal{O}_q(z_2, \bar{z}_2) \rangle_{S_2} = 0 \quad \text{unless } h_p = h_q, \ \tilde{h}_p = \tilde{h}_q . \tag{6.7.5}$$

Any three points $z_{1,2,3}$ can be brought to given positions by a Möbius transformation. For $n \geq 3$, Möbius invariance therefore reduces the expectation value from a function of n complex variables to a function of $n - 3$ complex variables. Again the result takes a simple form only for tensor fields. For example, for three tensor fields one finds

$$\left\langle \prod_{i=1}^3 \mathcal{O}_{p_i}(z_i, \bar{z}_i) \right\rangle_{S_2} = C_{p_1 p_2 p_3} \prod_{\substack{i,j=1 \\ i<j}}^3 z_{ij}^{h-2(h_i+h_j)} \bar{z}_{ij}^{\tilde{h}-2(\tilde{h}_i+\tilde{h}_j)} , \tag{6.7.6}$$

where $C_{p_1 p_2 p_3}$ is independent of position and $h = h_1 + h_2 + h_3$. For four primary fields,

$$\left\langle \prod_{i=1}^4 \mathcal{O}_{p_i}(z_i, \bar{z}_i) \right\rangle_{S_2} = C_{p_1 p_2 p_3 p_4}(z_c, \bar{z}_c)(z_{12} z_{34})^h (\bar{z}_{12} \bar{z}_{34})^{\tilde{h}} \times \prod_{\substack{i,j=1 \\ i<j}}^4 z_{ij}^{-h_i-h_j} \bar{z}_{ij}^{-\tilde{h}_i-\tilde{h}_j} , \tag{6.7.7}$$

where $h = \sum_i h_i$, $\tilde{h} = \sum_i \tilde{h}_i$, and $z_c = z_{12} z_{34} / z_{13} z_{24}$ is the Möbius invariant cross-ratio. The function $C_{p_1 p_2 p_3 p_4}(z_c, \bar{z}_c)$ is not determined by conformal invariance, so we are reduced from a function of four variables to an unknown function of one variable.

On the disk represented as the upper half-plane, only the Möbius transformations with $\alpha, \beta, \gamma, \delta$ real remain, forming the group $PSL(2, \mathbf{R})$. The extension of the above is left as an exercise. One learns much more by considering the full conformal algebra. We will see in chapter 15 that it determines all expectation values in terms of those of the tensor fields.

Path integrals and matrix elements

The path integrals we have considered can be related to operator expressions. Consider the path integral on the sphere with two operators, one at the origin and one at infinity:

$$\langle \mathscr{A}'_i(\infty, \infty) \mathscr{A}_j(0, 0) \rangle_{S_2} . \tag{6.7.8}$$

The prime indicates the u-frame, which we have to take for the operator at infinity; by a slight abuse of notation we still give the position in terms of z. Using the state–operator mapping we can replace the disk $|z| < 1$ containing $\mathscr{A}_j$ at $z = 0$ by the state $\Psi_{\mathscr{A}_j}$ on the circle $|z| = 1$. We can also replace the disk $|z| > 1$ ($|u| < 1$) containing $\mathscr{A}_i$ at $u = 0$ by the state $\Psi_{\mathscr{A}_i}$ on the circle $|z| = 1$. All that is left is the integral over the fields ϕ_b on the circle, so the expectation value (6.7.8) becomes

$$\int [d\phi_\text{b}] \Psi_{\mathscr{A}_i}[\phi_\text{b}^\Omega] \Psi_{\mathscr{A}_j}[\phi_\text{b}] \; ; \qquad (6.7.9)$$

here $\phi_\text{b}^\Omega(\sigma) = \phi_\text{b}(2\pi - \sigma)$, arising from the mapping $zu = 1$.

This convolution of wavefunctions resembles an inner product, so we define

$$\langle\!\langle i | j \rangle\!\rangle = \left\langle \mathscr{A}_i'(\infty, \infty) \mathscr{A}_j(0, 0) \right\rangle_{S_2} . \qquad (6.7.10)$$

This is essentially the inner product introduced by Zamolodchikov. For tensor operators in unitary theories, a Möbius transformation shows that this is the same as the coefficient of 1 in the OPE,

$$\mathcal{O}_i(z, \bar{z}) \mathcal{O}_j(0, 0) = \frac{\langle\!\langle i | j \rangle\!\rangle}{z^{h_i + h_j} \bar{z}^{\tilde{h}_i + \tilde{h}_j} \left\langle 1 \right\rangle_{S_2}} + \ldots . \qquad (6.7.11)$$

We have abbreviated $|\mathscr{A}_i\rangle$ to $|i\rangle$. The $\langle\!\langle \; | \; \rangle\!\rangle$ product is not the same as the quantum mechanical inner product $\langle \; | \; \rangle$ introduced in chapter 4. The latter is Hermitean, whereas the former includes no complex conjugation and so is bilinear, up to a sign if i and j are anticommuting. That is,

$$\langle\!\langle i | j \rangle\!\rangle = \pm \langle\!\langle j | i \rangle\!\rangle , \qquad (6.7.12)$$

since all we have done is to interchange the two operators and rename $z \leftrightarrow u$. There is a simple relation between the two inner products, which we will develop later.

We will sometimes write $\mathscr{G}_{ij}$ for $\langle\!\langle i | j \rangle\!\rangle$, and $\mathscr{G}^{ij}$ for the inverse matrix, where i, j run over a complete set. The matrices $\mathscr{G}_{ij}$ and $\mathscr{G}^{ij}$ will be used to raise and lower indices, $\mathscr{A}^i = \mathscr{G}^{ij} \mathscr{A}_j$, $\mathscr{A}_i = \mathscr{G}_{ij} \mathscr{A}^j$.

Operators in the path integral translate in the usual way into operators in Hilbert space. For example,

$$\left\langle \mathscr{A}_i'(\infty, \infty) \mathscr{A}_k(1, 1) \mathscr{A}_j(0, 0) \right\rangle_{S_2} = \langle\!\langle i | \hat{\mathscr{A}}_k(1, 1) | j \rangle\!\rangle , \qquad (6.7.13)$$

where we reintroduce the hat to emphasize that we are in a Hilbert space formalism. Using the OPE, the left-hand side becomes

$$\sum_l c^l_{kj} \left\langle \mathscr{A}_i'(\infty, \infty) \mathscr{A}_l(0, 0) \right\rangle_{S_2} = c_{ikj} . \qquad (6.7.14)$$

Thus the three-point expectation values on the sphere, the OPE coefficients with all indices lowered, and the matrix elements of general local operators are all the same thing. By a Möbius transformation (rescaling of z) we have also

$$\left\langle \mathscr{A}_i'(\infty, \infty) \mathscr{A}_k(z_1, \bar{z}_1) \mathscr{A}_j(0, 0) \right\rangle_{S_2} = z_1^{h_i - h_k - h_j} \bar{z}_1^{\tilde{h}_i - \tilde{h}_k - \tilde{h}_j} c_{ikj} . \qquad (6.7.15)$$

The four-point function translates into an operator expression

$$\langle \mathscr{A}'_i(\infty,\infty)\mathscr{A}_k(z_1,\bar{z}_1)\mathscr{A}_l(z_2,\bar{z}_2)\mathscr{A}_j(0,0) \rangle_{S_2} = \langle\langle i|\mathrm{T}[\hat{\mathscr{A}}_k(z_1,\bar{z}_1)\hat{\mathscr{A}}_l(z_2,\bar{z}_2)]|j\rangle ,$$
(6.7.16)

where T denotes radial ordering. Let $|z_1| > |z_2|$ and insert a complete set of states,

$$1 = |m\rangle\mathscr{G}^{mn}\langle\langle n| .$$
(6.7.17)

The four-point amplitude (6.7.16) becomes

$$\sum_m z_1^{h_i-h_k-h_m}\bar{z}_1^{\tilde{h}_i-\tilde{h}_k-\tilde{h}_m}z_2^{h_m-h_l-h_j}\bar{z}_2^{\tilde{h}_m-\tilde{h}_l-\tilde{h}_j}c_{ikm}c^m{}_{lj} .$$
(6.7.18)

Thus, the operator product coefficients determine not only the three-point expectation values on the sphere, but also the four-point and, by the same construction, arbitrary n-point amplitudes. For $|z_1 - z_2| > |z_1|$, which overlaps the region $|z_1| > |z_2|$ where the expansion (6.7.18) is valid, we can translate $\mathscr{A}_k$ to the origin and give a similar expansion in terms of $c_{ilm}c^m{}_{kj}$. The equality of these two expansions is associativity of the OPE, figure 2.8.

Operator calculations

The Hilbert space expressions give us one more way to calculate expectations values. We take as an example the case of four exponential operators

$$\left\langle :e^{ik_4\cdot X(\infty,\infty)}: \; :e^{ik_1\cdot X(z_1,\bar{z}_1)}: \; :e^{ik_2\cdot X(z_2,\bar{z}_2)}: \; :e^{ik_3\cdot X(0,0)}: \right\rangle_{S_2}$$
$$= \langle\langle 0;k_4|\mathrm{T}[\;{}^\circ_\circ e^{ik_1\cdot X_1}{}^\circ_\circ \; {}^\circ_\circ e^{ik_2\cdot X_2}{}^\circ_\circ]|0;k_3\rangle .$$
(6.7.19)

We have used the result (2.7.11) that : : ordered operators are the same as ${}^\circ_\circ$ ${}^\circ_\circ$ ordered operators for this CFT. We have abbreviated $X^\mu(z_i,\bar{z}_i)$ as X^μ_i and to avoid clutter are omitting the hats on operators. By definition

$${}^\circ_\circ e^{ik\cdot X}{}^\circ_\circ = e^{ik\cdot X_C}e^{ik\cdot X_A} ,$$
(6.7.20)

where

$$X^\mu_C(z,\bar{z}) = x^\mu - i\left(\frac{\alpha'}{2}\right)^{1/2}\sum_{m=1}^\infty \frac{1}{m}(\alpha^\mu_{-m}z^m + \tilde{\alpha}^\mu_{-m}\bar{z}^m) ,$$
(6.7.21a)

$$X^\mu_A(z,\bar{z}) = -i\frac{\alpha'}{2}p^\mu \ln|z|^2 + i\left(\frac{\alpha'}{2}\right)^{1/2}\sum_{m=1}^\infty \frac{1}{m}\left(\frac{\alpha^\mu_m}{z^m} + \frac{\tilde{\alpha}^\mu_m}{\bar{z}^m}\right) .$$
(6.7.21b)

For $|z_1| > |z_2|$ the matrix element (6.7.19) becomes

$$\langle\langle 0;k_4|e^{ik_1\cdot X_{1C}}e^{ik_1\cdot X_{1A}}e^{ik_2\cdot X_{2C}}e^{ik_2\cdot X_{2A}}|0;k_3\rangle .$$
(6.7.22)

To evaluate this use the Campbell–Baker–Hausdorff (CBH) formula

$$e^{ik_1 \cdot X_{1A}} e^{ik_2 \cdot X_{2C}} = e^{ik_2 \cdot X_{2C}} e^{ik_1 \cdot X_{1A}} e^{-[k_1 \cdot X_{1A}, k_2 \cdot X_{2C}]}$$
$$= e^{ik_2 \cdot X_{2C}} e^{ik_1 \cdot X_{1A}} |z_{12}|^{\alpha' k_1 \cdot k_2} . \tag{6.7.23}$$

The expectation value (6.7.22) becomes

$$|z_{12}|^{\alpha' k_1 \cdot k_2} \langle\!\langle 0; k_4 | e^{ik_1 \cdot X_{1C} + ik_2 \cdot X_{2C}} e^{ik_1 \cdot X_{1A} + ik_2 \cdot X_{2A}} |0; k_3 \rangle$$
$$= |z_{12}|^{\alpha' k_1 \cdot k_2} \langle\!\langle 0; k_4 | e^{i(k_1 + k_2) \cdot x} e^{\alpha'(k_1 \ln|z_1| + k_2 \ln|z_2|) \cdot p} |0; k_3 \rangle$$
$$= |z_{12}|^{\alpha' k_1 \cdot k_2} |z_1|^{\alpha' k_1 \cdot k_3} |z_2|^{\alpha' k_2 \cdot k_3} \langle\!\langle 0; k_1 + k_2 + k_4 |0; k_3 \rangle$$
$$= i C_{S_2}^X (2\pi)^d \delta^d(\textstyle\sum_i k_i) |z_{12}|^{\alpha' k_1 \cdot k_2} |z_1|^{\alpha' k_1 \cdot k_3} |z_2|^{\alpha' k_2 \cdot k_3} , \tag{6.7.24}$$

where we have used the two-point expectation value to normalize the last line. This is the familiar result (6.2.31), obtained by two other methods in section 6.2, after one includes in the latter a factor $|z_4|^{\alpha' k_4^2}$ from the change of frame and takes $z_4 \to \infty$. All other free-field results can be obtained by this same oscillator method.

Relation between inner products

Nondegenerate bilinear and Hermitean inner products can always be related to one another by an appropriate antilinear operation on the bra. Let us consider an example. From the free-field expectation values, we have

$$\langle\!\langle 0; k | 0; l \rangle = i C_{S_2}^X (2\pi)^d \delta^d(k + l) , \tag{6.7.25}$$

using the fact that $|0; k\rangle$ maps to $e^{ik \cdot X}$. Compare this with the inner product from the X^μ CFT,

$$\langle 0; k | 0; l \rangle = (2\pi)^d \delta^d(l - k) . \tag{6.7.26}$$

These differ only by $k \to -k$ from conjugating $e^{ik \cdot X}$, and normalization,

$$\langle\!\langle 0; k | = i C_{S_2}^X \langle 0; -k | . \tag{6.7.27}$$

For more general operators, there is a natural notion of conjugation in CFT. In Euclidean quantum mechanics, Hermitean conjugation inverts Euclidean time, eq. (A.1.37), so the natural operation of Euclidean conjugation is conjugation × time-reversal: an operator that is Hermitean in Minkowski space is also Hermitean under this combined operation. In CFT, we make the same definition, but also must include a time-reversal on the conformal frame,

$$\overline{\mathcal{A}(p)} = \mathcal{A}'(p')^\dagger . \tag{6.7.28}$$

Here p and p' are related by radial time-reversal, $z' = \bar{z}^{-1}$, and the unprimed operator is in the z-frame and the primed operator in the u-

frame. To see how this works, consider a holomorphic operator of weight h, whose Laurent expansion is

$$\mathcal{O}(z) = i^h \sum_{n=-\infty}^{\infty} \frac{\mathcal{O}_n}{z^{n+h}} . \tag{6.7.29}$$

Its simple adjoint is

$$\mathcal{O}(z)^\dagger = i^{-h} \sum_{n=-\infty}^{\infty} \frac{\mathcal{O}_n^\dagger}{\bar{z}^{n+h}} . \tag{6.7.30}$$

Then the Euclidean adjoint is

$$\overline{\mathcal{O}(z)} = i^{-h}(-z^{-2})^h \sum_{n=-\infty}^{\infty} \frac{\mathcal{O}_n^\dagger}{z^{-n-h}} = i^h \sum_{n=-\infty}^{\infty} \frac{\mathcal{O}_{-n}^\dagger}{z^{n+h}} . \tag{6.7.31}$$

For an operator that is Hermitean in Minkowski time, $\mathcal{O}_{-n}^\dagger = \mathcal{O}_n$, and so this operator is also Hermitean[3] under $\overline{\mathcal{O}}$. The Euclidean adjoint (6.7.28) conjugates all *explicit* factors of i, but leaves z and $\bar{z}$ indices unchanged. This is its whole effect, other than an overall factor $(-1)^{N_a(N_a-1)/2}$ from reversing the order of anticommuting fields; here N_a is the total number of anticommuting fields in the operator.

This is the natural conformally invariant operation of conjugation, so it must be that

$$\langle\!\langle \overline{\mathcal{A}_i} | = K \langle \mathcal{A}_i | \tag{6.7.32}$$

for some constant K. For a direct demonstration, the best we have come up with is

$$
\begin{aligned}
\langle\!\langle \bar{\imath} | j \rangle &\equiv \langle\!\langle 1 | \overline{\mathcal{A}_i(\infty, \infty)} | j \rangle = K \langle 1 | \overline{\mathcal{A}_i(\infty, \infty)} | j \rangle \\
&= K \langle 1 | \mathcal{A}_i(0,0)^\dagger | j \rangle = K \langle j | \mathcal{A}_i(0,0) | 1 \rangle^* \\
&= K \langle j | i \rangle^* = K \langle i | j \rangle .
\end{aligned} \tag{6.7.33}
$$

The only step here that is not either a definition or obvious is the assumption of proportionality, $\langle\!\langle 1 | = K \langle 1 |$. This must hold because $|1\rangle$ is the unique $SL(2,\mathbf{C})$-invariant state.

For the X CFT we have seen that $K^X = iC_{S_2}^X$. For the ghost CFT, the Laurent expansion of the amplitude (6.3.4) gives $\langle\!\langle 0 | \tilde{c}_0 c_0 | 0 \rangle = -C_{S_2}^g$, where $|0\rangle = \tilde{c}_1 c_1 |1\rangle$. The Hermitean inner product was defined in eq. (4.3.18) as $\langle 0 | \tilde{c}_0 c_0 | 0 \rangle = i$, so $K^g = iC_{S_2}^g$.

Except for the is, which are absent in a unitary CFT, one can set $K = 1$ by adding an Euler number term to the action. In a Hermitean basis of operators, the two inner products are then identical, and the distinction

---

[3] Because its Laurent expansion has no factor of i, the Faddeev–Popov c ghost is actually anti-Hermitean under the Euclidean adjoint. This is due to an inconvenient conflict between conventions, and has no significance.

between them is often ignored. Notice, however, that a vertex operator of definite nonzero momentum cannot be Hermitean.

   Again, this can all be extended to the open string, with the disk in place of the sphere.

Exercises

6.1 Verify that the expectation value (6.2.31) is smooth in the u-patch.

6.2 For the linear dilaton CFT, $\Phi = V_\mu X^\mu$, the X^μ zero-mode path integral diverges. This reflects the fact that the coupling is diverging in some direction. Consider therefore imaginary $\Phi = i b_\mu X^\mu$; this is unphysical but has technical applications (one could instead take complex momenta).
(a) Generalize the calculation that gave (6.2.17) to this case. Show that in the end the only changes are in the Weyl dependence (reflecting a change in the dimension of the exponential) and in the spacetime momentum conservation.
(b) Generalize the calculation of the same quantity using holomorphicity. Show that the result is the same as in the regular X^μ CFT but with an extra *background charge* operator $\exp(-2ib \cdot X)$ taken to infinity.

6.3 For one b and four c fields, show that expressions (6.3.5) and (6.3.8) are equal. Ignore the antiholomorphic fields.

6.4 (a) Write the amplitude for n open string tachyons on the disk as a generalization of eq. (6.4.5).
(b) Show by a change of variables that it is independent of the positions of the fixed vertex operators.

6.5 (a) Show that the residue of the pole in $I(s,t)$ at $\alpha' s = J - 1$ is a polynomial in $u - t$ of degree J, corresponding to intermediate particles of spin up to J.
(b) Consider the amplitude $I(s,t)$ without assuming $D = 26$. In the center-of-mass frame, write the residue of the pole at $\alpha' s = 1$ in terms of

$$k_1^i k_1^j P_{ij,kl}^J k_3^k k_3^l \,,$$

where P^0 projects onto spin 0 (the unit matrix) and P^2 projects onto spin 2 (traceless symmetric tensors). Show that the coefficient of P^0 is positive for $D < 26$, zero at $D = 26$, and negative at $D > 26$. Compare with the expected string spectrum at this level, and with the discussion of this level in section 4.1.

6.6 (a) In the hard scattering limit, evaluate the integral in the Veneziano amplitude (6.4.5) using the saddle point approximation. Take the naive saddle point; the justification requires consideration of the analytic continuations needed to define the integral.
(b) Do the same for the Virasoro–Shapiro amplitude.

6.7 (a) Derive the amplitude (6.5.12).
(b) Verify the relation (6.5.14) between vertex operator normalizations. Show that this agrees with the state–operator mapping.

6.8 Derive the three-gauge-boson amplitude (6.5.15).

6.9 (a) Obtain the four-gauge-boson amplitude. To reduce the rather extensive algebra, consider only those terms in which the polarizations appear as $e_1 \cdot e_2 \, e_3 \cdot e_4$.
(b) Compare the behavior at small $\alpha' k_i \cdot k_j$ with the same amplitude in Yang–Mills theory.

6.10 Carry out the integral in eq. (6.6.5) to obtain the Virasoro–Shapiro amplitude (6.6.10). The relation

$$|z|^{-2a} = \frac{1}{\Gamma(a)} \int_0^\infty dt \, t^{a-1} \exp(-tz\bar{z})$$

is useful. (Reference: Green, Schwarz, & Witten (1987), section 7.2; note that their d^2z is $\frac{1}{2}$ times ours.)

6.11 (a) Verify the amplitude (6.6.14).
(b) Verify the relation (6.6.15) between vertex operator normalizations.
(c) Complete the field theory calculation and verify the relation (6.6.18) between g_c and the gravitational coupling.

6.12 Calculate the disk amplitude with one closed string tachyon and two open string tachyons.

6.13 (a) Find the $PSL(2, \mathbf{C})$ transformation that takes three given points $z_{1,2,3}$ into chosen positions $\hat{z}_{1,2,3}$.
(b) Verify the Möbius invariance results (6.7.3)–(6.7.7). Show that to derive (6.7.5) it is sufficient that L_1 and $\tilde{L}_1$ annihilate the operators.

6.14 (a) Find the $PSL(2, \mathbf{R})$ transformation that takes three given points $y_{1,2,3}$ into chosen positions $\hat{y}_{1,2,3}$.
(b) Generalize the Möbius invariance results (6.7.3)–(6.7.7) to operators on the boundary of a disk.
(c) Find similar results for the disk with one interior, two interior, and one interior and one boundary operators.

6.15 (a) Generalize the operator calculation (6.7.24) to a product of n exponentials.
(b) Generalize it to the expectation value (6.2.25).

7

One-loop amplitudes

After describing the relevant surfaces, we focus on the torus, first on the CFTs on this surface and then on the string amplitudes. We then generalize to the open and unoriented theories. The most important issue to be understood is the absence of short-distance (UV) divergences.

7.1 Riemann surfaces

There are four Riemann surfaces with Euler number zero.

The torus

The torus T^2, discussed in section 5.1, is the only closed oriented surface with Euler number zero. We describe it as the complex plane with metric $ds^2 = dw d\bar{w}$ and identifications

$$w \cong w + 2\pi \cong w + 2\pi\tau . \tag{7.1.1}$$

There are two moduli, the real and imaginary parts of $\tau = \tau_1 + i\tau_2$, and two CKVs, the translations. In terms of the real coordinates $w = \sigma^1 + i\sigma^2$,

$$(\sigma^1, \sigma^2) \cong (\sigma^1 + 2\pi, \sigma^2) \cong (\sigma^1 + 2\pi\tau_1, \sigma^2 + 2\pi\tau_2) , \tag{7.1.2}$$

so that one can think of the torus as a cylinder of circumference 2π and length $2\pi\tau_2$ with the ends rotated by an angle of $2\pi\tau_1$ and then sewn together.

In terms of the coordinate $z = \exp(-iw)$, the identification $w \cong w + 2\pi$ is automatic, while $w \cong w + 2\pi\tau$ becomes

$$z \cong z \exp(-2\pi i\tau) . \tag{7.1.3}$$

A fundamental region is the annulus

$$1 \leq |z| \leq \exp(2\pi\tau_2) . \tag{7.1.4}$$

The torus is formed by rotating the outer circle by $2\pi\tau_1$ and then sewing the inner and outer circles together. We will use the w-coordinate unless otherwise noted.

The cylinder (annulus)

The cylinder C_2 is the region

$$0 \le \mathrm{Re}\, w \le \pi , \quad w \cong w + 2\pi i t . \tag{7.1.5}$$

That is, it is a strip of width π and length $2\pi t$, with the ends joined. There is a single modulus t, which runs over the whole range $0 < t < \infty$. Unlike the torus there is no modular group, the long-cylinder limit $t \to 0$ being quite different from the long-strip limit $t \to \infty$. There is a single CKV, the translation parallel to the boundary.

The cylinder can be obtained from the torus with imaginary $\tau = it$ by identifying under the involution

$$w' = -\bar{w} , \tag{7.1.6}$$

which is a reflection through the imaginary axis. The lines $\sigma^1 = 0, \pi$ are fixed by this reflection and so become boundaries. The other coordinate is periodically identified,

$$(\sigma^1, 0) \cong (\sigma^1, 2\pi t) . \tag{7.1.7}$$

The Klein bottle

The Klein bottle K_2 can be regarded as the complex plane with identification

$$w \cong w + 2\pi \cong -\bar{w} + 2\pi i t \tag{7.1.8}$$

or

$$(\sigma^1, \sigma^2) \cong (\sigma^1 + 2\pi, \sigma^2) \cong (-\sigma^1, \sigma^2 + 2\pi t) . \tag{7.1.9}$$

This is a cylinder of circumference 2π and length $2\pi t$, with the ends sewn together after a parity-reversal Ω. The single modulus t runs over $0 < t < \infty$, and there is no modular group. Translation in the σ^2 direction is the only CKV. The Klein bottle can be obtained from the torus with modulus $\tau = 2it$ by identifying under

$$w' = -\bar{w} + 2\pi i t . \tag{7.1.10}$$

The Klein bottle can be thought of as a sphere with two cross-caps, but we postpone this description until section 7.4.

The Möbius strip

The Möbius strip M_2 is sewn from a strip with a twist by Ω,

$$0 \leq \operatorname{Re} w \leq \pi , \quad w \cong -\bar{w} + \pi + 2\pi i t . \qquad (7.1.11)$$

Again the modulus t runs over $0 < t < \infty$ and the only CKV is the σ^2-translation. It can be obtained from the torus with $\tau = 2it$ by identifying under the *two* involutions

$$w' = -\bar{w} \quad \text{and} \quad w' = w + \pi(2it + 1) . \qquad (7.1.12)$$

The Möbius strip can be thought of as a disk with a cross-cap, but again we defer this description.

7.2 CFT on the torus

Scalar correlators

As in the case of the sphere, we start with the Green's function (6.2.7), satisfying

$$\frac{2}{\alpha'} \bar{\partial} \partial G'(w, \bar{w}; w', \bar{w}') = -2\pi \delta^2(w - w') + \frac{1}{4\pi \tau_2} . \qquad (7.2.1)$$

The Green's function is periodic in both directions on the torus, and aside from the source and the background charge term it would be the sum of holomorphic and antiholomorphic functions. These properties identify it as being associated with the theta functions, whose properties are given at the end of this section. In particular we guess

$$G'(w, \bar{w}; w', \bar{w}') \sim -\frac{\alpha'}{2} \ln \left| \vartheta_1 \left(\frac{w - w'}{2\pi} \bigg| \tau \right) \right|^2 . \qquad (7.2.2)$$

The function ϑ_1 vanishes linearly when its first argument goes to zero, giving the correct behavior as $w \to w'$. Elsewhere the logarithm is the sum of holomorphic and antiholomorphic functions and so is annihilated by $\bar{\partial}\partial$. However, this function is not quite doubly periodic owing to the quasiperiodicity (7.2.32b) — it changes by $-\alpha'[\operatorname{Im}(w - w') + \pi\tau_2]$ under $w \to w + 2\pi\tau$. Also, the background charge is missing. Both properties are easily corrected:

$$G'(w, \bar{w}; w', \bar{w}') = -\frac{\alpha'}{2} \ln \left| \vartheta_1 \left(\frac{w - w'}{2\pi} \bigg| \tau \right) \right|^2 + \alpha' \frac{[\operatorname{Im}(w - w')]^2}{4\pi\tau_2} + k(\tau, \bar{\tau}) .$$

$$(7.2.3)$$

The function $k(\tau, \bar{\tau})$ is determined by orthogonality to X_0, but as in the case of the sphere it drops out due to spacetime momentum conservation.

The expectation value of a product of vertex operators is given, in parallel to the earlier result (6.2.13) on the sphere, by

$$
\left\langle \prod_{i=1}^{n} :e^{ik_i \cdot X(z_i,\bar{z}_i)}: \right\rangle_{T^2} = iC_{T^2}^X(\tau)(2\pi)^d \delta^d(\textstyle\sum_i k_i)
$$

$$
\times \prod_{i<j} \left| \frac{2\pi}{\partial_v \vartheta_1(0|\tau)} \vartheta_1\left(\frac{w_{ij}}{2\pi}\Big|\tau\right) \exp\left[-\frac{(\mathrm{Im}\, w_{ij})^2}{4\pi\tau_2}\right] \right|^{\alpha' k_i \cdot k_j}. \qquad (7.2.4)
$$

The factor $2\pi/\partial_v \vartheta_1$ comes from the renormalized self-contractions.

The scalar partition function

The overall normalization of the sphere was absorbed into the string coupling constant, but we cannot do this for the torus, in particular because there is a nontrivial τ-dependence. In fact, we will see that there is a great deal of physics in the amplitude with no vertex operators at all. Consider the path integral with no vertex operators, $\langle 1 \rangle_{T^2(\tau)} \equiv Z(\tau)$. We can think of the torus with modulus τ as formed by taking a field theory on a circle, evolving for Euclidean time $2\pi\tau_2$, translating in σ^1 by $2\pi\tau_1$, and then identifying the ends. In operator language this gives a trace,

$$
Z(\tau) = \mathrm{Tr}\left[\exp(2\pi i \tau_1 P - 2\pi\tau_2 H)\right]
$$

$$
= (q\bar{q})^{-d/24} \mathrm{Tr}\left(q^{L_0} \bar{q}^{\tilde{L}_0}\right). \qquad (7.2.5)
$$

Here $q = \exp(2\pi i \tau)$, the momentum $P = L_0 - \tilde{L}_0$ generates translations of σ^1, and the Hamiltonian $H = L_0 + \tilde{L}_0 - \frac{1}{24}(c + \tilde{c})$ generates translations of σ^2 as in eq. (2.6.10). Such a trace, weighted by the exponential of the Hamiltonian and other conserved quantities, is termed a *partition function* as in statistical mechanics.

The trace breaks up into a sum over occupation numbers $N_{\mu n}$ and $\tilde{N}_{\mu n}$ for each μ and n and an integral over momentum k^μ. The Virasoro generators similarly break up into a sum, and $Z(\tau)$ becomes

$$
V_d (q\bar{q})^{-d/24} \int \frac{d^d k}{(2\pi)^d} \exp(-\pi\tau_2\alpha' k^2) \prod_{\mu,n} \sum_{N_{\mu n}, \tilde{N}_{\mu n}=0}^{\infty} q^{nN_{\mu n}} \bar{q}^{n\tilde{N}_{\mu n}}. \qquad (7.2.6)
$$

The factor of spacetime volume V_d comes from the continuum normalization of the momentum, $\sum_k$ becoming $V_d(2\pi)^{-d} \int d^d k$. The various sums are geometric,

$$
\sum_{N=0}^{\infty} q^{nN} = (1 - q^n)^{-1}, \qquad (7.2.7)
$$

and so we obtain

$$Z(\tau) = iV_d Z_X(\tau)^d , \tag{7.2.8}$$

where

$$Z_X(\tau) = (4\pi^2\alpha'\tau_2)^{-1/2}|\eta(\tau)|^{-2} . \tag{7.2.9}$$

Here

$$\eta(\tau) = q^{1/24} \prod_{n=1}^{\infty}(1 - q^n) \tag{7.2.10}$$

is the Dedekind eta function, discussed further at the end of the section. The i comes from the rotation $k^0 \to ik^d$ needed to define the divergent integral. Also, the earlier $C_{T^2}^X(\tau)$ is $Z_X(\tau)^d$.

The result (7.2.8) is obviously invariant under $\tau \to \tau + 1$. Invariance of Z_X under $\tau \to -1/\tau$ follows from eq. (7.2.44), $\eta(-1/\tau) = (-i\tau)^{1/2}\eta(\tau)$, thus generating the full modular group. Similarly the expectation value with vertex operators is modular-covariant, taking into account the Weyl transformations of the operators.

We have evaluated the path integral by translating into an operator expression, but as on the sphere there are two other methods. We leave to the reader the direct path integral evaluation. Fourier transforming in σ^1, it just breaks up into a product of harmonic oscillator path integrals.

The evaluation using holomorphicity is less direct. We include it for illustration because it is used at higher genus and for more complicated CFTs. The idea is to get a differential equation with respect to τ. Consider the torus with modulus τ and make a small change in the metric, $\delta g_{ww} = \epsilon^*$. The new metric is

$$ds^2 = dwd\bar{w} + \epsilon^* dw^2 + \epsilon d\bar{w}^2$$
$$= (1 + \epsilon^* + \epsilon) d[w + \epsilon(\bar{w} - w)] d[\bar{w} + \epsilon^*(w - \bar{w})] + O(\epsilon^2) . \tag{7.2.11}$$

That is, the metric is Weyl-equivalent to $dw'd\bar{w}'$ with $w' = w + \epsilon(\bar{w} - w)$. This has the periodicities

$$w' \cong w' + 2\pi \cong w' + 2\pi(\tau - 2i\tau_2\epsilon) . \tag{7.2.12}$$

This change in the metric is therefore equivalent to a change

$$\delta\tau = -2i\tau_2\epsilon \tag{7.2.13}$$

in the modulus.

Under a change in the metric, the change in the path integral is

$$\delta Z(\tau) = -\frac{1}{2\pi} \int d^2w \left[\delta g_{\bar{w}\bar{w}} \langle T_{ww}(w)\rangle + \delta g_{ww} \langle T_{\bar{w}\bar{w}}(\bar{w})\rangle\right]$$
$$= -2\pi i\left[\delta\tau \langle T_{ww}(0)\rangle - \delta\bar{\tau} \langle T_{\bar{w}\bar{w}}(0)\rangle\right] . \tag{7.2.14}$$

In the second line we have used translation invariance to carry out the integral over w. All expectation values are understood to be on the torus with modulus τ. To get the expectation value of the energy-momentum tensor, we use the OPE

$$\partial_w X^\mu(w)\partial_w X_\mu(0) = -\frac{\alpha'd}{2w^2} - \alpha' T_{ww}(0) + O(w) . \qquad (7.2.15)$$

Now,

$$Z(\tau)^{-1}\langle \partial_w X^\mu(w)\partial_w X_\mu(0)\rangle = d\partial_w\partial_{w'}G'(w,\bar{w};w',\bar{w}')|_{w'=0}$$

$$= \frac{\alpha'd}{2}\frac{\vartheta_1\partial_w^2\vartheta_1 - \partial_w\vartheta_1\partial_w\vartheta_1}{\vartheta_1^2} + \frac{\alpha'd}{8\pi\tau_2} , \qquad (7.2.16)$$

where the arguments of all the theta functions are $(w/2\pi, \tau)$. This does indeed have a double pole at $w = 0$. Carefully expanding both numerator and denominator to order w^2 (this is simplified by the fact that ϑ_1 is odd), the order w^0 term in the expectation value (7.2.16) is

$$\frac{\alpha'd}{6}\frac{\partial_w^3\vartheta_1}{\partial_w\vartheta_1} + \frac{\alpha'd}{8\pi\tau_2} . \qquad (7.2.17)$$

The OPE (7.2.15) then gives

$$\langle T_{ww}(0)\rangle = \left(-\frac{d}{6}\frac{\partial_w^3\vartheta_1(0|\tau)}{\partial_w\vartheta_1(0|\tau)} - \frac{d}{8\pi\tau_2}\right)Z(\tau) , \qquad (7.2.18)$$

and the variation (7.2.14) gives the differential equation

$$\partial_\tau \ln Z(\tau) = \frac{\pi i d}{3}\frac{\partial_w^3\vartheta_1(0|\tau)}{\partial_w\vartheta_1(0|\tau)} + \frac{id}{4\tau_2} . \qquad (7.2.19)$$

To proceed further, we use the fact, which is easily verified from the infinite sum, that the theta functions satisfy

$$\partial_w^2\vartheta_1\left(\frac{w}{2\pi}\bigg|\tau\right) = \frac{i}{\pi}\partial_\tau\vartheta_1\left(\frac{w}{2\pi}\bigg|\tau\right) . \qquad (7.2.20)$$

Rewrite (7.2.19) as

$$\partial_\tau \ln Z(\tau) = -\frac{d}{3}\partial_\tau \ln \partial_w\vartheta_1(0|\tau) + \frac{id}{4\tau_2} . \qquad (7.2.21)$$

Together with the conjugate equation, this determines

$$Z(\tau) \propto |\partial_w\vartheta_1(0|\tau)|^{2d/3}\tau_2^{-d/2} \qquad (7.2.22)$$

up to a numerical coefficient that must be fixed by other means. Noting the relation (7.2.43), this is the same as the earlier result (7.2.8).

The bc CFT

For the ghost CFT, the partition function is again determined easily in terms of a trace over states. There are two sets of raising operators b_{-n} and c_{-n} on both the right and left, but now these are anticommuting so the occupation numbers can only be zero or one — a single oscillator at mode n thus gives $(1 + q^n)$. In addition there are four ground states, giving

$$\text{Tr}\left[\exp(2\pi i\tau_1 P - 2\pi\tau_2 H)\right] = (q\bar{q})^{13/12}\text{Tr}\left(q^{L_0}\bar{q}^{\bar{L}_0}\right)$$

$$= 4(q\bar{q})^{1/12}\prod_{n=1}^{\infty}|1 + q^n|^4 . \qquad (7.2.23)$$

Recall, however, that for anticommuting fields the simple trace corresponds to a path integral with antiperiodic boundary conditions in the time direction. In order to calculate the Faddeev–Popov determinant, the ghosts must have the same periodicity as the original coordinate transformations, which were periodic. Thus we need

$$Z(\tau) = \text{Tr}\left[(-1)^F \exp(2\pi i\tau_1 P - 2\pi\tau_2 H)\right]$$
$$= 0 , \qquad (7.2.24)$$

where $(-1)^F$ anticommutes with all the ghost fields. The trace vanishes because the states $|\downarrow\downarrow\rangle$ and $|\uparrow\uparrow\rangle$ have opposite ghost number mod 2 from $|\uparrow\downarrow\rangle$ and $|\downarrow\uparrow\rangle$. From the path integral point of view, $Z(\tau)$ vanishes due to ghost zero modes from the moduli and CKVs. These must be saturated with appropriate insertions, the simplest nonvanishing amplitude (and the one we will need) being

$$\left\langle c(w_1)b(w_2)\tilde{c}(\bar{w}_3)\tilde{b}(\bar{w}_4) \right\rangle . \qquad (7.2.25)$$

In the operator calculation we write each field in terms of its mode expansion. Only the $n = 0$ terms contribute, the expectation values of the others vanishing as above. Then (7.2.25) becomes

$$\text{Tr}\left[(-1)^F c_0 b_0 \tilde{c}_0 \tilde{b}_0 \exp(2\pi i\tau_1 P - 2\pi\tau_2 H)\right] . \qquad (7.2.26)$$

The operator $c_0 b_0 \tilde{c}_0 \tilde{b}_0$ projects onto the single ground state $|\uparrow\uparrow\rangle$, so the result is

$$(q\bar{q})^{1/12}\prod_{n=1}^{\infty}|1 - q^n|^4 = |\eta(\tau)|^4 . \qquad (7.2.27)$$

Note the sign change in the infinite product, from the $(-1)^F$ in the trace. The result (7.2.27) is independent of the positions of the ghost fields because the CKV and quadratic differentials on the torus are constants.

Again, we do not try to keep track of the overall sign, setting it to be positive.

General CFTs

For general CFTs,

$$Z(\tau) = \sum_i q^{h_i - c/24} \bar{q}^{\tilde{h}_i - \tilde{c}/24} (-1)^{F_i} , \qquad (7.2.28)$$

where i runs over all states of the CFT, and F_i is again the world-sheet fermion number. Invariance under $\tau \to \tau + 1$ requires $h_i - \tilde{h}_i - \frac{1}{24}(c - \tilde{c})$ to be an integer. Looking at the unit operator, this requires $c - \tilde{c}$ to be a multiple of 24. For general operators one then has the requirement that the spin be an integer,

$$h_i - \tilde{h}_i \in \mathbf{Z} . \qquad (7.2.29)$$

Incidentally, CFTs for which $c - \tilde{c}$ is not a multiple of 24 are certainly of interest. These cannot be separately modular-invariant, but with appropriate projections can be combined into a modular-invariant theory.

Invariance under $\tau \to -1/\tau$ puts a further constraint on the spectrum. Let us here note just one general consequence, which relates the states of large h to those of lowest h. Consider the partition function for $\tau = i\ell$ as $\ell \to 0$. The convergence factor $q = \exp(-2\pi\ell)$ is approaching unity, so the partition function is determined by the density of states at high weight. Using modular invariance, this is equal to the partition function at $\tau = i/\ell$. In the latter case, $q = \exp(-2\pi/\ell)$ is going to zero, so the state of lowest weight dominates the sum. This is the unit state, with $L_0 = \tilde{L}_0 = 0$, giving

$$Z(i\ell) \overset{\ell \to 0}{\approx} \exp\left[\frac{\pi(c + \tilde{c})}{12\ell}\right] . \qquad (7.2.30)$$

The density of states at high weight is thus determined by the central charge. This generalizes the free-boson result, where $c = \tilde{c} = d$ counts the number of free bosons.[1]

---

[1] To be precise, the result (7.2.30) holds in compact unitary CFTs. In nonunitary theories there are usually operators of negative weight, whose contribution will dominate that of the unit operator. In the linear dilaton theory, the central charge depends on V^μ but the density of states does not: here the exception occurs because of complications due to the noncompactness of X^μ combined with lack of translation invariance.

Theta functions

The basic *theta function* is

$$\vartheta(v,\tau) = \sum_{n=-\infty}^{\infty} \exp(\pi i n^2 \tau + 2\pi i n v) . \qquad (7.2.31)$$

It has the periodicity properties

$$\vartheta(v+1,\tau) = \vartheta(v,\tau) , \qquad (7.2.32a)$$
$$\vartheta(v+\tau,\tau) = \exp(-\pi i \tau - 2\pi i v)\vartheta(v,\tau) , \qquad (7.2.32b)$$

and these in fact determine it up to normalization. Under modular transformations,

$$\vartheta(v,\tau+1) = \vartheta(v+1/2,\tau) , \qquad (7.2.33a)$$
$$\vartheta(v/\tau,-1/\tau) = (-i\tau)^{1/2} \exp(\pi i v^2/\tau)\vartheta(v,\tau) . \qquad (7.2.33b)$$

The theta function has a unique zero, up to the periodicity (7.2.32), at $v = \frac{1}{2}(1+\tau)$. It can also be written as an infinite product,

$$\vartheta(v,\tau) = \prod_{m=1}^{\infty} (1-q^m)(1+zq^{m-1/2})(1+z^{-1}q^{m-1/2}) , \qquad (7.2.34)$$

in terms of

$$q = \exp(2\pi i \tau), \qquad z = \exp(2\pi i v) . \qquad (7.2.35)$$

We often need the asymptotic behavior of the theta function as $q \to 0$ or as $q \to 1$. The $q \to 0$ behavior can be read immediately from either the infinite sum or infinite product form. The $q \to 1$ behavior is not manifest in either form, because an infinite number of terms contribute. It can be obtained from the $\tau \to -1/\tau$ modular transformation, which relates these two limits by $q \to \exp(4\pi^2/\ln q)$.

It is also useful to define the *theta function with characteristics,*

$$\vartheta\begin{bmatrix} a \\ b \end{bmatrix}(v,\tau) = \exp\left[\pi i a^2 \tau + 2\pi i a(v+b)\right]\vartheta(v+a\tau+b,\tau)$$

$$= \sum_{n=-\infty}^{\infty} \exp\left[\pi i(n+a)^2\tau + 2\pi i(n+a)(v+b)\right] . \qquad (7.2.36)$$

Other common notations are

$$\vartheta_{00}(v,\tau) = \vartheta_3(v|\tau) = \vartheta\begin{bmatrix} 0 \\ 0 \end{bmatrix}(v,\tau) = \sum_{n=-\infty}^{\infty} q^{n^2/2} z^n , \qquad (7.2.37a)$$

$$\vartheta_{01}(v,\tau) = \vartheta_4(v|\tau) = \vartheta\begin{bmatrix} 0 \\ 1/2 \end{bmatrix}(v,\tau) = \sum_{n=-\infty}^{\infty} (-1)^n q^{n^2/2} z^n , \qquad (7.2.37b)$$

$$\vartheta_{10}(v,\tau) = \vartheta_2(v|\tau) = \vartheta\begin{bmatrix} 1/2 \\ 0 \end{bmatrix}(v,\tau) = \sum_{n=-\infty}^{\infty} q^{(n-1/2)^2/2} z^{n-1/2} \,,$$

$$\text{(7.2.37c)}$$

$$\vartheta_{11}(v,\tau) = -\vartheta_1(v|\tau) = \vartheta\begin{bmatrix} 1/2 \\ 1/2 \end{bmatrix}(v,\tau)$$

$$= -i \sum_{n=-\infty}^{\infty} (-1)^n q^{(n-1/2)^2/2} z^{n-1/2} \,. \qquad \text{(7.2.37d)}$$

These have the product representations

$$\vartheta_{00}(v,\tau) = \prod_{m=1}^{\infty} (1 - q^m)(1 + z q^{m-1/2})(1 + z^{-1} q^{m-1/2}) \,, \qquad \text{(7.2.38a)}$$

$$\vartheta_{01}(v,\tau) = \prod_{m=1}^{\infty} (1 - q^m)(1 - z q^{m-1/2})(1 - z^{-1} q^{m-1/2}) \,, \qquad \text{(7.2.38b)}$$

$$\vartheta_{10}(v,\tau) = 2 \exp(\pi i \tau/4) \cos \pi v \prod_{m=1}^{\infty} (1 - q^m)(1 + z q^m)(1 + z^{-1} q^m) \,,$$

$$\text{(7.2.38c)}$$

$$\vartheta_{11}(v,\tau) = -2 \exp(\pi i \tau/4) \sin \pi v \prod_{m=1}^{\infty} (1 - q^m)(1 - z q^m)(1 - z^{-1} q^m) \,.$$

$$\text{(7.2.38d)}$$

Their modular transformations are

$$\vartheta_{00}(v, \tau+1) = \vartheta_{01}(v,\tau) \,, \qquad \text{(7.2.39a)}$$

$$\vartheta_{01}(v, \tau+1) = \vartheta_{00}(v,\tau) \,, \qquad \text{(7.2.39b)}$$

$$\vartheta_{10}(v, \tau+1) = \exp(\pi i/4)\vartheta_{10}(v,\tau) \,, \qquad \text{(7.2.39c)}$$

$$\vartheta_{11}(v, \tau+1) = \exp(\pi i/4)\vartheta_{11}(v,\tau) \,, \qquad \text{(7.2.39d)}$$

and

$$\vartheta_{00}(v/\tau, -1/\tau) = (-i\tau)^{1/2} \exp(\pi i v^2/\tau)\vartheta_{00}(v,\tau) \,, \qquad \text{(7.2.40a)}$$

$$\vartheta_{01}(v/\tau, -1/\tau) = (-i\tau)^{1/2} \exp(\pi i v^2/\tau)\vartheta_{10}(v,\tau) \,, \qquad \text{(7.2.40b)}$$

$$\vartheta_{10}(v/\tau, -1/\tau) = (-i\tau)^{1/2} \exp(\pi i v^2/\tau)\vartheta_{01}(v,\tau) \,, \qquad \text{(7.2.40c)}$$

$$\vartheta_{11}(v/\tau, -1/\tau) = -i(-i\tau)^{1/2} \exp(\pi i v^2/\tau)\vartheta_{11}(v,\tau) \,. \qquad \text{(7.2.40d)}$$

We will encounter these functions more extensively in the superstring.

The theta function satisfies Jacobi's 'abstruse identity,' a special case of a quartic identity of Riemann,

$$\vartheta_{00}^4(0,\tau) - \vartheta_{01}^4(0,\tau) - \vartheta_{10}^4(0,\tau) = 0 \,. \qquad \text{(7.2.41)}$$

Note also that

$$\vartheta_{11}(0,\tau) = 0 \,. \qquad \text{(7.2.42)}$$

Finally, the Dedekind eta function is

$$\eta(\tau) = q^{1/24} \prod_{m=1}^{\infty} (1 - q^m) = \left[\frac{\partial_\nu \vartheta_{11}(0, \tau)}{-2\pi} \right]^{1/3} . \qquad (7.2.43)$$

It has the modular transformations

$$\eta(\tau + 1) = \exp(i\pi/12)\, \eta(\tau) , \qquad (7.2.44a)$$

$$\eta(-1/\tau) = (-i\tau)^{1/2}\, \eta(\tau) . \qquad (7.2.44b)$$

7.3   The torus amplitude

We now apply to the torus the general result (5.3.9) expressing the string amplitude as an integral with ghost and vertex operator insertions. The two CKVs require one vertex operator to be fixed, so

$$S_{T^2}(1; 2; \ldots; n)$$

$$= \frac{1}{2} \int_{F_0} d\tau d\bar{\tau} \left\langle B\tilde{B}\, \tilde{c}c\mathscr{V}_1(w_1, \bar{w}_1) \prod_{i=2}^{n} \int dw_i d\bar{w}_i\, \mathscr{V}_i(w_i, \bar{w}_i) \right\rangle_{T^2} . \qquad (7.3.1)$$

The fundamental region F_0 and the factor of $\frac{1}{2}$ from $w \to -w$ were discussed in section 5.1. As usual for complex variables, $d\tau d\bar{\tau} = 2d\tau_1 d\tau_2$, where $\tau = \tau_1 + i\tau_2$. The ghost insertion for $d\tau$ is

$$B = \frac{1}{4\pi}(b, \partial_\tau g) = \frac{1}{2\pi} \int d^2w\, b_{ww}(w)\partial_\tau g_{\bar{w}\bar{w}}$$

$$= \frac{i}{4\pi\tau_2} \int d^2w\, b_{ww}(w)$$

$$\to 2\pi i b_{ww}(0) . \qquad (7.3.2)$$

We have used the result (7.2.13) for the τ-derivative of the metric. In the final line we have used the fact that the ghost path integral (7.2.27) is independent of position to carry out the w-integration. (Recall that $d^2w = 2d\sigma^1 d\sigma^2$ and that $g_{w\bar{w}} = \frac{1}{2}$.)

The CKG consists of the translations of the torus. Since this group has finite volume, we need not fix it: we can rewrite the amplitude so that all vertex operators are integrated and the volume of the CKG is divided out. It is easy to do this by hand here. The CKVs are constant, so the expectation value in eq. (7.3.1) is independent of where the c ghosts are placed: we can move them away from w_1, putting them at some fixed position. For the vertex operators, translation invariance implies that the amplitude is unchanged if we replace $w_i \to w_i + w$. Average over translations

$$\int \frac{dw d\bar{w}}{2(2\pi)^2 \tau_2} , \qquad (7.3.3)$$

the denominator being the area of the torus, which is the volume of the CKG. Using eqs. (7.3.2) and (7.3.3), the amplitude becomes

$$S_{T^2}(1; 2; \ldots; n)$$

$$= \int_{F_0} \frac{d\tau d\bar{\tau}}{4\tau_2} \left\langle b(0)\tilde{b}(0)\tilde{c}(0)c(0) \prod_{i=1}^{n} \int dw_i d\bar{w}_i \, \mathcal{V}_i(w_i, \bar{w}_i) \right\rangle_{T^2} . \tag{7.3.4}$$

All vertex operators are now on an equal footing.

The form (7.3.4) could also have been derived directly, without the intermediate step of fixing a vertex operator, and is valid even without vertex operators,

$$Z_{T^2} = \int_{F_0} \frac{d\tau d\bar{\tau}}{4\tau_2} \left\langle b(0)\tilde{b}(0)\tilde{c}(0)c(0) \right\rangle_{T^2} . \tag{7.3.5}$$

This vacuum amplitude is quite interesting and will be our main focus. Not only does it reveal the essential difference between the ultraviolet behaviors of string theory and field theory, but it has an important physical interpretation of its own.

For 26 flat dimensions, the matter and ghost path integrals (7.3.5) were evaluated in the previous section, with the result

$$Z_{T^2} = iV_{26} \int_{F_0} \frac{d\tau d\bar{\tau}}{4\tau_2} (4\pi^2\alpha'\tau_2)^{-13} |\eta(\tau)|^{-48} . \tag{7.3.6}$$

The amplitude has the important property of modular invariance. The product $\tau_2|\eta(\tau)|^4$ is invariant by eq. (7.2.44), and it is easy to check that

$$\frac{d\tau d\bar{\tau}}{\tau_2^2} \tag{7.3.7}$$

is modular-invariant as well. Note the exponent -48, from the 24 left- and right-moving oscillators. The contribution of the ghosts cancels two sets of oscillators, leaving only the contribution of the transverse modes. The amplitude with n tachyons is given by including the vertex operator expectation value (7.2.4) and integrating over positions.

For a general CFT the path integral on the torus is expressed as a trace in eq. (7.2.28). As long as there are $d \geq 2$ noncompact flat dimensions the ghosts still cancel two sets of bosonic operators, and the vacuum amplitude is

$$Z_{T^2} = V_d \int_{F_0} \frac{d\tau d\bar{\tau}}{4\tau_2} \int \frac{d^d k}{(2\pi)^d} \exp(-\pi\tau_2\alpha' k^2) \sum_{i \in \mathcal{H}^\perp} q^{h_i-1} \bar{q}^{\tilde{h}_i-1} \tag{7.3.8a}$$

$$= iV_d \int_{F_0} \frac{d\tau d\bar{\tau}}{4\tau_2} (4\pi^2\alpha'\tau_2)^{-d/2} \sum_{i \in \mathcal{H}^\perp} q^{h_i-1} \bar{q}^{\tilde{h}_i-1} . \tag{7.3.8b}$$

Here $\mathscr{H}^{\perp}$ is the closed string Hilbert space excluding the ghosts, the $\mu = 0, 1$ oscillators, and the noncompact momenta.

In order to understand the physics of these amplitudes it is useful to compare with the corresponding quantity in field theory, the sum over all particle paths with the topology of a circle. This is given by

$$
Z_{S_1}(m^2) = V_d \int \frac{d^d k}{(2\pi)^d} \int_0^\infty \frac{dl}{2l} \exp[-(k^2 + m^2)l/2]
$$
$$
= i V_d \int_0^\infty \frac{dl}{2l} (2\pi l)^{-d/2} \exp(-m^2 l/2) . \tag{7.3.9}
$$

This can be derived by gauge-fixing the point-particle path integral, which was given as exercise 5.2. The result is intuitive: l is the modulus for the circle, $\frac{1}{2}(k^2 + m^2)$ is the world-line Hamiltonian, and the $2l$ in the denominator removes the overcounting from translation and reversal of the world-line coordinate.

Let us now take this point-particle result, which is for a particle of given m^2, and sum over the actual spectrum of the string. As shown in chapter 4, the physical spectrum is in one-to-one correspondence with $\mathscr{H}^{\perp}$, where the mass is related to the transverse weights by

$$
m^2 = \frac{2}{\alpha'}(h + \tilde{h} - 2) \tag{7.3.10}
$$

and with the constraint $h = \tilde{h}$. It is useful to write the latter constraint in integral form,

$$
\delta_{h,\tilde{h}} = \int_{-\pi}^{\pi} \frac{d\theta}{2\pi} \exp\left[i(h - \tilde{h})\theta\right] . \tag{7.3.11}
$$

In writing this we are assuming that the difference $h - \tilde{h}$ is an integer. This is true in 26 flat dimensions, and as we have discussed in the previous section it is necessary in general for modular invariance. Then

$$
\sum_{i \in \mathscr{H}^{\perp}} Z_{S_1}(m_i^2) = i V_d \int_0^\infty \frac{dl}{2l} \int_{-\pi}^{\pi} \frac{d\theta}{2\pi} (2\pi l)^{-d/2}
$$
$$
\times \sum_{i \in \mathscr{H}^{\perp}} \exp\left[-(h_i + \tilde{h}_i - 2)l/\alpha' + i(h_i - \tilde{h}_i)\theta\right]
$$
$$
= i V_d \int_R \frac{d\tau d\bar{\tau}}{4\tau_2} (4\pi^2 \alpha' \tau_2)^{-d/2} \sum_{i \in \mathscr{H}^{\perp}} q^{h_i - 1} \bar{q}^{\tilde{h}_i - 1} , \tag{7.3.12}
$$

where $\theta + il/\alpha' = 2\pi\tau$. Here the region of integration R is

$$
R : \quad \tau_2 > 0, \quad |\tau_1| < \frac{1}{2} . \tag{7.3.13}
$$

Now let us interpret these results. The amplitude (7.3.9) for a single point particle diverges as $l \to 0$. This is the usual UV divergence of quantum field theory (note that l has units of spacetime length-squared). Summing this over the string spectrum as in eq. (7.3.12) only makes things worse, as all the states of the string contribute with the same sign. However, compare the sum (7.3.12) to the actual string amplitude (7.3.8). There is great similarity but also one remarkable difference. The integrands are identical but the regions of integration are different: in the string it is the fundamental region

$$F_0: \quad |\tau| > 1, \quad |\tau_1| < \frac{1}{2}, \quad \tau_2 > 0 \qquad (7.3.14)$$

and in field theory the larger region R. In the string amplitude, the UV divergent region is simply *absent*. We can also see this from the point of view of the momentum integral (7.3.8a). Since τ_2 is bounded below, this integral (after contour rotation) is a convergent Gaussian.

Another possible divergence comes from the limit $\tau_2 \to \infty$, where the torus becomes very long. In this region, the string amplitude in 26 flat dimensions has the expansion

$$iV_{26} \int^{\infty} \frac{d\tau_2}{2\tau_2} (4\pi^2 \alpha' \tau_2)^{-13} \left[\exp(4\pi\tau_2) + 24^2 + \dots \right], \qquad (7.3.15)$$

the asymptotic behavior being controlled by the lightest string states. The first term in the series diverges, and the interpretation is clear from the field theory point of view. The series is in order of increasing mass-squared. The first term is from the tachyon and diverges due to the positive exponential in the path sum (7.3.9). This divergence is therefore an artifact of a theory with a tachyon, and will not afflict more realistic string theories. Incidentally, this path sum can be defined by analytic continuation from positive mass-squared, but there is still a pathology — the continued energy density is complex. This signifies an instability, the tachyon field rolling down its inverted potential. With a general CFT, eq. (7.3.12) becomes

$$iV_d \int^{\infty} \frac{d\tau_2}{2\tau_2} (4\pi^2 \alpha' \tau_2)^{-d/2} \sum_i \exp(-\pi\alpha' m_i^2 \tau_2). \qquad (7.3.16)$$

Because the tachyon is always present in bosonic string theory this again diverges; in the absence of tachyons it will converge.

The torus illustrates a general principle, which holds for all string amplitudes: there is no UV region of moduli space that might give rise to high energy divergences. All limits are controlled by the lightest states, the long distance physics. For the torus with vertex operators the τ-integration is still cut off as above, but there are more limits to consider as vertex operators approach one another. The same general principle applies to these as well, as we will develop further in chapter 9.

One might try to remove the UV divergence from field theory by cutting off the l integral. Similarly, one might try to make an analogous modification of string theory, for example replacing the usual fundamental region F_0 with the region $|\tau_1| \leq \frac{1}{2}$, $\tau_2 > 1$ whose lower edge is straight. However, in either case this would spoil the consistency of the theory: unphysical negative norm states would no longer decouple. We have seen in the general discussion in section 5.4 that the coupling of a BRST-null state is proportional to a total derivative on moduli space. For the fundamental region F_0, the apparent boundaries I and I' and II and II' respectively (shown in fig. 5.2) are identified and the surface terms cancel. Modifying the region of integration introduces real boundaries. Surface terms on the modified moduli space will no longer cancel, so there is nonzero amplitude for null states and an inconsistent quantum theory. This is directly analogous to what happens in gravity if we try to make the theory finite by some brute force cutoff on the short-distance physics: it is exceedingly difficult to do this without spoiling the local spacetime symmetry and making the theory inconsistent. String theory manages in a subtle way to soften the short-distance behavior and eliminate the divergences without loss of spacetime gauge invariance.

Physics of the vacuum amplitude

Besides serving as a simple illustration of the behavior of string amplitudes, the one-loop vacuum amplitude has an interesting physical interpretation. In point-particle theory, 'vacuum' paths consist of any number n of disconnected circles. Including a factor of $1/n!$ for permutation symmetry and summing in n leads to

$$Z_{\text{vac}}(m^2) = \exp\left[Z_{S_1}(m^2)\right]. \tag{7.3.17}$$

Translating into canonical field theory,

$$Z_{\text{vac}}(m^2) = \langle 0| \exp(-iHT)|0\rangle$$
$$= \exp(-i\rho_0 V_d), \tag{7.3.18}$$

where ρ_0 is the vacuum energy density:

$$\rho_0 = \frac{i}{V_d} Z_{S_1}(m^2). \tag{7.3.19}$$

The l integral in $Z_{S_1}(m^2)$ diverges as $l \to 0$. In renormalizable field theory this divergence is canceled by counterterms or supersymmetry. To get a rough insight into the physics we will cut off the integral at $l > \epsilon$ and then take $\epsilon \to 0$, but drop divergent terms. With this prescription,

$$\int_0^\infty \frac{dl}{2l} \exp[-(k^2 + m^2)l/2] \to -\frac{1}{2}\ln(k^2 + m^2) \tag{7.3.20}$$

and

$$i \int_0^\infty \frac{dl}{2l} \int_{-\infty}^\infty \frac{dk^0}{2\pi} \exp[-(k^2 + m^2)l/2] \rightarrow \frac{\omega_{\mathbf{k}}}{2}, \qquad (7.3.21)$$

where $\omega_{\mathbf{k}}^2 = \mathbf{k} \cdot \mathbf{k} + m^2$. Using the latter form, the vacuum energy density becomes

$$\rho_0 = \int \frac{d^{d-1}\mathbf{k}}{(2\pi)^{d-1}} \frac{\omega_{\mathbf{k}}}{2}, \qquad (7.3.22)$$

which is just the sum of the zero-point energies of the modes of the field. This is in spacetime, of course — we encountered the same kind of sum but on the world-sheet in section 1.3.

The description of a quantum field theory as a sum over particle paths is not as familiar as the description in terms of a sum over field histories, but it is equivalent. In particular, the free path integral in field theory can be found in most modern field theory textbooks:

$$\ln Z_{\text{vac}}(m^2) = -\frac{1}{2} \text{Tr} \ln(-\partial^2 + m^2)$$

$$= -\frac{V_d}{2} \int \frac{d^d k}{(2\pi)^d} \ln(k^2 + m^2). \qquad (7.3.23)$$

Using eq. (7.3.20), this is the same as the path sum result (7.3.9).

In older quantum field theory texts, one often reads that vacuum amplitudes such as eq. (7.3.18) are irrelevant because they give an overall phase that divides out of any expectation value. This is true if one is considering scattering experiments in a fixed background and ignoring gravity, but there are at least two circumstances in which the vacuum energy density is quite important. The first is in comparing the energy densities of *different* states to determine which is the true ground state of a theory. It is very likely, for example, that the breaking of the electroweak $SU(2) \times U(1)$ symmetry is determined in part by such quantum corrections to the vacuum energy density. This was the original motivation that led to the Coleman–Weinberg formula (7.3.23). The generalization to a theory with particles of arbitrary spin is

$$\rho_0 = \frac{i}{V_d} \sum_i (-1)^{\mathbf{F}_i} Z_{S_i}(m_i^2). \qquad (7.3.24)$$

The sum runs over all physical particle states. Each polarization counts separately, giving a factor of $2s_i + 1$ for a particle of spin s_i in four dimensions. Here $\mathbf{F}_i$ is the spacetime fermion number, defined mod 2, so fermions contribute with the opposite sign.

The second circumstance is when one considers the coupling to gravity. The vacuum energy gives a source term in Einstein's equation, the cosmological constant, and so has observable effects. In fact, this cosmological constant is a great challenge, because the fact that spacetime is approximately flat and static means that its value is very small,

$$|\rho_0| \lesssim 10^{-44} \, \mathrm{GeV}^4 \,. \tag{7.3.25}$$

If one considers only the contributions to the vacuum energy from the vacuum fluctuations of the known particles up to the currently explored energy (roughly the electroweak scale m_{ew}), the zero-point energy is already of order

$$m_{\mathrm{ew}}^4 \approx 10^8 \, \mathrm{GeV}^4 \,, \tag{7.3.26}$$

52 orders of magnitude too large. The Higgs field potential energy and the QCD vacuum energy are also far too large. Finding a mechanism that would account for the cancellation of the net cosmological constant to great accuracy has proven very difficult. For example, in a supersymmetric theory, the contributions of degenerate fermions and bosons in the sum (7.3.24) cancel. But supersymmetry is not seen in nature, so it must be a broken symmetry. The cancellation is then imperfect, leaving again a remainder of at least m_{ew}^4. The cosmological constant problem is one of the most nagging difficulties, and therefore probably one of the best clues, in trying to find a unified theory with gravity.

What about the cosmological constant problem in string theory? At string tree level we had a consistent theory with a flat metric, so the cosmological constant was zero. In fact we arranged this by hand, when we took 26 dimensions. One sees from the spacetime action (3.7.20) that there would otherwise be a tree-level potential energy proportional to $D - 26$. The one-loop vacuum energy density in bosonic string theory is nonzero, and is necessarily of the order of the string scale (ignoring the tachyon divergence). In four dimensions this would correspond to $10^{72} \, \mathrm{GeV}^4$, which is again far too large. In a supersymmetric string theory, there will be a certain amount of cancellation, but again one expects a remainder of at least m_{ew}^4 in a realistic theory with supersymmetry breaking.

The cosmological constant problem is telling us that there is something that we still do not understand about the vacuum, in field theory and string theory equally. We will return to this problem at various points.

7.4 Open and unoriented one-loop graphs

The cylinder

The results for the torus are readily extended to the other surfaces of Euler number zero. For example, the vacuum amplitude from the cylinder

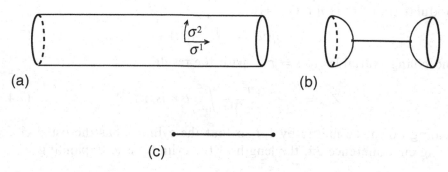

(c)

Fig. 7.1. (a) Cylinder in the limit of small t. (b) The amplitude separated into disk tadpole amplitudes and a closed string propagator. (c) Analogous field theory graph. The heavy circles represent the tadpoles.

in the oriented theory is

$$
\begin{aligned}
Z_{C_2} &= \int_0^\infty \frac{dt}{2t} \operatorname{Tr}'_o[\exp(-2\pi t L_0)] \\
&= iV_d \int_0^\infty \frac{dt}{2t} (8\pi^2 \alpha' t)^{-d/2} \sum_{i \in \mathscr{H}_o^\perp} \exp[-2\pi t(h_i - 1)] \\
&\to iV_{26}\, n^2 \int_0^\infty \frac{dt}{2t} (8\pi^2 \alpha' t)^{-13} \eta(it)^{-24} .
\end{aligned}
\tag{7.4.1}
$$

This can be obtained either by working out the path integral measure in terms of the ghost zero modes, as we did for the torus, or by guessing (correctly) that we should again sum the point-particle result (7.3.9) over the open string spectrum. In the first line, the trace is over the full open string CFT, except that for omission of the ghost zero modes as denoted by the prime. In the final line the trace has been carried out for the case of 26 flat dimensions, with n Chan–Paton degrees of freedom. Including tachyon vertex operators is straightforward.

The $t \to \infty$ limit of the cylinder is much like the $\tau_2 \to \infty$ limit of the torus. The cylinder looks like a long strip, and the leading asymptotics are given by the lightest open string states. As in the closed string there is a divergence, but only from the open string tachyon.

The $t \to 0$ limit is rather interesting. Unlike the case of the torus, there is no modular group acting to cut off the range of integration, and so the UV divergence of field theory still seems to be present. However, we will see that, as with all divergences in string theory, this should actually be interpreted as a *long-distance* effect. In the $t \to 0$ limit the cylinder is very long as shown in figure 7.1(a). That is, it looks like a closed string appearing from the vacuum, propagating a distance, and then disappearing again into the vacuum. To make this clear, use the modular transformation (7.2.44)

$$\eta(it) = t^{-1/2}\eta(i/t) \tag{7.4.2}$$

and change variables to $s = \pi/t$, with the result

$$Z_{C_2} = i\frac{V_{26}n^2}{2\pi(8\pi^2\alpha')^{13}} \int_0^\infty ds\,\eta(is/\pi)^{-24} . \tag{7.4.3}$$

Scaling down the metric by $1/t$ so that the cylinder has the usual closed string circumference 2π, the length of the cylinder is s. Expanding

$$\eta(is/\pi)^{-24} = \exp(2s) \prod_{n=1}^\infty [1 - \exp(-2ns)]^{-24}$$
$$= \exp(2s) + 24 + O(\exp(-2s)) , \tag{7.4.4}$$

one sees the expected asymptotics from expanding in a complete set of closed string states. In other words, if we think of σ^2 as world-sheet time and σ^1 as world-sheet space, figure 7.1(a) is a very short open string loop. If we reverse the roles of world-sheet space and time, it is a very long closed string world-line beginning and ending on boundary loops. In the Euclidean path integral, either description can be used, and each is useful in a different limit of moduli space.

The leading divergence in the vacuum amplitude is from the closed string tachyon. This is uninteresting and can be defined by analytic continuation,

$$\int_0^\infty ds\,\exp(\beta s) \equiv -\frac{1}{\beta} . \tag{7.4.5}$$

The second term, from the massless closed string states, gives a divergence of the form $1/0$ even with the continuation. To see the origin of this divergence, imagine separating the process as shown in figure 7.1(b). As we have discussed in section 6.6, there is a nonzero amplitude with one closed string vertex operator on the disk. This tadpole corresponds to a closed string appearing from or disappearing into the vacuum. In between the two tadpoles is the closed string propagator. In momentum space the massless propagator is proportional to $1/k^2$. Here, momentum conservation requires that the closed string appear from the vacuum with zero momentum, so the propagator diverges.

This same kind of divergence occurs in quantum field theory. Consider a massless scalar field ϕ with a term linear in ϕ in the Lagrangian. There is then a vertex which connects to just a single propagator, and the graph shown in figure 7.1(c) exists. This diverges because the intermediate propagator is

$$\frac{1}{k^2}\bigg|_{k^\mu=0} . \tag{7.4.6}$$

This is a long-distance (IR) divergence, because poles in propagators are associated with propagation over long spacetime distances.

UV and IR divergences in quantum field theory have very different physical origins. UV divergences usually signify a breakdown of the theory, the need for new physics at some short distance. IR divergences generally mean that we have asked the wrong question, or expanded in the wrong way. So it is here. In normal perturbation theory we expand around $\phi(x) = 0$, or some other constant configuration. For the action

$$-\frac{1}{g^2} \int d^d x \left(\frac{1}{2} \partial_\mu \phi \partial^\mu \phi + g \Lambda \phi\right) , \qquad (7.4.7)$$

the equation of motion

$$\partial^2 \phi = g \Lambda \qquad (7.4.8)$$

does not allow $\phi(x) = 0$ as a solution. Instead we must expand around a solution to the equation (7.4.8); any solution is necessarily position-dependent. The resulting amplitudes are free of divergences. This is true even though the right-hand side is a perturbation (we have included the factor of g appropriate to the disk), because it is a singular perturbation. In particular, the corrected background breaks some of the Poincaré symmetry of the zeroth order solution.

The situation is just the same in string theory. The disk tadpole is a source

$$-\Lambda \int d^{26} x (-G)^{1/2} e^{-\Phi} \qquad (7.4.9)$$

for both the dilaton and metric. Expanding around a solution to the corrected field equations (with the metric and dilaton no longer constant) leads to finite amplitudes. The details are a bit intricate, and will be discussed somewhat further in chapter 9. Incidentally, in supersymmetric string theories if the tree-level background is invariant under supersymmetry, it usually receives no loop corrections.

The pole (7.4.6) is the same kind of divergence encountered in the tree-level amplitudes, a resonance corresponding to propagation over long spacetime distances. If we add open string vertex operators to each end of the cylinder so that it represents a one-loop open string scattering amplitude, then the momentum k^μ flowing from one boundary to the other is in general nonzero. The large-s limit (7.4.4) then includes a factor of

$$\exp(-\alpha' k^2 s/2) \qquad (7.4.10)$$

and the divergence becomes a momentum pole representing scattering of open strings into an intermediate closed string. Thus, as claimed in chapter 3, an open string theory must include closed strings as well. It is curious that the mechanism that removes the UV divergence is different in the closed and open string cases. In the closed string it is an effective cutoff on the modular integration; in the open string it is a reinterpretation of the dangerous limit of moduli space in terms of long distances.

In eq. (7.4.1) the path integral on the cylinder has been related to a trace over the open string spectrum by cutting open the path integral with σ^2 treated as time. It can also be obtained in terms of the closed string Hilbert space by treating σ^1 as time. Let σ^2 be periodic with period 2π and the boundaries be at $\sigma^1 = 0, s$. The closed string appears in some state $|B\rangle$ at $\sigma^1 = 0$ and then disappears in the same way at $\sigma^1 = s$. Including the measure insertions, the path integral is then proportional to

$$\langle B|c_0 b_0 \exp[-s(L_0 + \tilde{L}_0)]|B\rangle. \tag{7.4.11}$$

The *boundary state* $|B\rangle$ is determined by the condition that $\partial_1 X^\mu$, c^1, and b_{12} vanish on the boundary. In the Hamiltonian form these must annihilate $|B\rangle$, which in terms of the Laurent coefficients is

$$(\alpha_n^\mu + \tilde{\alpha}_{-n}^\mu)|B\rangle = (c_n + \tilde{c}_{-n})|B\rangle = (b_n - \tilde{b}_{-n})|B\rangle = 0 , \quad \text{all } n . \tag{7.4.12}$$

This determines

$$|B\rangle \propto (c_0 + \tilde{c}_0) \exp\left[-\sum_{n=1}^{\infty}(n^{-1}\alpha_{-n}\cdot\tilde{\alpha}_{-n} + b_{-n}\tilde{c}_{-n} + \tilde{b}_{-n}c_{-n})\right]|0;0\rangle . \tag{7.4.13}$$

Using this in (7.4.11) gives the result (7.4.3), except that the normalization of $|B\rangle$ is undetermined. This representation is useful in analyzing the $t \to 0$ limit and the closed string poles.

By comparing the string and field theory calculations we can determine the disk tadpole Λ, but it will be more convenient to do this in the next chapter as a special case of a more general result.

The Klein bottle

The vacuum amplitude from the Klein bottle is

$$Z_{K_2} = \int_0^\infty \frac{dt}{4t} \,\text{Tr}'_c\left\{\Omega \exp[-2\pi t(L_0 + \tilde{L}_0)]\right\}$$

$$= iV_d \int_0^\infty \frac{dt}{4t}(4\pi^2\alpha't)^{-d/2} \sum_{i\in\mathscr{H}_c^\perp} \Omega_i \exp[-2\pi t(h_i + \tilde{h}_i - 2)] , \tag{7.4.14}$$

where the notation follows the cylinder (7.4.1). Relative to the torus in the oriented theory there is an extra factor of $\frac{1}{2}$ from the projection operator $\frac{1}{2}(1 + \Omega)$. For the same reason there is an extra $\frac{1}{2}$ in both the torus and cylinder amplitudes in the unoriented theory. One can also think of this as coming from the extra gauge invariance $w \to \bar{w}$. To evaluate the trace for 26 flat dimensions, note that the only diagonal elements of Ω are those

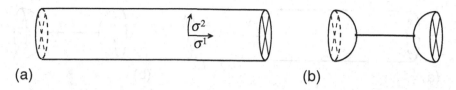

(a) (b)

Fig. 7.2. (a) Klein bottle in the limit of small t as a cylinder capped by cross-caps. (b) The amplitude separated into RP_2 tadpole amplitudes and a closed string propagator.

for which the right- and left-movers are in the same state, and these states contribute with $\Omega = +1$. The trace is then effectively over only one side, which is the same as the open string spectrum, except that the weight is doubled because the right- and left-movers make equal contributions. The result is

$$Z_{K_2} \to iV_{26} \int_0^\infty \frac{dt}{4t} (4\pi^2\alpha' t)^{-13} \eta(2it)^{-24} . \qquad (7.4.15)$$

The modulus t runs over the same range $0 < t < \infty$ as for the cylinder, so the $t \to 0$ divergences are the same. Again the $1/0$ pole has a long-distance interpretation in terms of a closed string pole. To see this consider the regions

$$0 \le \sigma^1 \le 2\pi , \quad 0 \le \sigma^2 \le 2\pi t , \qquad (7.4.16a)$$
$$0 \le \sigma^1 \le \pi , \quad 0 \le \sigma^2 \le 4\pi t , \qquad (7.4.16b)$$

each of which is a fundamental region for the identification

$$w \cong w + 2\pi \cong -\bar{w} + 2\pi it . \qquad (7.4.17)$$

In (7.4.16a), the left- and right-hand edges are periodically identified while the upper and lower edges are identified after a parity-reversal, giving an interpretation as a closed string loop weighted by Ω. For (7.4.16b), note that the identifications (7.4.17) imply also that

$$w \cong w + 4\pi it , \quad w + \pi \cong -(\bar{w} + \pi) + 2\pi it . \qquad (7.4.18)$$

It follows that the upper and lower edges of the region (7.4.16b) are periodically identified. The left-hand edge is identified with itself after translation by half its length and reflection through the edge, and similarly the right-hand edge: this is the definition of a cross-cap. Thus we have the representation in figure 7.2(a) as a cylinder capped by two cross-caps. Rescaling by $1/2t$, the cylinder has circumference 2π and length $s = \pi/2t$. After a modular transformation, the amplitude becomes

$$Z_{K_2} = i\frac{2^{26}V_{26}}{4\pi(8\pi^2\alpha')^{13}} \int_0^\infty ds \, \eta(is/\pi)^{-24} . \qquad (7.4.19)$$

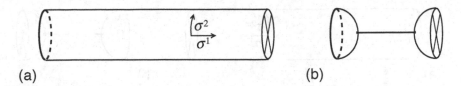

(a)　　　　　　　　　　　　　　　　(b)

Fig. 7.3. (a) Möbius in the limit of small t as a cylinder capped by one cross-cap. (b) The amplitude separated into D_2 and RP_2 tadpole amplitudes and a closed string propagator.

All the discussion of the cylinder divergence now applies, except that the tadpole is from the projective plane rather than the disk.

The Möbius strip

For the Möbius strip,

$$Z_{M_2} = iV_d \int_0^\infty \frac{dt}{4t} (8\pi^2\alpha't)^{-d/2} \sum_{i\in\mathcal{H}_0^\perp} \Omega_i \exp[-2\pi t(h_i - 1)] \qquad (7.4.20)$$

This differs from the result on the cylinder only by the Ω and the factor of $\frac{1}{2}$ from the projection operator. In 26 flat dimensions, the effect of the operator Ω in the trace is an extra -1 at the even mass levels plus an appropriate accounting of the Chan–Paton factors. The oscillator trace is thus

$$\exp(2\pi t) \prod_{n=1}^\infty [1 - (-1)^n \exp(-2\pi nt)]^{-24} = \vartheta_{00}(0, 2it)^{-12} \eta(2it)^{-12} . \qquad (7.4.21)$$

For the $SO(n)$ theory the $\frac{1}{2}n(n + 1)$ symmetric states have $\Omega = +1$ while the $\frac{1}{2}n(n - 1)$ antisymmetric states have $\Omega = -1$ for a net contribution of n. For the $Sp(k)$ theory these degeneracies are reversed, giving $-n$ (recall that in our notation $Sp(k)$ corresponds to $n = 2k$ Chan–Paton states). The amplitude is then

$$Z_{M_2} = \pm inV_{26} \int_0^\infty \frac{dt}{4t} (8\pi^2\alpha't)^{-13} \vartheta_{00}(0, 2it)^{-12} \eta(2it)^{-12} . \qquad (7.4.22)$$

By the same construction as for the Klein bottle, the Möbius strip can be represented as a cylinder with a boundary at one end and a cross-cap on the other, as in figure 7.3(a). The length of the cylinder is now $s = \pi/4t$. By a modular transformation, the amplitude becomes

$$Z_{M_2} = \pm 2in\frac{2^{13}V_{26}}{4\pi(8\pi^2\alpha')^{13}} \int_0^\infty ds\, \vartheta_{00}(0, 2is/\pi)^{-12} \eta(2is/\pi)^{-12} . \qquad (7.4.23)$$

As for the annulus this can be written as an operator expression, one of the boundary states in eq. (7.4.11) being replaced by an analogous cross-cap state $|C\rangle$. There is again a $t \to 0$ divergence; it corresponds to the process of figure 7.3(a) with one tadpole from the disk and one from the projective plane.

In the unoriented theory the divergences on these three surfaces combine into

$$i\frac{24V_{26}}{4\pi(8\pi^2\alpha')^{13}}(2^{13} \mp n)^2 \int_0^\infty ds \,. \qquad (7.4.24)$$

That is, the total tadpole is proportional to $2^{13} \mp n$. For the gauge group $SO(2^{13}) = SO(8192)$ this vanishes, the tadpoles from the disk and projective plane canceling. For the bosonic string this probably has no special significance, but the analogous cancellation for $SO(32)$ in superstring theory does.

Exercises

7.1 Fill in the steps leading to the expectation value (7.2.4) of exponential operators on the torus. Show that it has the correct transformation under $\tau \to -1/\tau$.

7.2 Derive the torus vacuum amplitude (7.3.6) by regulating and evaluating the determinants, as is done for the harmonic operator in appendix A. Show that a modular transformation just permutes the eigenvalues. [Compare exercise A.3.] (Reference: Polchinski (1986).)

7.3 Derive the Green's function (7.2.3) by carrying out the eigenfunction sum (6.2.7).

7.4 Evaluate $\langle \partial_w X^\mu(w)\partial_w X_\mu(0) \rangle$ on the torus by representing it as a trace. Show that the result agrees with eq. (7.2.16).

7.5 (a) Obtain the leading behavior of the theta functions (7.2.37) as $\mathrm{Im}\,\tau \to \infty$.

(b) By using the modular transformations, do the same as $\tau \to 0$ along the imaginary axis.

(c) A more exotic question: what if τ approaches a nonzero point on the real axis, along a path parallel to the imaginary axis? [Hint: the answer depends crucially on whether the point is rational or irrational.]

7.6 (a) Verify eqs. (7.3.20) and (7.3.21).

(b) Evaluate $Z_{S_1}(m^2)$ with a cutoff on the l-integral in any dimension d. Show that the counterterms are analytic in m^2. Give the nonanalytic part for $d = 4$. (Reference: Coleman & Weinberg (1973).)

7.7 Obtain the amplitude for n open string tachyons on the cylinder in a form analogous to the torus amplitude (eq. (7.3.6) combined with eq. (7.2.4)). The necessary Green's function can be obtained from that on the torus by the method of images, using the representation of the cylinder as an involution of the torus.

7.8 Consider the amplitude from the previous problem, in the case that there are some vertex operators on each of the two boundaries. Show from the explicit result of the previous problem that the $t \to 0$ limit gives a series of closed string poles as a function of the momentum flowing from one boundary to the other.

7.9 If you have done both exercises 6.12 and 7.8, argue that unitarity relates the square of the former to the tachyon pole of the latter. Use this to find the numerical relation between g_c and g_o^2. We will do something similar to this, in a more roundabout way, in the next chapter.

7.10 Argue that unitarity of the cylinder amplitude with respect to intermediate open strings requires the Chan–Paton factors to satisfy

$$\text{Tr}\,(\lambda^a \lambda^b)\text{Tr}\,(\lambda^c \lambda^d) = \text{Tr}\,(\lambda^a \lambda^b \lambda^e \lambda^c \lambda^d \lambda^e).$$

Show that the gauge group $U(n_1) \times U(n_2) \times \ldots \times U(n_k)$ is consistent with unitarity at tree level, but that this additional condition singles out $U(n)$.

7.11 This exercise reproduces the first appearance of $D = 26$ in string theory (Lovelace, 1971). Calculate the cylinder vacuum amplitude from the Coleman–Weinberg formula in the form (7.3.9), assuming D spacetime dimensions and D' net sets of oscillators. Fix the constant in the open string mass-squared so that the tachyon remains at $m^2 = -1/\alpha'$, as required by conformal invariance of the vertex operator. Insert an additional factor of $\exp(-\alpha' k^2 s/2)$ as in eq. (7.4.10) to simulate a scattering amplitude. Show that unless $D = 26$ and $D' = 24$, the amplitude has branch cuts (rather than poles) in k^2.

7.12 Repeat the previous exercise for the torus, showing in this case that the vacuum amplitude is modular-invariant only for $D = 26$ and $D' = 24$ (Shapiro, 1972).

7.13 Carry out the steps, parallel to those used to derive the torus amplitude (7.3.6), to obtain the vacuum Klein bottle amplitude (7.4.15) from the string path integral.

7.14 Show that the amplitude (7.4.11) with the boundary state (7.4.13) reproduces the cylinder amplitude (7.4.3).

7.15 (a) Find the state $|C\rangle$ corresponding to a cross-cap.
(b) By formulae analogous to eq. (7.4.11), obtain the Klein bottle and Möbius strip vacuum amplitudes.

8
Toroidal compactification and T-duality

Realistic compactifications of string theory will be the subject of the later chapters of volume two, but it is interesting to look now at the simplest compactification of string theory, in which one or more dimensions are periodically identified. We first consider the same compactification in field theory, encountering in particular the Kaluza–Klein unification of gauge interactions and gravity. We then extend this to string theory, where several new and intrinsically stringy phenomena arise: winding states, enhanced gauge symmetries, T-duality, and D-branes. We are also led to consider slightly more complicated compactifications, namely orbifolds and orientifolds.

8.1 Toroidal compactification in field theory

In general relativity, the geometry of spacetime is dynamical. The three spatial dimensions we see are expanding and were once highly curved. It is a logical possibility that there are additional dimensions that remain small. In fact, this was put forward as early as 1914 as a means of unifying the electromagnetic and gravitational fields as components of a single higher-dimensional field. Consider the case of a five-dimensional theory, with x^4 periodic,

$$x^4 \cong x^4 + 2\pi R , \qquad (8.1.1)$$

and with x^μ noncompact for $\mu = 0, \dots, 3$. This is *toroidal compactification*. The five-dimensional metric separates into $G_{\mu\nu}$, $G_{\mu 4}$, and G_{44}. From the four-dimensional point of view, these are a metric, a vector, and a scalar.

Let us see this in detail, taking the more general case of $D = d + 1$ spacetime dimensions with x^d periodic. Since we are still in field theory we leave D arbitrary. Parameterize the metric as

$$ds^2 = G^D_{MN} dx^M dx^N = G_{\mu\nu} dx^\mu dx^\nu + G_{dd}(dx^d + A_\mu dx^\mu)^2 . \qquad (8.1.2)$$

231

Indices M, N run over all dimensions $0, \ldots, d$ and indices μ, ν run only over noncompact dimensions $0, \ldots, d-1$. We designate the full D-dimensional metric by G_{MN}^D. One should note that $G_{\mu\nu} \neq G_{\mu\nu}^D$; in d-dimensional actions we raise and lower indices with $G_{\mu\nu}$. For now the fields $G_{\mu\nu}$, G_{dd}, and A_μ are allowed to depend only on the noncompact coordinates x^μ. The form (8.1.2) is the most general metric invariant under translations of x^d. This form still allows d-dimensional reparameterizations $x'^\mu(x^\nu)$ and also reparameterizations

$$x'^d = x^d + \lambda(x^\mu) . \tag{8.1.3}$$

Under the latter,

$$A'_\mu = A_\mu - \partial_\mu \lambda , \tag{8.1.4}$$

so gauge transformations arise as part of the higher-dimensional coordinate group. This is the *Kaluza–Klein* mechanism.

To see the effect of x^d-dependence, consider a massless scalar ϕ in D dimensions, where for simplicity the metric in the compact direction is $G_{dd} = 1$. The momentum in the periodic dimension is quantized, $p_d = n/R$. Expand the x^d-dependence of ϕ in a complete set,

$$\phi(x^M) = \sum_{n=-\infty}^{\infty} \phi_n(x^\mu) \exp(inx^d/R) . \tag{8.1.5}$$

The D-dimensional wave equation $\partial_M \partial^M \phi = 0$ becomes

$$\partial_\mu \partial^\mu \phi_n(x^\mu) = \frac{n^2}{R^2} \phi_n(x^\mu) . \tag{8.1.6}$$

The modes ϕ_n of the D-dimensional field thus become an infinite tower of d-dimensional fields, labeled by n. The d-dimensional mass-squared

$$-p^\mu p_\mu = \frac{n^2}{R^2} \tag{8.1.7}$$

is nonzero for all fields with nonvanishing p_d. At energies small compared to R^{-1}, only the x^d-independent fields remain and the physics is d-dimensional. At energies above R^{-1}, one sees the tower of Kaluza–Klein states.

The charge corresponding to the Kaluza–Klein gauge invariance (8.1.3) is the p_d-momentum. In this simple example, all fields carrying the Kaluza–Klein charge are massive. More generally, with higher spin fields and curved backgrounds, there can be *massless* charged fields.

The effective action for the massless fields is always an important object to consider. Define $G_{dd} = e^{2\sigma}$. The Ricci scalar for the metric (8.1.2) is

$$R = R_d - 2e^{-\sigma}\nabla^2 e^{\sigma} - \frac{1}{4}e^{2\sigma}F_{\mu\nu}F^{\mu\nu} , \tag{8.1.8}$$

where $\boldsymbol{R}$ is constructed from G_{MN}^D and $\boldsymbol{R}_d$ from $G_{\mu\nu}$. The graviton-dilaton action (3.7.20) becomes

$$
\begin{aligned}
S_1 &= \frac{1}{2\kappa_0^2} \int d^D x \, (-G_D)^{1/2} e^{-2\Phi} (\boldsymbol{R} + 4\nabla_\mu \Phi \nabla^\mu \Phi) \\
&= \frac{\pi R}{\kappa_0^2} \int d^d x \, (-G_d)^{1/2} e^{-2\Phi + \sigma} \\
&\quad \times \left(\boldsymbol{R}_d - 4\partial_\mu \Phi \partial^\mu \sigma + 4\partial_\mu \Phi \partial^\mu \Phi - \frac{1}{4} e^{2\sigma} F_{\mu\nu} F^{\mu\nu} \right) \\
&= \frac{\pi R}{\kappa_0^2} \int d^d x \, (-G_d)^{1/2} e^{-2\Phi_d} \\
&\quad \times \left(\boldsymbol{R}_d - \partial_\mu \sigma \partial^\mu \sigma + 4\partial_\mu \Phi_d \partial^\mu \Phi_d - \frac{1}{4} e^{2\sigma} F_{\mu\nu} F^{\mu\nu} \right) , \quad (8.1.9)
\end{aligned}
$$

giving kinetic terms for all the massless fields. Here G_d denotes the determinant of $G_{\mu\nu}$, and in the last line we have introduced the effective d-dimensional dilaton, $\Phi_d = \Phi - \sigma/2$. The apparent wrong sign of the dilaton kinetic term is illusory because mixing of the graviton with the trace of the metric must also be taken into account. This is most easily done by means of a Weyl transformation as in eq. (3.7.25), and the resulting kinetic term has the correct sign.

The field equations do not determine the radius of the compact dimension:[1] a flat metric and constant dilaton are a solution for any values of Φ and σ. In other words, there is no potential energy for Φ or for σ, and so these fields are massless, much like Goldstone bosons. The different values of Φ and σ label degenerate configurations (or states in the quantum theory), and a state in which these fields are slowly varying has energy only from the gradient. The difference from the Goldstone phenomenon is that the degenerate states are not related to one another by any symmetry, and in fact the physics depends on Φ and σ. In this bosonic theory the degeneracy is accidental and the one-loop energy, discussed in the previous chapter, breaks the degeneracy. In supersymmetric theories, the existence of physically inequivalent but degenerate vacua is quite common and plays an important role in understanding the dynamics. The massless fields, which label the inequivalent vacua, are called *moduli*. In nature,

---

[1] Of course, only the invariant radius

$$\rho = Re^\sigma$$

distinguishes physically inequivalent solutions. We could set R to some convenient value, say $\alpha'^{1/2}$, or to unity if we use dimensionless variables, but it is convenient to leave it general; often it will be more convenient to set G_{dd} to unity instead. Similarly we could set κ_0^2 to be the appropriate power of α' by an additive shift of Φ, but again it does not hurt to leave it general and we will see in chapter 13 that it is natural to fix the additive normalization of Φ in a different way.

supersymmetry breaking must almost surely give mass to all moduli, else they would mediate infinite-range interactions of roughly gravitational strength. We will see in chapter 14 that in some string theories the dilaton itself is the radius of a hidden dimension, one that is not evident in string perturbation theory.

Defining $A_\mu = R\tilde{A}_\mu$, the covariant derivative is

$$\partial_\mu + ip_d A_\mu = \partial_\mu + in\tilde{A}_\mu \,, \tag{8.1.10}$$

so that the charges are integers. The d-dimensional gauge and gravitational couplings are conventionally defined as follows. The coefficient of $\tilde{F}_{\mu\nu}\tilde{F}^{\mu\nu}$ in the Lagrangian density is defined to be $-1/4g_d^2$, and that of R_d is defined to be $1/2\kappa_d^2$. The gauge coupling is thus determined in terms of the gravitational coupling,

$$g_d^2 = \frac{\kappa_0^2 e^{2\Phi_d}}{\pi R^3 e^{2\sigma}} = \frac{2\kappa_d^2}{\rho^2} \,. \tag{8.1.11}$$

The d-dimensional and D-dimensional gravitational couplings are related

$$\frac{1}{\kappa_d^2} = \frac{2\pi\rho}{\kappa^2} \,, \tag{8.1.12}$$

$2\pi\rho$ being the volume of the compactified dimension.

The antisymmetric tensor also gives rise to a gauge symmetry by a generalization of the Kaluza–Klein mechanism. Separating B_{MN} into $B_{\mu\nu}$ and $A'_\mu = B_{d\mu}$, the gauge parameter ζ_M defined in eq. (3.7.7) separates into a d-dimensional antisymmetric tensor transformation ζ_μ and an ordinary gauge invariance ζ_d. The gauge field is $B_{d\mu}$ and the field strength $H_{d\mu\nu}$. The antisymmetric tensor action becomes

$$S_2 = -\frac{1}{24\kappa_0^2} \int d^D x (-G_D)^{1/2} e^{-2\Phi} H_{MNL} H^{MNL}$$

$$= -\frac{\pi R}{12\kappa_0^2} \int d^d x (-G_d)^{1/2} e^{-2\Phi_d} \left(\tilde{H}_{\mu\nu\lambda} \tilde{H}^{\mu\nu\lambda} + 3e^{-2\sigma} H_{d\mu\nu} H_d^{\mu\nu} \right) \,. \tag{8.1.13}$$

We have defined

$$\tilde{H}_{\mu\nu\lambda} = (\partial_\mu B_{\nu\lambda} - A_\mu H_{d\nu\lambda}) + \text{cyclic permutations} \,. \tag{8.1.14}$$

The term proportional to the vector potential arises from the inverse metric G^{MN}. It is known as a *Chern–Simons term*, this signifying the antisymmetrized combination of one gauge potential and any number of field strengths. Such terms appear in many places in supersymmetric theories, and are associated with interesting physical effects. Notice in particular that $\tilde{H}_{\mu\nu\lambda}$ is gauge-invariant because the variation (8.1.4) of A_μ is canceled by a variation

$$B'_{\nu\lambda} = B_{\nu\lambda} - \lambda H_{d\nu\lambda} \ . \tag{8.1.15}$$

There is no way to couple the potential B_{MN} minimally to other fields, and so unlike the Kaluza–Klein case there are no fields charged under the antisymmetric tensor gauge symmetry; this will be different in string theory.

8.2 Toroidal compactification in CFT

Now we consider the conformal field theory of a single periodic scalar field,

$$X \cong X + 2\pi R \ . \tag{8.2.1}$$

To keep the equations uncluttered we drop the superscript from X^d and set $G_{dd} = 1$. The world-sheet action is as in the noncompact theory, $\int d^2z \, \partial X \bar{\partial} X / 2\pi\alpha'$, so the equations of motion, operator products, and energy-momentum tensor are unchanged. In particular, the theory is still conformally invariant. The periodicity has two effects. First, string states must be single-valued under the identification (8.2.1). That is, the operator $\exp(2\pi i R p)$ which translates strings once around the periodic dimension must leave states invariant, so the center-of-mass momentum is quantized

$$k = \frac{n}{R} \ , \quad n \in \mathbf{Z} \ . \tag{8.2.2}$$

This is just as in field theory.

The second effect is special to string theory. A closed string may now wind around the compact direction,

$$X(\sigma + 2\pi) = X(\sigma) + 2\pi R w \ , \qquad w \in \mathbf{Z} \ . \tag{8.2.3}$$

The integer w is the *winding number*. States of winding numbers $+1$, 0, and -1 are shown in figure 8.1(a). From the point of view of the world-sheet field theory, strings of nonzero winding number are *topological solitons*, states with a topologically nontrivial field configuration. A consistent string theory must include the winding number states: by the usual splitting–joining process a $w = 0$ string can turn into a $w = +1$, $w = -1$ pair as shown in figure 8.1(b). It is easy to see that winding number is always conserved as in this example.

To describe the states of the closed string CFT, consider the Laurent

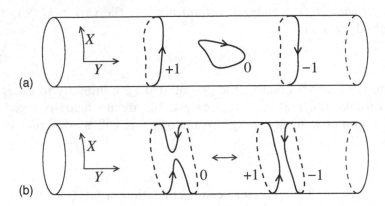

(a)

(b)

Fig. 8.1. (a) Closed oriented strings of winding number $w = +1, 0, -1$. For illustration, one compact dimension X and one noncompact dimension Y are shown. (b) Transition of a $w = 0$ string into $w = +1$ and $w = -1$ strings.

expansions

$$\partial X(z) = -i\left(\frac{\alpha'}{2}\right)^{1/2} \sum_{m=-\infty}^{\infty} \frac{\alpha_m}{z^{m+1}} \; , \qquad \bar{\partial} X(\bar{z}) = -i\left(\frac{\alpha'}{2}\right)^{1/2} \sum_{m=-\infty}^{\infty} \frac{\tilde{\alpha}_m}{\bar{z}^{m+1}} \; . \quad (8.2.4)$$

The total change in the coordinate X in going around the string is

$$2\pi R w = \oint (dz\, \partial X + d\bar{z}\, \bar{\partial} X) = 2\pi(\alpha'/2)^{1/2}(\alpha_0 - \tilde{\alpha}_0) \; . \quad (8.2.5)$$

The total Noether momentum is

$$p = \frac{1}{2\pi\alpha'} \oint (dz\, \partial X - d\bar{z}\, \bar{\partial} X) = (2\alpha')^{-1/2}(\alpha_0 + \tilde{\alpha}_0) \; . \quad (8.2.6)$$

For a noncompact dimension this would give the usual relation $\alpha_0 = \tilde{\alpha}_0 = p(\alpha'/2)^{1/2}$, but for the periodic dimension we have

$$p_L \equiv (2/\alpha')^{1/2} \alpha_0 = \frac{n}{R} + \frac{wR}{\alpha'} \; , \quad (8.2.7a)$$

$$p_R \equiv (2/\alpha')^{1/2} \tilde{\alpha}_0 = \frac{n}{R} - \frac{wR}{\alpha'} \; . \quad (8.2.7b)$$

The Virasoro generators are

$$L_0 = \frac{\alpha' p_L^2}{4} + \sum_{n=1}^{\infty} \alpha_{-n}\alpha_n \; , \quad (8.2.8a)$$

$$\tilde{L}_0 = \frac{\alpha' p_R^2}{4} + \sum_{n=1}^{\infty} \tilde{\alpha}_{-n}\tilde{\alpha}_n \; . \quad (8.2.8b)$$

The partition function

The partition function for X is now

$$(q\bar{q})^{-1/24}\text{Tr}\left(q^{L_0}\bar{q}^{\tilde{L}_0}\right)$$

$$= |\eta(\tau)|^{-2}\sum_{n,w=-\infty}^{\infty}q^{\alpha' p_L^2/4}\bar{q}^{\alpha' p_R^2/4}$$

$$= |\eta(\tau)|^{-2}\sum_{n,w=-\infty}^{\infty}\exp\left[-\pi\tau_2\left(\frac{\alpha' n^2}{R^2}+\frac{w^2 R^2}{\alpha'}\right)+2\pi i\tau_1 nw\right]. \quad (8.2.9)$$

The oscillator sum is the same as in the noncompact case, while the momentum integration is replaced by a sum over n and w. Modular invariance is not manifest, but can be made so by using the Poisson resummation formula,

$$\sum_{n=-\infty}^{\infty}\exp(-\pi a n^2+2\pi ibn)=a^{-1/2}\sum_{m=-\infty}^{\infty}\exp\left[-\frac{\pi(m-b)^2}{a}\right]. \quad (8.2.10)$$

The partition function becomes

$$2\pi R Z_X(\tau)\sum_{m,w=-\infty}^{\infty}\exp\left(-\frac{\pi R^2|m-w\tau|^2}{\alpha'\tau_2}\right). \quad (8.2.11)$$

Here $Z_X(\tau)$ is the modular-invariant expression (7.2.9) from the noncompact theory. The sum is obviously invariant under $\tau \to \tau+1$, by making a change of variables $m \to m+w$. It is also invariant under $\tau \to -1/\tau$ with $m \to -w$, $w \to m$.

The expression (8.2.11) has a simple path integral interpretation. Summing over all genus one world-sheets in a periodic spacetime, each nontrivial closed curve on the world-sheet can be wound around the compact direction,

$$X(\sigma^1+2\pi,\sigma^2)=X(\sigma^1,\sigma^2)+2\pi wR, \quad (8.2.12a)$$

$$X(\sigma^1+2\pi\tau_1,\sigma^2+2\pi\tau_2)=X(\sigma^1,\sigma^2)+2\pi mR. \quad (8.2.12b)$$

That is, the path integral breaks up into topologically distinct sectors labeled by w and m. The Gaussian path integration can be done by the usual strategy of writing X as a classical solution of the correct periodicity,

$$X_{\text{cl}}=\sigma^1 wR+\sigma^2(m-w\tau_1)R/\tau_2, \quad (8.2.13)$$

plus a quantum piece with periodic boundary conditions. The action separates into a classical piece plus a quantum piece. The path integral over the quantum piece is just as in the noncompact case, while the classical action appears as the exponent in eq. (8.2.11). The effect of the modular transformations on the periodicities is then simply to change the summation variables m and w.

Vertex operators

In order to create a winding state, which has $\alpha_0 \neq \tilde{\alpha}_0$, we need independent variables x_L and x_R, with

$$[x_L, p_L] = [x_R, p_R] = i . \tag{8.2.14}$$

The field X then splits into holomorphic and antiholomorphic parts,

$$X(z, \bar{z}) = X_L(z) + X_R(\bar{z}) , \tag{8.2.15}$$

with

$$X_L(z) = x_L - i\frac{\alpha'}{2}p_L \ln z + i\left(\frac{\alpha'}{2}\right)^{1/2} \sum_{\substack{m=-\infty \\ m \neq 0}}^{\infty} \frac{\alpha_m}{mz^m} , \tag{8.2.16a}$$

$$X_R(\bar{z}) = x_R - i\frac{\alpha'}{2}p_R \ln \bar{z} + i\left(\frac{\alpha'}{2}\right)^{1/2} \sum_{\substack{m=-\infty \\ m \neq 0}}^{\infty} \frac{\tilde{\alpha}_m}{m\bar{z}^m} . \tag{8.2.16b}$$

If we restrict attention to states with $k_L = k_R$ and to operators constructed only from the sum $X_L(z) + X_R(\bar{z})$, this reduces to the CFT for a noncompact dimension.

Let us now discuss the OPE and vertex operators, first in a slightly heuristic way and then filling in the details. It is easy to guess what the operator products should be, since the usual XX operator product $-(\alpha'/2)\ln(z_{12}\bar{z}_{12})$ separates into a sum of holomorphic plus antiholomorphic functions. Thus,

$$X_L(z_1)X_L(z_2) \sim -\frac{\alpha'}{2}\ln z_{12} , \quad X_R(\bar{z}_1)X_R(\bar{z}_2) \sim -\frac{\alpha'}{2}\ln \bar{z}_{12} , \tag{8.2.17a}$$

$$X_L(z_1)X_R(\bar{z}_2) \sim 0 . \tag{8.2.17b}$$

The operator corresponding to the state $|0; k_L, k_R\rangle$ would be

$$\mathcal{V}_{k_L k_R}(z, \bar{z}) = :e^{ik_L X_L(z) + ik_R X_R(\bar{z})} : \tag{8.2.18}$$

with OPE

$$\mathcal{V}_{k_L k_R}(z_1, \bar{z}_1)\mathcal{V}_{k'_L k'_R}(z_2, \bar{z}_2) \sim z_{12}^{\alpha' k_L k'_L/2}\bar{z}_{12}^{\alpha' k_R k'_R/2}\mathcal{V}_{(k+k')_L (k+k')_R}(z_2, \bar{z}_2) . \tag{8.2.19}$$

One encounters branch cuts in various expressions, which is not surprising because the field X is no longer single-valued. It is important, however, that the full vertex operator OPE is single-valued: the net phase when z_1 circles z_2 is

$$\exp[\pi i\alpha'(k_L k'_L - k_R k'_R)] = \exp[2\pi i(nw' + wn')] = 1 . \tag{8.2.20}$$

This is as it must be in order for the string amplitudes to be well defined.

A technicality

The above is essentially correct, but we have been cavalier with the location of the branch cut in the logarithm. One can see a problem in the OPE (8.2.19) if one interchanges the points z_1 and z_2 and the momenta k and k'. The left-hand side is symmetric, but the right-hand side changes by $\exp[\pi i(nw' + wn')]$ and so changes sign if $nw' + wn'$ is odd. One can also see the problem as follows. From the mode expansion one may derive the equal time ($|z_1| = |z_2|$) commutator

$$[X_L(z_1), X_L(z_2)] = \frac{\pi i \alpha'}{2}\mathrm{sign}(\sigma_1^1 - \sigma_2^1) . \qquad (8.2.21)$$

The CBH formula then shows that if we define the operators simply as creation–annihilation ordered, then for $nw' + wn'$ odd the operators $\mathscr{V}_{k_L k_R}$ and $\mathscr{V}_{k'_L k'_R}$ anticommute rather than commute; the invisible branch cut (8.2.20) is separated into two visible ones. A correct oscillator expression for the vertex operator (there is some arbitrariness in the phase) is

$$\mathscr{V}_{k_L k_R}(z, \bar{z}) = \exp[\pi i(k_L - k_R)(p_L + p_R)\alpha'/4] \, {}^\circ_\circ e^{ik_L X_L(z) + ik_R X_R(\bar{z})} {}^\circ_\circ , \qquad (8.2.22)$$

where as usual the ps are momentum operators and the ks are numbers (the momenta carried by the given vertex operator). When $\mathscr{V}_{k_L k_R}$ and $\mathscr{V}_{k'_L k'_R}$ are commuted past each other, the additional factors, known as *cocycles*, give an additional phase

$$\exp\left\{ \pi i\left[(k_L - k_R)(k'_L + k'_R) - (k'_L - k'_R)(k_L + k_R)\right]\alpha'/4\right\}$$
$$= \exp[\pi i(nw' - wn')] , \qquad (8.2.23)$$

removing the branch cuts. For most purposes one can ignore this complication and work with the simpler expressions of the previous paragraph; the cocycle affects only the relative signs of certain amplitudes.

The general X^μ path integral (6.2.18) factorizes in an obvious way into holomorphic times antiholomorphic, so that one can simply replace

$$\prod_{\substack{i,j=1 \\ i<j}}^{n} |z_{ij}|^{\alpha' k_i k_j} \rightarrow \prod_{\substack{i,j=1 \\ i<j}}^{n} z_{ij}^{\alpha' k_{Li} k_{Lj}/2} \bar{z}_{ij}^{\alpha' k_{Ri} k_{Rj}/2} , \qquad (8.2.24)$$

and also replace the $2\pi\delta(\sum k)$ from the noncompact integral over x_0 by

$$2\pi R \delta_{\Sigma_i n_i, 0}\delta_{\Sigma_i w_i, 0} . \qquad (8.2.25)$$

Again this is naive; the precise expression (8.2.22) gives some additional signs. The reader can readily calculate an amplitude with four vertex operators, which is similar to the Virasoro–Shapiro amplitude.

DDF operators

We have just seen that the exponential operator can be separated into holomorphic and antiholomorphic parts. This has many applications. We now use it to tie up a loose end from the end of chapter 4, the DDF operators.

We will work with light-cone coordinates $X^{\pm} = 2^{-1/2}(X^0 \pm X^1)$, whose holomorphic parts have the OPE (we drop the subscript L)

$$X^+(z)X^-(0) \sim \frac{\alpha'}{2} \ln z \;, \quad X^+(z)X^+(0) \sim X^-(z)X^-(0) \sim 0 \;. \tag{8.2.26}$$

It follows that the operators

$$V^i(nk_0, z) = \partial X^i(z)e^{ink_0 X^+(z)}(2/\alpha')^{1/2} \tag{8.2.27}$$

with i transverse are $(1,0)$ primary fields with OPE

$$V^i(nk_0, z)\, V^j(mk_0, 0) \sim -\frac{\delta^{ij}}{z^2} e^{i(n+m)k_0 X^+(0)}$$
$$-\frac{ink_0\delta^{ij}}{z}\partial X^+(0)e^{i(n+m)k_0 X^+(0)} \;. \tag{8.2.28}$$

Define the DDF operators by

$$A^i_n = \oint \frac{dz}{2\pi} V^i(nk_0, z) \;. \tag{8.2.29}$$

The residue of the z^{-1} term in the OPE (8.2.28) is a total derivative except when $n + m = 0$, and so the DDF operators satisfy an oscillator algebra

$$[A^i_m, A^j_n] = m\delta^{ij}\delta_{m,-n}\frac{\alpha' k_0 p^+}{2} \;, \tag{8.2.30}$$

which is an oscillator algebra, up to normalization.

Consider the action of the DDF operators on a state of given momentum q. The OPE of $V^i(nk_0, z)$ with the vertex operator for this state involves $z^{-\alpha' nk_0 q^-/2}$ and so is single-valued if we fix $k_0 = 2/\alpha' q^-$. The contour integral (8.2.29) then defines a sensible operator in this sector. The DDF operators, being integrals of $(1,0)$ tensors, commute with the Virasoro generators and so take physical states into physical states. It follows that we can build physical states by acting with the DDF operators on the oscillator ground state. The states obtained in this way are in one-to-one correspondence with the light-cone states, and so are a complete set of physical states. The DDF operators do not involve X^- and so these states have no α^-_{-m} excitations. They are therefore the same states given by the more general but less explicit construction (4.4.19).

8.3 Closed strings and T-duality

Now that we are looking at string theory we take $D = 26$, with only X^{25} periodic at first. The mass-shell (L_0 and $\tilde{L}_0$) conditions are

$$m^2 = -k^\mu k_\mu = (k_L^{25})^2 + \frac{4}{\alpha'}(N - 1)$$

$$= (k_R^{25})^2 + \frac{4}{\alpha'}(\tilde{N} - 1) \,, \tag{8.3.1}$$

or

$$m^2 = \frac{n^2}{R^2} + \frac{w^2 R^2}{\alpha'^2} + \frac{2}{\alpha'}(N + \tilde{N} - 2) \,, \tag{8.3.2a}$$

$$0 = nw + N - \tilde{N} \,. \tag{8.3.2b}$$

One recognizes four contributions to the mass-squared: those from the compact momentum, the potential energy of the winding string, the oscillators, and the zero-point energy. The discussion of the physical spectrum in chapter 4 was valid for an arbitrary compactification, so we get the correct counting of states by considering only the transverse oscillators $M = 2, \ldots, 25$.

Let us first recover the field theory results for the massless spectrum. At a generic value of R, the only way a state can be massless is if $n = w = 0$ and $N = \tilde{N} = 1$. These are the same 24^2 states as in the noncompact theory, but it is useful now to separate them according to whether the oscillations are in the spacetime directions μ or the internal direction 25:

$$\alpha_{-1}^\mu \tilde{\alpha}_{-1}^\nu |0;k\rangle \,, \quad (\alpha_{-1}^\mu \tilde{\alpha}_{-1}^{25} + \alpha_{-1}^{25} \tilde{\alpha}_{-1}^\mu)|0;k\rangle \,,$$

$$(\alpha_{-1}^\mu \tilde{\alpha}_{-1}^{25} - \alpha_{-1}^{25} \tilde{\alpha}_{-1}^\mu)|0;k\rangle \,, \quad \alpha_{-1}^{25} \tilde{\alpha}_{-1}^{25}|0;k\rangle \,. \tag{8.3.3}$$

The first of these breaks up further into a 25-dimensional graviton plus dilaton plus antisymmetric tensor. The second, a graviton with one spacetime and one internal index, is the Kaluza–Klein vector. The third state is the vector from the antisymmetric tensor. The final state is a scalar, the modulus for the radius of the compact direction. Its vertex operator, $:\partial X^{25} \bar{\partial} X^{25} e^{ik \cdot X}:$, is a perturbation of the metric $G_{25,25}$. This is the same massless spectrum as found by considering the low energy field theory.

It is interesting to consider the charges of massive states under the $U(1) \times U(1)$ symmetry. Again the Kaluza–Klein gauge symmetry is from X^{25}-translations, so the corresponding charge is the compact momentum p_{25}. For the antisymmetric tensor gauge symmetry, consider the zero-momentum vertex operator, which measures the charge of any state to which it couples. It is proportional to

$$\partial X^\mu \bar{\partial} X^{25} - \partial X^{25} \bar{\partial} X^\mu = \bar{\partial}(X^{25} \partial X^\mu) - \partial(X^{25} \bar{\partial} X^\mu) \,. \tag{8.3.4}$$

This is a total derivative, but it does not integrate to zero in a winding state because X^{25} is not single-valued. So the $B_{\mu,25}$ charge is the winding number. This is our first of several examples of 'stringy' physics in this compactification: in field theory there were no states carrying this charge.

Let us check the gauge couplings explicitly from the string three-point coupling. The gauge boson vertex operators are

$$\frac{2^{1/2}g_{c,25}}{\alpha'} :(\partial X^\mu \bar\partial X^{25} \pm \partial X^{25} \bar\partial X^\mu)e^{ik\cdot X}: . \tag{8.3.5}$$

We have defined the 25-dimensional string coupling $g_{c,25} = g_c(2\pi R)^{-1/2}$. The factor $(2\pi R)^{-1/2}$ is from normalizing the zero-mode wavefunction. The vertex operator for a tachyon with general compact momentum and winding is

$$g_{c,25} :e^{ik_L\cdot X_L(z)+ik_R\cdot X_R(\bar z)}: . \tag{8.3.6}$$

The three-point amplitude for one gauge boson and two tachyons, with the tachyons having nonzero compact momenta and winding, is similar to eq. (6.6.14),

$$-2^{-1/2}\pi i g_{c,25}(2\pi)^{25}\delta^{25}(\textstyle\sum_i k_i)k_{23}^\mu(k_{L23}^{25} \pm k_{R23}^{25})$$
$$\to -2^{3/2}\pi i g_{c,25}(2\pi)^{25}\delta^{25}(\textstyle\sum_i k_i)k_2^\mu(k_{L2}^{25} \pm k_{R2}^{25}) . \tag{8.3.7}$$

In the second line we have taken the gauge boson momentum $k_1 \to 0$, which defines the gauge coupling. The two gauge bosons thus couple respectively to $k_{L2}^{25}\pm k_{R2}^{25}$, the compact momentum and winding, as expected. By considering also the amplitude with a graviton replacing the gauge boson, one can recover the same relations (8.1.11), (8.1.12) between the various couplings as found from the effective action.

Enhanced gauge symmetries

Everything so far is almost exactly as it would be in field theory with one compact dimension. The 26-dimensional graviton gives rise to a 25-dimensional graviton plus vector plus modulus, and so on. The one stringy effect is that winding states carry the $B_{\mu,25}$ charge.

Further and more striking effects arise at special compactification radii. Our discussion of the massless spectrum omitted states that are massless only for special values of the radius R. The richest case is $R = \alpha'^{1/2}$, where $k_{L,R}^{25} = (n \pm w)\alpha'^{-1/2}$ and the condition for a massless state is

$$(n+w)^2 + 4N = (n-w)^2 + 4\tilde N = 4 . \tag{8.3.8}$$

In addition to the generic solution $n = w = 0$, $N = \tilde N = 1$, there are now also

$$n = w = \pm 1, \; N = 0, \; \tilde N = 1 , \quad n = -w = \pm 1, \; N = 1, \; \tilde N = 0 , \tag{8.3.9}$$

and

$$n = \pm 2, \; w = N = \tilde{N} = 0 \,, \quad w = \pm 2, \; n = N = \tilde{N} = 0 \,. \tag{8.3.10}$$

The states (8.3.9) include four new gauge bosons, with vertex operators

$$:\bar{\partial}X^\mu e^{ik\cdot X} \exp[\pm 2i\alpha'^{-1/2}X_L^{25}]: \,, \quad :\partial X^\mu e^{ik\cdot X} \exp[\pm 2i\alpha'^{-1/2}X_R^{25}]: \,. \tag{8.3.11}$$

The precise definition of the exponential operators is as discussed in the previous section. These states have internal momentum and winding number, so they carry the Kaluza–Klein and antisymmetric tensor gauge charges. The only consistent theory of charged massless vectors is non-Abelian gauge theory, so the new gauge bosons must combine with the old into a non-Abelian theory. It is now useful to work with the basis $\partial X^{25}\bar{\partial}X^\mu$ and $\partial X^\mu \bar{\partial}X^{25}$ for the earlier vectors. The first of these couples to k_L^{25}, under which the first pair of states in (8.3.11) carries charges ± 1 and the second pair is neutral. The second similarly couples to k_R^{25} and so to the other pair. This identifies the gauge group as $SU(2) \times SU(2)$, with the three vectors involving $\bar{\partial}X^\mu$ forming one $SU(2)$ and the three involving ∂X^μ forming the other.

To exhibit the $SU(2) \times SU(2)$ define the three $(1,0)$ currents

$$j^1(z) = :\cos[2\alpha'^{-1/2}X_L^{25}(z)]: \,, \tag{8.3.12a}$$

$$j^2(z) = :\sin[2\alpha'^{-1/2}X_L^{25}(z)]: \,, \tag{8.3.12b}$$

$$j^3(z) = i\partial X_L^{25}(z)/\alpha'^{1/2} \,. \tag{8.3.12c}$$

These have been normalized to have the OPE

$$j^i(z)j^j(0) \sim \frac{\delta^{ij}}{2z^2} + i\frac{\epsilon^{ijk}}{z}j^k(0) \,. \tag{8.3.13}$$

There is of course also a similar OPE of $(0,1)$ currents $\tilde{j}^i$. The single pole term implies that the corresponding charges form an $SU(2)$ algebra. In fact, since the currents are holomorphic there is an infinite-dimensional algebra formed by the Laurent coefficients

$$j^i(z) = \sum_{m=-\infty}^{\infty} \frac{j_m^i}{z^{m+1}} \,, \tag{8.3.14a}$$

$$[j_m^i, j_n^j] = \frac{m}{2}\delta_{m,-n}\delta^{ij} + i\epsilon^{ijk}j_{m+n}^k \,, \tag{8.3.14b}$$

known variously as a *current algebra*, *affine Lie algebra*, or *Kac–Moody algebra*. We will encounter such algebras extensively beginning in chapter 11. It will be shown that the coefficient of the z^{-2} term is quantized, with the value in (8.3.13) being the minimum, so this is known as a *level one* $SU(2)$ current algebra.

The existence of this $SU(2) \times SU(2)$ symmetry is the first indication that string theory sees spacetime geometry differently from what we are used to in field theory, where only a $U(1) \times U(1)$ symmetry is evident at any radius. The j^3 and $j^{1,2}$ currents appear to have very different origins, as oscillator excitations and as winding–momentum excitations respectively, and their expressions in terms of X^{25} are very different, but in their action on the string spectrum they are related to one another by symmetry. The extra vectors are massless because the energies of internal momentum and winding are offset by the negative zero-point energy. The reader might therefore suspect that these could arise only in theories with tachyons, but we will see otherwise in volume two.

Scales and couplings

At the $SU(2) \times SU(2)$ radius, the relation (8.1.11) becomes[2]

$$g_{25}^2 = 2\kappa_{25}^2/\alpha' . \tag{8.3.15}$$

This relation continues to hold if we compactify more dimensions, each side scaling the same way, so in particular in four dimensions

$$g_4^2 = 2\kappa_4^2/\alpha' . \tag{8.3.16}$$

In nature, the non-Abelian gauge couplings g_4^2 are within an order of magnitude of unity, so this implies that the string length $\alpha'^{1/2}$ is not too far from the gravitational length κ_4.

We can also think about this as follows. The four-dimensional gauge coupling is dimensionless, but the four-dimensional gravitational coupling is dimensional. Define an effective dimensionless gravitational coupling depending on the energy scale E,

$$g_{G,4}^2(E) = \kappa_4^2 E^2 , \tag{8.3.17}$$

which increases quadratically with energy. Of course, the 'dimensionless' gauge couplings run with energy, but this is slow compared to the dimensional scaling of the gravitational coupling. Then by eq. (8.3.16), the string mass scale is where the gauge and gravitational couplings are roughly equal, $g_{G,4}^2(E) \approx g_4^2$.

It is also interesting to consider the possibility that in compactifying down to four dimensions, some of the compactified dimensions are larger than others. This can best be analyzed by starting at low energy and working upwards. At low energy there is a dimensionless gauge coupling g_4^2. At an energy ρ_5^{-1}, one compact dimension becomes visible and the physics

---

[2] To be precise, this was the coupling of an $(n, w) = (1, 0)$ state. The non-Abelian gauge bosons have $(|n|, |w|) = (1, 1)$ and so their coupling, which is the conventional particle physics definition of the $SU(2)$ coupling, is $g_{SU(2),25}^2 = 4\kappa_{25}^2/\alpha'$. We will see in chapter 18 that this holds for all level one current algebras.

becomes five-dimensional (for simplicity we consider only one dimension at this scale; with more dimensions the conclusion is strengthened). The effective five-dimensional coupling is

$$g_5^2 = 2\pi\rho_5 g_4^2 . \tag{8.3.18}$$

This follows from the effective action, as in eq. (8.1.12). The coupling g_5^2 has units of length. That is, five-dimensional Yang–Mills theory, like four-dimensional gravity, is a nonrenormalizable theory. Correspondingly, the effective dimensionless coupling

$$\hat{g}_5^2 = g_5^2 E = 2\pi\rho_5 E g_4^2 \tag{8.3.19}$$

grows linearly with energy. However, g_4^2 is not much less than one in nature, so the coupling quickly becomes strong at high energy; presumably string theory is then needed to make the short-distance behavior sensible. So the compactification scale must be close to the string scale, which as we have already learned is close to the gravitational scale. This analysis is modified somewhat in open string theory, and when the recent understanding of strongly coupled string theory is taken into account; it is possible, but unlikely, that string or Kaluza–Klein excitations lie far below the Planck scale.

Higgs mechanism

It is instructive to consider what happens as we move R away from the $SU(2) \times SU(2)$ radius. The extra gauge bosons now acquire masses

$$m = \frac{|R^2 - \alpha'|}{R\alpha'} \approx \frac{2}{\alpha'}\left|R - \alpha'^{1/2}\right| , \tag{8.3.20}$$

the approximation being for radii close to the $SU(2) \times SU(2)$ radius $\alpha'^{1/2}$. Close to this radius the mass is much less than the string scale and we should be able to understand the physics in terms of low energy field theory.

There is only one way to give such a mass to a gauge boson, spontaneous symmetry breaking, and that is just what is happening here. At $R = \alpha'^{1/2}$ there are ten massless scalars: the dilaton, the modulus $G_{25,25}$, the four states in (8.3.9) where a 25-oscillator is excited, and the four states (8.3.10). The last nine are created by the vertex operators

$$:j^i(z)\tilde{j}^j(\bar{z})e^{ik\cdot X(z,\bar{z})}: . \tag{8.3.21}$$

The index i is a vector under the left-moving $SU(2)$ and the index j is a vector under the right-moving $SU(2)$. That is, these transform as $(3,3)$ under $SU(2) \times SU(2)$. In particular, the modulus for the radius is $j^3\tilde{j}^3$. Moving away from the $SU(2) \times SU(2)$ radius gives an expectation value to this field and so breaks the gauge symmetry down to the $U(1) \times U(1)$ that leaves the 3-axes invariant. Indeed, this is the gauge group at generic radii. Near the $SU(2)$ radius, the mass (8.3.20) is linear in the magnitude $|R - \alpha'^{1/2}|$ of the modulus, as usual in spontaneous breaking.

Let us label the spacetime fields corresponding to the nine massless scalars as M_{ij}. A change in the modulus is $M_{33} \neq 0$, but it is instructive to look for more general backgrounds near the $SU(2) \times SU(2)$ point. The interactions of these massless fields will be described by some spacetime action, which will include a potential $U(M)$. These fields are massless at the symmetric point so there is no term of order M^2 in U, but an $(SU(2) \times SU(2))$-invariant cubic term is possible,

$$U(M) \propto \epsilon^{ijk} \epsilon^{i'j'k'} M_{ii'} M_{jj'} M_{kk'} = \det M \ . \tag{8.3.22}$$

It is not hard to verify from the string three-point amplitude that this term does appear. Being cubic in bosons it is unbounded below, but like the tachyon this is an irrelevant artifact of bosonic string theory.

It is interesting to look for static classical solutions, ignoring the fact that they are unstable to small variations. In order to have a static background solution, the graviton and dilaton field equations require the potential to vanish and the M_{ij} field equations require the potential to be stationary,

$$U(M) = \frac{\partial U(M)}{\partial M_{ij}} = 0 \ . \tag{8.3.23}$$

Using an $SU(2) \times SU(2)$ rotation to diagonalize M, the conditions (8.3.23) become

$$M_{11}M_{22}M_{33} = M_{11}M_{22} = M_{11}M_{33} = M_{22}M_{33} = 0 \ . \tag{8.3.24}$$

Thus, only one of the diagonal components can be nonvanishing. By an $SU(2) \times SU(2)$ rotation we can take this to be M_{33}. So we do not construct any physically new string backgrounds in this way.

A continuous family of static background solutions is known as a *flat direction*. Here there are nine massless scalars, which by an $SU(2) \times SU(2)$ rotation are reduced to the three diagonal fields, but there is only one flat direction (gauge-equivalent directions not being counted). Of course, we have only computed the cubic term in the potential, and in a general situation this could vanish and higher powers still be nonzero. To establish an exactly flat direction one needs some special argument. Here the special argument is that we can explicitly construct the free CFT for any value of R. Notice that if we treat $j^1 \bar{\jmath}^1$ or $j^2 \bar{\jmath}^2$ as the modulus the world-sheet action looks very complicated, with cosines and sines of X^{25}, but the world-sheet theory is solvable because it is equivalent to the free theory obtained by varying R.

T-duality

From the mass formula

$$m^2 = \frac{n^2}{R^2} + \frac{w^2 R^2}{\alpha'^2} + \frac{2}{\alpha'}(N + \tilde{N} - 2), \qquad (8.3.25)$$

we see that as $R \to \infty$ winding states become infinitely massive, while the compact momenta go over to a continuous spectrum. This is as would be expected with a noncompact dimension. Looking at the opposite limit $R \to 0$, however, one gets a surprise. The states with compact momentum become infinitely massive, but the spectrum of winding states now approaches a continuum — it does not cost much energy to wrap a string around a small circle. Thus as the radius goes to zero the spectrum again seems to approach that of a *noncompact* dimension. This is quite different from the behavior in field theory, where there is compact momentum n but no winding number w, and no states become light as $R \to 0$.

In fact, the $R \to 0$ and $R \to \infty$ limits are physically identical. The spectrum (8.3.25) is invariant under

$$R \to R' = \frac{\alpha'}{R}, \qquad n \leftrightarrow w. \qquad (8.3.26)$$

This equivalence extends to the interactions as well. Notice that reversing n and w is the same as

$$p_L^{25} \to p_L^{25}, \qquad p_R^{25} \to -p_R^{25}. \qquad (8.3.27)$$

Consider the theory at radius R. Recalling the separation $X^{25}(z, \bar{z}) = X_L^{25}(z) + X_R^{25}(\bar{z})$, define

$$X'^{25}(z, \bar{z}) = X_L^{25}(z) - X_R^{25}(\bar{z}). \qquad (8.3.28)$$

The field X'^{25} has the same OPEs and energy momentum tensor as X^{25}, the minus signs always entering in pairs. The only change in the CFT from using X'^{25} in place of X^{25} is that it includes the sign change (8.3.27), which changes the spectrum from that of the theory with radius R to that of the theory with radius R'. That is, these are the same theory, one written in terms of X^{25} and one in terms of X'^{25}. It should be noted that the relation between $X^{25}(z, \bar{z})$ and $X'^{25}(z, \bar{z})$ is nonlocal on the world-sheet, because the separate left- and right-moving fields are nonlocal operators.

This equivalence is known as *T-duality*. That the $R \to 0$ limit is physically the same as the $R \to \infty$ limit is completely unlike the behavior of point particles, and is one more sign that strings see geometry very differently at short distances. The space of inequivalent theories is the half-line $R \geq \alpha'^{1/2}$. We could take instead the range $0 \leq R \leq \alpha'^{1/2}$ but it is more natural to think in terms of the larger of the two equivalent radii:

momentum continua are more familiar than winding number continua, and in particular questions of locality are clearer in the larger-R picture. Thus, there is no radius smaller than the self-dual radius

$$R_{\text{self-dual}} = R_{SU(2) \times SU(2)} = \alpha'^{1/2} \,. \tag{8.3.29}$$

This indication of a minimum distance scale of the order of the string length will recur elsewhere in string perturbation theory, but we will see that nonperturbatively there is structure at shorter distance.

It is important in many applications that T-duality acts nontrivially on the string dilaton Φ. Consider the scattering amplitude for gravitons without winding or compact momentum. These states are invariant under T-duality, and so must be the amplitude as well. The latter can be read from the low energy action (8.1.9), and so the 25-dimensional coupling κ_{25} must be invariant under duality. Since the 26-dimensional coupling is related to this by $\kappa = (2\pi\rho)^{1/2}\kappa_{25}$, duality must act as

$$\rho' = \frac{\alpha'}{\rho} \,, \quad \kappa' = \frac{\alpha'^{1/2}}{\rho}\kappa \,. \tag{8.3.30}$$

Since $\kappa \propto e^{\Phi}$, this implies

$$e^{\Phi'} = \frac{\alpha'^{1/2}}{\rho}e^{\Phi} \,. \tag{8.3.31}$$

T-duality is a symmetry, relating different states (backgrounds) in a single theory, and in fact it is a gauge symmetry. We saw that the modulus δR is the 33-component of a $(\mathbf{3},\mathbf{3})$ field. A rotation by π around the 1-axis in *one* of the $SU(2)$s reverses the sign of this modulus, so decreasing R from the $SU(2) \times SU(2)$ radius is gauge-equivalent to increasing it. Thus the $\mathbf{Z}_2$ symmetry of duality is a small part of the $SU(2) \times SU(2)$ gauge symmetry. This implies further that duality is a symmetry not only of string perturbation theory but of the *exact* theory. If we have a massless gauge boson to leading approximation, even small explicit violations of the symmetry lead to inconsistency (spontaneous violation is not an issue, as T-duality is already spontaneously broken away from the self-dual radius).

This last argument illustrates an important idea, the complementary use of stringy and spacetime arguments. The existence of the enhanced gauge symmetry is a purely stringy phenomenon. We do not have a nonperturbative understanding of string theory, but we do know a great deal about low energy field theory, even nonperturbatively, and we can use this.

8.4 Compactification of several dimensions

It is useful to generalize the analysis to k periodic dimensions,

$$X^m \cong X^m + 2\pi R , \quad 26 - k \leq m \leq 25 . \tag{8.4.1}$$

Let $d = 26 - k$ be the number of noncompact dimensions. Spacetime is now $M^d \times T^k$. The coordinate periodicity (8.4.1) is being held fixed, but the actual geometry of the k-torus depends on internal metric G_{mn}. With more than one compact dimension, the antisymmetric tensor also has scalar components B_{mn}. The total number of scalars from both sources is k^2. There are also Kaluza–Klein gauge bosons A^m_μ and antisymmetric tensor gauge bosons $B_{m\mu}$. The low energy action, obtained as in section 8.1, is

$$
\begin{aligned}
S = \frac{(2\pi R)^k}{2\kappa_0^2} \int d^d x \, (-G_d)^{1/2} e^{-2\Phi_d} \Big[& R_d + 4\partial_\mu \Phi_d \partial^\mu \Phi_d \\
& - \frac{1}{4} G^{mn} G^{pq} (\partial_\mu G_{mp} \partial^\mu G_{nq} + \partial_\mu B_{mp} \partial^\mu B_{nq}) \\
& - \frac{1}{4} G_{mn} F^m_{\mu\nu} F^{n\mu\nu} - \frac{1}{4} G^{mn} H_{m\mu\nu} H^{\mu\nu}_n - \frac{1}{12} H_{\mu\nu\lambda} H^{\mu\nu\lambda} \Big] ,
\end{aligned} \tag{8.4.2}
$$

where $\Phi_d = \Phi - \frac{1}{4} \ln \det G_{mn}$.

The string spectrum

The main new issue is the antisymmetric tensor background B_{mn}. The contribution of this term to the world-sheet Lagrangian density is proportional to

$$B_{mn} \partial_a (g^{1/2} \epsilon^{ab} X^m \partial_b X^n) , \tag{8.4.3}$$

which is a total derivative for constant B_{mn}. It thus has no effect locally and the world-sheet theory is still a CFT, a good string background. It does, however, change the spectrum. We will see this in two ways, canonically and via the path integral.

In the canonical approach, focus on the zero-mode contribution to the world-sheet action. Inserting

$$X^m(\sigma^1, \sigma^2) = x^m(\sigma^2) + w^m R \sigma^1 \tag{8.4.4}$$

into the world-sheet action gives

$$L = \frac{1}{2\alpha'} G_{mn} (\dot{x}^m \dot{x}^n + w^m w^n R^2) - \frac{i}{\alpha'} B_{mn} \dot{x}^m w^n R . \tag{8.4.5}$$

A dot indicates a derivative with respect to world-sheet time σ^2. The canonical momenta are

$$p_m = -\frac{\partial L}{\partial v^m} = \frac{1}{\alpha'} (G_{mn} v^n + B_{mn} w^n R) , \tag{8.4.6}$$

where $v^m = i\dot{x}^m$. Some unfamiliar signs and is appear because we are using Euclidean world-sheet time. Continuing to Minkowski time, v^m becomes the velocity $\partial_0 x^m$. The periodicity of the wavefunction implies the quantization of the canonical momenta, $p_m = n_m/R$, and so

$$v_m = \alpha' \frac{n_m}{R} - B_{mn} w^n R \ . \tag{8.4.7}$$

The zero-mode contribution to the world-sheet Hamiltonian is

$$\frac{1}{2\alpha'} G_{mn}(v^m v^n + w^m w^n R^2) \ , \tag{8.4.8}$$

and the closed string mass is

$$m^2 = \frac{1}{2\alpha'^2} G_{mn}(v_L^m v_L^n + v_R^m v_R^n) + \frac{2}{\alpha'}(N + \tilde{N} - 2) \ , \tag{8.4.9a}$$

$$v_{L,R}^m = v^m \pm w^m R \ . \tag{8.4.9b}$$

The B_{mn} background thus shifts the masses of the winding states through the dependence of v^m on B_{mn}. The $L_0 - \tilde{L}_0$ constraint is

$$\begin{aligned}
0 &= G_{mn}(v_L^m v_L^n - v_R^m v_R^n) + 4\alpha'(N - \tilde{N}) \\
&= 4\alpha'(n_m w^m + N - \tilde{N}) \ . \tag{8.4.10}
\end{aligned}$$

Alternatively, consider the torus path integral for the partition function. The antisymmetric tensor term is locally a total derivative and so depends only on the topology of the configuration. Consider the configuration

$$X^m = (w_1^m \sigma^1 + w_2^m \sigma^2) R \ . \tag{8.4.11}$$

The spatial direction on the torus winds w_1^m times around each periodic spacetime dimension x^m, and the timelike dimension w_2^m times. The B_{mn} world-sheet action is

$$2\pi i b_{mn} w_1^m w_2^n \ , \tag{8.4.12}$$

where $b_{mn} = B_{mn} R^2/\alpha'$. One can check that the sum over path integral sectors with phase (8.4.12) is related by Poisson resummation to the partition function for the shifted spectrum (8.4.9).

In the description above the metric G_{mn} appears in the action for the world-sheet fields. For vertex operator calculations it is more convenient to keep the usual action. Write the metric in terms of a tetrad e_m^r,

$$G_{mn} = e_m^r e_n^r \ , \tag{8.4.13}$$

with tangent space indices $r, s, \ldots$. The coordinates $X^r = e_m^r X^m$ then have the standard OPEs. The vertex operator momenta are

$$k_{rL} = e_r^m \frac{v_{mL}}{\alpha'} \ , \qquad k_{rR} = e_r^m \frac{v_{mR}}{\alpha'} \ , \tag{8.4.14}$$

with the inverse tetrad $e_r{}^m$. The mass-shell conditions become

$$m^2 = \frac{1}{2}(k_{rL}k_{rL} + k_{rR}k_{rR}) + \frac{2}{\alpha'}(N + \tilde{N} - 2) , \qquad (8.4.15a)$$

$$0 = \alpha'(k_{rL}k_{rL} - k_{rR}k_{rR}) + 4(N - \tilde{N}) . \qquad (8.4.15b)$$

We use the coordinates X^r implicitly in any discussion of vertex operators, such as the one that now follows.

Narain compactification

There is an elegant description of the general toroidal compactification. Consider the winding state vertex operators $e^{ik_L \cdot X_L + ik_R \cdot X_R}$. For any given compactification, the spectrum of momenta (k_{rL}, k_{rR}) forms a lattice in a $2k$-dimensional momentum space $\mathbf{R}^{2k}$. That is, the momentum spectrum consists of all integer linear combinations of $2k$ linearly independent basis vectors. Let us work with the dimensionless momenta $l_{L,R} = k_{L,R}(\alpha'/2)^{1/2}$ and call the corresponding lattice Γ. The OPE of two vertex operators is

$$: e^{ik_L \cdot X_L(z) + ik_R \cdot X_R(\bar{z})} : \; : e^{ik'_L \cdot X_L(0) + ik'_R \cdot X_R(0)} :$$
$$\sim z^{l_L \cdot l'_L} \bar{z}^{l_R \cdot l'_R} \; : e^{i(k_L + k'_L) \cdot X_L(0) + i(k_R + k'_R) \cdot X_R(0)} : . \qquad (8.4.16)$$

As one vertex operator circles the other, the product picks up the phase $\exp[2\pi i(l_L \cdot l'_L - l_R \cdot l'_R)]$. Single-valuedness of the operator product thus requires

$$l_L \cdot l'_L - l_R \cdot l'_R \equiv l \circ l' \in \mathbf{Z} \qquad (8.4.17)$$

for all l and l' in Γ. We have defined the product $\circ$, which has signature (k, k) in $\mathbf{R}^{2k}$. The dual lattice Γ^* is defined to be the set of points in $\mathbf{R}^{2k}$ that have integer $\circ$ product with all points in Γ. The single-valuedness condition (8.4.17) is then the statement that

$$\Gamma \subset \Gamma^* . \qquad (8.4.18)$$

Modular invariance also constrains Γ. In section 7.2 it was shown that invariance under $\tau \to \tau + 1$ requires that $L_0 - \tilde{L}_0$ be an integer for all states in the string CFT. Then

$$l \circ l \in 2\mathbf{Z} \quad \text{for all } l \in \Gamma . \qquad (8.4.19)$$

The condition (8.4.19) actually implies condition (8.4.17), given closure of the OPE

$$2l \circ l' = (l + l') \circ (l + l') - l \circ l - l' \circ l' \in 2\mathbf{Z} . \qquad (8.4.20)$$

Modular invariance under $\tau \to -1/\tau$ requires a more detailed calculation. The partition function for the compact dimensions is

$$Z_\Gamma(\tau) = |\eta(\tau)|^{-2k} \sum_{l \in \Gamma} \exp(\pi i \tau l_L^2 - \pi i \bar{\tau} l_R^2) . \qquad (8.4.21)$$

To determine the transformation property of Z_Γ use again Poisson resummation. Write

$$\sum_{l' \in \Gamma} \delta(l - l') = V_\Gamma^{-1} \sum_{l'' \in \Gamma^*} \exp(2\pi i l'' \circ l) .$$ (8.4.22)

Here, V_Γ is the volume of a unit cell of the lattice Γ. Eq. (8.4.22) is familiar from Fourier series: the phase averages to zero except when $l \in \Gamma$, and the normalization is fixed by integrating over a unit cell. Using the representation (8.4.22), the sum (8.4.21) can be written

$$Z_\Gamma(\tau) = V_\Gamma^{-1} |\eta(\tau)|^{-2k} \sum_{l'' \in \Gamma^*} \int d^{2k}l \, \exp(2\pi i l'' \circ l + \pi i \tau l_L^2 - \pi i \bar{\tau} l_R^2)$$

$$= V_\Gamma^{-1} (\tau \bar{\tau})^{-k/2} |\eta(\tau)|^{-2k} \sum_{l'' \in \Gamma^*} \exp(-\pi i l_L''^2/\tau + \pi i l_R''^2/\bar{\tau})$$

$$= V_\Gamma^{-1} Z_{\Gamma^*}(-1/\tau) ,$$ (8.4.23)

where in the last line we have used the modular transformation (7.2.44) of the eta function. Then the condition

$$\Gamma = \Gamma^*$$ (8.4.24)

is sufficient for modular invariance, because $V_\Gamma = V_{\Gamma^*}^{-1}$ and so is equal to unity if (8.4.24) holds. A little thought shows that this is also necessary if modular invariance is to hold for all τ.

The consistency conditions can be summarized as the requirement that Γ be an even self-dual lattice of signature (k, k), *even* referring to the property (8.4.19) and *self-dual* to the property (8.4.24).

All such lattices have been classified. Note that the consistency conditions (8.4.19) and (8.4.24) depend on the momenta l only through the $\circ$ product, which is invariant under Lorentz boosts of the $2k$-dimensional space, $O(k, k, \mathbf{R})$ transformations. If Γ is an even self-dual lattice then so is

$$\Gamma' = \Lambda \Gamma ,$$ (8.4.25)

which is obtained by rotating all the points of the lattice by a common $O(k, k, \mathbf{R})$ transformation Λ. Now, it is important that $O(k, k, \mathbf{R})$ is *not* a symmetry of the theory. The mass-shell condition and operator products involve the separate dot products $l_L \cdot l_L'$ and $l_R \cdot l_R'$ and so are invariant only under the $O(k, \mathbf{R}) \times O(k, \mathbf{R})$ that acts on l_L and l_R separately. Thus, most $O(k, k, \mathbf{R})$ transformations produce inequivalent theories.

Consider for example $k = 1$, where

$$l_{L,R} = \frac{n}{r} \pm \frac{mr}{2}$$ (8.4.26)

in terms of the dimensionless radius $r = R(2/\alpha')^{1/2}$. This is indeed an even self-dual lattice. The boost

$$l'_L = l_L \cosh \lambda + l_R \sinh \lambda \,, \quad l'_R = l_L \sinh \lambda + l_R \cosh \lambda \qquad (8.4.27)$$

changes this to the lattice for $r' = re^{-\lambda}$.

All even self-dual lattices of given Lorentzian signature can be obtained from any single lattice by $O(k, k, \mathbf{R})$ transformations. Start with a given solution, for example all compact dimensions orthogonal and at the $SU(2)$ radius with $B_{mn} = 0$. The corresponding momentum lattice Γ_0 is known as kP_2 and gives the gauge group $SU(2)^{2k}$. As discussed above, two $O(k, k, \mathbf{R})$ transformations Λ and $\Lambda'\Lambda$ give equivalent string theories if $\Lambda' \in O(k, \mathbf{R}) \times O(k, \mathbf{R})$, so the space of inequivalent theories obtained in this way is

$$\frac{O(k, k, \mathbf{R})}{O(k, \mathbf{R}) \times O(k, \mathbf{R})} \,, \qquad (8.4.28)$$

up to some discrete identifications to be discussed shortly. This is equivalent to the earlier description in terms of the backgrounds of G_{mn} and B_{mn}. In particular, the number of parameters in the coset (8.4.28) is

$$\frac{2k(2k-1)}{2} - k(k-1) = k^2 \,, \qquad (8.4.29)$$

the same as the number of components in G_{mn} and B_{mn}.

The T-duality symmetry is much enlarged from the case of one compactified dimension. In the abstract description, there is some discrete subgroup of $O(k, k, \mathbf{R})$ that takes the initial lattice Γ_0 into itself (though it permutes the individual points of the lattice). This subgroup is conventionally but somewhat inaccurately denoted $O(k, k, \mathbf{Z})$. If Λ'' is in this group, then clearly $\Lambda\Gamma_0$ and $\Lambda\Lambda''\Gamma_0$ are the same lattice. In all we have the equivalence

$$\Lambda\Gamma_0 \cong \Lambda'\Lambda\Lambda''\Gamma_0 \,, \qquad (8.4.30a)$$
$$\Lambda' \in O(k, \mathbf{R}) \times O(k, \mathbf{R}) \,, \quad \Lambda'' \in O(k, k, \mathbf{Z}) \,. \qquad (8.4.30b)$$

The space of inequivalent lattices, and inequivalent backgrounds, is then

$$\frac{O(k, k, \mathbf{R})}{O(k, \mathbf{R}) \times O(k, \mathbf{R}) \times O(k, k, \mathbf{Z})} \,. \qquad (8.4.31)$$

One should remember that the continuous denominator group acts by left-multiplication and the discrete group by right-multiplication, as in eq. (8.4.30a).

In terms of the backgrounds, the T-duality group $O(k, k, \mathbf{Z})$ contains transformations of several types. One is $R \to \alpha'/R$ dualities on the individual axes. Another is large spacetime coordinate transformations respecting the periodicity,

$$x'^m = L^m{}_n x^n \qquad (8.4.32)$$

with $L^m{}_n$ integers and $\det L = 1$ for invertibility. This is the group $SL(k, \mathbf{Z})$. Finally there are shifts of the antisymmetric tensor background,

$$b_{mn} \to b_{mn} + N_{mn} \qquad (8.4.33)$$

with N_{mn} integers. From either the canonical result (8.4.7) or the path integral phase (8.4.12) it follows that this leaves the spectrum invariant. Together these generate the whole T-duality group.

Identifying first by the left-multiplication produces the space (8.4.28), and one can think of the discrete right-multiplication as acting on this covering space. For generic Λ, there are no $\Lambda_{1,2}$ in the respective denominator groups such that $\Lambda_1 \Lambda \Lambda_2 = \Lambda$, so that there are no T-dualities that leave this point on the covering space fixed. At special Λ there are solutions to $\Lambda_1 \Lambda \Lambda_2 = \Lambda$, and these will be fixed points of the T-duality element Λ_2. In the one-dimensional case we had such a fixed point, which was also a point of enhanced gauge symmetry. In the higher-dimensional case, besides products of $SU(2)$s, larger gauge groups arise at various points. We will defer further discussion of this to chapter 11.

When there is a moduli space parameterized by scalar fields ϕ^i, the kinetic term

$$-\frac{1}{2} g_{ij}(\phi) \partial_\mu \phi^i \partial^\mu \phi^j \qquad (8.4.34)$$

defines a natural *metric on moduli space*. To be precise, an unambiguous definition requires that one first perform a spacetime Weyl transformation as in eq. (3.7.25) so as to make the coefficient of the gravitational action moduli-independent and eliminate mixing between the moduli and the spacetime metric. In the low energy action (8.4.2), the resulting scalar kinetic terms are proportional to

$$\frac{16}{d-2} \partial_\mu \Phi \partial^\mu \Phi + G^{mn} G^{pq} (\partial_\mu G_{mp} \partial^\mu G_{nq} + \partial_\mu B_{mp} \partial^\mu B_{nq}) . \qquad (8.4.35)$$

The coset space (8.4.31) has a unique $O(k, k, \mathbf{R})$-invariant metric, which is just the second term in eq. (8.4.35). This $O(k, k, \mathbf{R})$ is not a symmetry of the full theory; only its discrete T-duality subgroup $O(k, k, \mathbf{Z})$ is. The difference between the two is not visible at low energy — it comes from the quantization of the string zero modes and so affects the massive spectrum. Thus $O(k, k, \mathbf{R})$ is an accidental symmetry of the low energy theory. In this example the moduli space is a product of the dilaton moduli space and the moduli space of compactifications.

An example

The case of two compact dimensions is a good example. The four moduli $G_{24,24}$, $G_{24,25}$, $G_{25,25}$, and $B_{24,25}$ are usefully joined into two complex fields $\tau = \tau_1 + i\tau_2$ and $\rho = \rho_1 + i\rho_2$. Define

$$\rho = \frac{R^2}{\alpha'}\left(B_{24,25} + i \det{}^{1/2}G_{mn}\right)$$

$$= b_{24,25} + i\frac{V}{4\pi^2\alpha'}, \qquad (8.4.36)$$

where V is the volume of the compact two-torus. Parameterize the compact metric as

$$ds^2 = \frac{\alpha'\rho_2}{R^2\tau_2}\left|dX^{24} + \tau dX^{25}\right|^2. \qquad (8.4.37)$$

The coordinate transformations (8.4.32) act on the spacetime two-torus just as the modular group acts on the world-sheet two-torus, generating a $PSL(2,\mathbf{Z})$ on τ and leaving ρ invariant. The antisymmetric tensor shift (8.4.33) is $\rho \to \rho + N_{24,25}$ with τ invariant. A simultaneous T-duality on $X^{24,25}$ acts as $\rho \to -1/\rho$ with τ invariant. Together the last two transformations generate a second $PSL(2,\mathbf{Z})$, acting on ρ. In addition, duality on X^{24} alone takes $(\tau, \rho) \to (\rho, \tau)$ and spacetime parity $X^{24} \to -X^{24}$ takes $(\tau, \rho) \to (-\bar\tau, -\bar\rho)$. The full T-duality group is then

$$PSL(2,\mathbf{Z}) \times PSL(2,\mathbf{Z}) \rtimes \mathbf{Z}_2^2. \qquad (8.4.38)$$

A further equivalence comes from world-sheet parity, which takes $(\tau, \rho) \to (\tau, -\bar\rho)$. It reverses the sign of the $\circ$ product.

The moduli space is therefore two copies of the moduli space of the torus, together with some discrete identifications. The kinetic terms are proportional to

$$\frac{\partial_\mu\tau\partial^\mu\bar\tau}{\tau_2^2} + \frac{\partial_\mu\rho\partial^\mu\bar\rho}{\rho_2^2}. \qquad (8.4.39)$$

The modular-invariant combination $d^2\tau/\tau_2^2$ was encountered in the discussion of the genus-zero amplitude. The continuous group $PSL(2,\mathbf{R})$ carries any τ in the upper half-plane into any other, and a $U(1)$ subgroup leaves each τ invariant. Thus the upper half-plane can be regarded as $PSL(2,\mathbf{R})/U(1)$, and the full moduli space is

$$\frac{PSL(2,\mathbf{R}) \times PSL(2,\mathbf{R})}{U(1) \times U(1) \times PSL(2,\mathbf{Z}) \times PSL(2,\mathbf{Z}) \times \mathbf{Z}_2^2}. \qquad (8.4.40)$$

This is the same as the earlier (8.4.31). In particular, $PSL(2,\mathbf{R}) \times PSL(2,\mathbf{R})$ is locally the same as $O(2,2,\mathbf{R})$, in parallel to the fact that $SU(2) \times SU(2)$ is locally the same as $O(4,\mathbf{R})$.

8.5   Orbifolds

Rather than the periodic identification of X^{25}, consider instead identifying points under the reflection

$$X^{25} \cong -X^{25} . \tag{8.5.1}$$

Now a fundamental region is the half-line $X^{25} \geq 0$, and the hyperplane $X^{25} = 0$ becomes a boundary. This hyperplane is distinguished by the fact that its points are fixed under the reflection (8.5.1). More generally we could identify spacetime under the simultaneous reflection of k coordinates

$$X^m \to -X^m , \quad 26 - k \leq m \leq 25 . \tag{8.5.2}$$

Again there is a space of fixed points, at $X^{26-k} = \ldots = X^{25} = 0$. For $k \geq 2$ this space is a singularity. In particular, for $k = 2$ it is a conic singularity with deficit angle π.

One can form a compact space by combining the reflection (8.5.1) with the periodic identification $X^{25} \cong X^{25} + 2\pi R$; we will call these r and t respectively. One can think of first identifying by t to form the circle S^1, and then identifying points on the circle under r. The resulting compact space is a line segment, with $0 \leq X^{25} \leq \pi R$ as a fundamental region. There are fixed points at both ends of the segment, the point πR being fixed by tr. In all, one is identifying the real line under the group

$$t^m : \quad X^{25} \cong X^{25} + 2\pi R m , \tag{8.5.3a}$$

$$t^m r : \quad X^{25} \cong 2\pi R m - X^{25} , \tag{8.5.3b}$$

for any integer m.

Similarly one could identify the k-torus under the reflection (8.5.2). In that case there are 2^k distinct fixed points, with each of the X^m being either 0 or πR. The singular spaces obtained in this way are known as *orbifolds*. The space obtained by the identification (8.5.2) in noncompact space is $\mathbf{R}^k/\mathbf{Z}_2$, while the compact space obtained by the same identification on the k-torus is $T^k/\mathbf{Z}_2$.

It is not obvious that string theory on such singular spaces should make sense, but we will see that it does. It is not so different from toroidal compactification, but it breaks more symmetry (and, later on, supersymmetry) and so will be of interest in constructing string models.

The identification (8.5.1) or (8.5.2) has two effects. The first is that the string wavefunction must be invariant under the reflection, equal at identified points. Second, there is a new sector in the closed string spectrum, in which

$$X^{25}(\sigma^1 + 2\pi) = -X^{25}(\sigma^1) , \tag{8.5.4}$$

since these are the same point in the identified spacetime. Strings in this sector are known as *twisted states*. This is directly parallel to periodic identification. There, invariance of the wavefunction made the momentum discrete, while strings that closed only up to the periodic identification (winding strings) appeared. One can again make a pictorial argument, analogous to figure 8.1, to show that twisted strings can be produced from untwisted ones. We will also see below that modular invariance requires them in the spectrum.

We will focus on the compact one-dimensional orbifold $S^1/\mathbf{Z}_2$. Examining first the untwisted sector, the spectrum of the periodic theory is reduced by the projection onto invariant states. The effect of r on a general state is

$$|N, \tilde{N}; k^\mu, n, w\rangle \rightarrow (-1)^{\sum_{m=1}^\infty (N_m^{25} + \tilde{N}_m^{25})} |N, \tilde{N}; k^\mu, -n, -w\rangle , \qquad (8.5.5)$$

in particular reversing the compact winding and momentum. We must form linear combinations that are invariant under this. The states that are massless at generic R have $n = w = 0$, so the projection simply requires that the number of 25-excitations be even. The spacetime graviton, antisymmetric tensor, dilaton, and tachyon survive this projection. The modulus $\alpha_{-1}^{25} \tilde{\alpha}_{-1}^{25} |0; k^\mu, 0, 0\rangle$ also survives — this could have been anticipated, since the radius R can take any value. However, the Kaluza–Klein gauge bosons are no longer in the spectrum. The fate of the extra massless states at the self-dual point is quite interesting, and we will discuss it in detail later.

In the sector twisted by r, X^{25} is antiperiodic and so has a half-integral mode expansion

$$X^{25}(z, \bar{z}) = i \left(\frac{\alpha'}{2}\right)^{1/2} \sum_{m=-\infty}^\infty \frac{1}{m + 1/2} \left(\frac{\alpha_{m+1/2}^{25}}{z^{m+1/2}} + \frac{\tilde{\alpha}_{m+1/2}^{25}}{\bar{z}^{m+1/2}}\right) . \qquad (8.5.6)$$

The antiperiodicity forbids any center-of-mass coordinate or momentum, so the string cannot move away from the $X^{25} = 0$ fixed point: to the extent that the oscillations are small, one has $X^{25}(z, \bar{z}) \approx 0$. There are also twisted strings localized at the other fixed point,

$$X^{25}(\sigma^1 + 2\pi) = 2\pi R - X^{25}(\sigma^1) . \qquad (8.5.7)$$

The fields change by tr as one circles the loop. The mode expansion is as above with an additional constant term πR. All other fixed points $n\pi R$ are images of one of these under t.

According to the zero-point mnemonic from section 2.9, the zero-point energy of X^{25} is shifted from $-\frac{1}{24}$ for a periodic boson to $\frac{1}{48}$ for an antiperiodic boson, for a net $+\frac{1}{16}$. The mass-shell conditions in the twisted sectors are then

$$m^2 = \frac{4}{\alpha'}\left(N - \frac{15}{16}\right), \quad N = \tilde{N}. \tag{8.5.8}$$

Also, the 25-oscillators make half-integral contributions to the level N. The r-projection again requires the total number of 25-excitations to be even, which can also be deduced from the level-matching condition $N = \tilde{N}$. In the twisted sectors the ground states are $|T_{1,2}\rangle$, the subscript referring to the two fixed points. They are tachyonic, as are the first excited states $\alpha^{25}_{-1/2}\tilde{\alpha}^{25}_{-1/2}|T_{1,2}\rangle$. There are no massless twisted states. The extension to the more general case (8.5.2) is straightforward. The quantization is perfectly consistent in spite of the singularities in spacetime.

The one complication with orbifolds is that the twisted state vertex operators are not as simple as those for winding states; there is no such explicit formula as the exponential of a free field. For two twisted strings and any number of untwisted ones, the tree-level amplitude can be written in operator form $\langle T | \mathcal{V}_U \mathcal{V}_U \ldots | T \rangle$, which uses only untwisted vertex operators. For four or more twisted vertex operators (the number must be even for the path integral to make sense) two methods have been used. One is the stress tensor method, which we used in section 7.2 to get the partition function on the torus. The other is to go to the covering space on which X^{25} is single-valued (this is a genus g surface for $2g + 2$ twisted vertex operators on the sphere) and evaluate the path integral.

There is a simple result for the tree-level couplings of untwisted states. On the sphere, with all external states untwisted, the twists and projections do not enter into the calculation at all and the amplitudes are the same as in the untwisted theory. For example, the low energy effective action for the massless untwisted fields is the same as we found for toroidal compactification, with the omission of the vector fields. This property is often useful in analyzing the physics of orbifold theories; it is called the *inheritance principle*.

Consider now the partition function of the X^{25} CFT. In the untwisted sector, one is to project the spectrum onto $r = +1$ states,

$$(q\bar{q})^{-1/24}\mathrm{Tr}_U\left(\frac{1+r}{2}q^{L_0}\bar{q}^{\tilde{L}_0}\right). \tag{8.5.9}$$

The first term in the projection operator just gives half the partition function (8.2.9) for toroidal compactification. In the term with r in the trace, the diagonal elements of r must have $n = w = 0$, as one sees from eq. (8.5.5). The $(-1)^{N^{25}+\tilde{N}^{25}}$ changes the oscillator sum to $1 - q + q^2 - q^3 + \ldots = (1+q)^{-1}$. The partition function for the untwisted states is then

$$\frac{1}{2}Z_{\mathrm{tor}}(R, \tau) + \frac{1}{2}(q\bar{q})^{-1/24}\prod_{m=1}^{\infty}|1 + q^m|^{-2}. \tag{8.5.10}$$

The contribution of the twisted sectors is from the sum over half-integer moded oscillations,

$$(q\bar{q})^{1/48}\text{Tr}_T\left(\frac{1+r}{2}q^{L_0}\bar{q}^{\bar{L}_0}\right)$$

$$= (q\bar{q})^{1/48}\left[\prod_{m=1}^{\infty}\left|1-q^{m-1/2}\right|^{-2} + \prod_{m=1}^{\infty}\left|1+q^{m-1/2}\right|^{-2}\right], \quad (8.5.11)$$

the factor of 2 from the number of twisted sectors offsetting the $\frac{1}{2}$ in the projection operator.

In all, the partition function can be written as

$$Z_{\text{orb}}(R,\tau) = \frac{1}{2}Z_{\text{tor}}(R,\tau) + \left|\frac{\eta(\tau)}{\vartheta_{10}(0,\tau)}\right| + \left|\frac{\eta(\tau)}{\vartheta_{01}(0,\tau)}\right| + \left|\frac{\eta(\tau)}{\vartheta_{00}(0,\tau)}\right|. \quad (8.5.12)$$

The first term is already known to be modular-invariant, and it follows from the modular transformation properties given in section 7.2 that the sum of the final three is invariant as well. In terms of the path integral on the torus, $Z_{\text{tor}}(R,\tau)$ is from fields that are periodic up to translations. The term with ϑ_{ab} is from the path integral with

$$X^{25}(\sigma^1 + 2\pi, \sigma^2) = (-1)^{a+1}X^{25}(\sigma^1, \sigma^2), \quad (8.5.13a)$$

$$X^{25}(\sigma^1 + 2\pi\tau_1, \sigma^2 + 2\pi\tau_2) = (-1)^{b+1}X^{25}(\sigma^1, \sigma^2). \quad (8.5.13b)$$

For example, the ϑ_{10} term is from the untwisted sector with r in the trace, that is, from fields that are periodic in the spatial direction and antiperiodic in the time direction. The ϑ_{01} term, from the twisted sector with 1 in the trace, has the opposite boundary conditions. These are interchanged by $\tau \to -1/\tau$. Thus, the twisted sectors are necessary for modular invariance.

Twisting

Toroidal and orbifold compactifications are two examples of a useful construction, known as *twisting*, for constructing new string theories from old. In a given CFT whose symmetries include some discrete group H, we can form a new CFT in two steps. First add *twisted sectors*, in which the closed string world-sheet fields are periodic only up to some transformation $h \in H$,

$$\phi(\sigma^1 + 2\pi) = h \cdot \phi(\sigma^1), \quad (8.5.14)$$

where ϕ denotes a generic world-sheet field. Then, restrict the spectrum to H-invariant states. Although the fields (8.5.14) are no longer periodic, the vertex operators are periodic because of the projection. So the OPE is still well defined, and it closes because the product of H-invariant operators is H-invariant.

In the partition function, we sum over twists h_1 in the spatial direction and include a projection operator onto invariant states,

$$P_H = \frac{1}{\text{order}(H)} \sum_{h_2 \in H} \hat{h}_2 \ . \tag{8.5.15}$$

The operator $\hat{h}_2$ in the trace makes the fields in the time direction aperiodic by h_2, so in the end we are summing over twists in both the spatial and time directions on the torus,

$$Z = \frac{1}{\text{order}(H)} \sum_{h_1, h_2 \in H} Z_{h_1, h_2} \ . \tag{8.5.16}$$

The sum over boundary conditions is modular-invariant after a shift of summation variables. The modular transformation $\tau \to -1/\tau$ takes (h_1, h_2) to (h_2, h_1^{-1}), and $\tau \to \tau + 1$ takes (h_1, h_2) to $(h_1, h_1 h_2)$. Thus, if we project onto H-invariant states by summing over h_2, we must also sum over the twisted sectors. In more general situations, in particular right–left asymmetric theories, there will be an issue of phases possibly spoiling the naive modular invariance of the sum over boundary conditions, but we will deal with that in chapters 10, 11, and 16.

In the case of toroidal compactification, the original CFT is the non-compact theory and H consists of $t^m = \exp(2\pi i R m p)$ for all integer m. This happens to have infinite order, but we can regulate the sum by putting the system in a large box. For the orbifold, we can think of starting from the toroidal theory and twisting by r, or from the noncompact theory and twisting by r and t.

When H is non-Abelian, the path integral boundary conditions are inconsistent for h_1 and h_2 noncommuting and the path integral vanishes. This also follows from the canonical expression (8.5.15). If ϕ has spatial twist h_1, then

$$\phi'(\sigma^1) = h_2 \cdot \phi(\sigma^1) \tag{8.5.17}$$

has a different spatial twist

$$\phi'(\sigma^1 + 2\pi) = h_1' \cdot \phi'(\sigma^1) \ , \tag{8.5.18}$$

where

$$h_1' = h_2 h_1 h_2^{-1} \tag{8.5.19}$$

and so the diagonal matrix elements of $\hat{h}_2$ are zero. We also see that the sectors twisted by h_1 and h_1' are not independent — they are related by the projection onto h_2-invariant states. Thus the independent twisted sectors are in one-to-one correspondence with the *conjugacy classes* of H.

Twisting can also be thought of as *gauging* the discrete group H. It may not be obvious that this makes sense, as the discreteness forces the gauge parameter to be constant. In particular, there can be no associated gauge field. The point is that on a world-sheet with nontrivial closed paths, only gauge-invariant quantities need to be periodic, so the fields can be aperiodic by an H transformation. Gauging a discrete symmetry adds in these path integral sectors and correspondingly projects onto invariant states. This is just the twisting construction. It is also possible to gauge a continuous world-sheet symmetry to produce a new CFT. This is more involved and will be discussed in chapter 15.

The inheritance principle holds for general twisted theories: the path integral with untwisted vertex operators on the sphere is unaffected by the twist.

c = 1 CFTs

We have found two families of compact $c = 1$ CFTs. These are toroidal compactification, restricting the radius to $R \geq \alpha'^{1/2}$ so as to count only distinct theories, and orbifold compactification with the same range of R. These two families are connected. Starting from the $SU(2) \times SU(2)$ radius $R = \alpha'^{1/2}$, twist by

$$r' : \quad X^{25} \rightarrow X^{25} + \pi\alpha'^{1/2} . \tag{8.5.20}$$

This produces the toroidal theory at $R = \alpha'^{1/2}/2$, which by T-duality is the same as at $R = 2\alpha'^{1/2}$. In terms of the $SU(2)$ currents (8.3.12), r' flips the signs of $j^{1,2}$ and $\bar{j}^{1,2}$ and so is a rotation by π around the 3-axis in each $SU(2)$. The orbifold r flips the signs of $j^{2,3}$ and $\bar{j}^{2,3}$ and is a rotation by π around the 1-axis in each $SU(2)$. But of course these rotations are equivalent under the $SU(2) \times SU(2)$, so the resulting theories are the same, the $R = 2\alpha'^{1/2}$ torus and the $R = \alpha'^{1/2}$ orbifold. The corresponding partition functions are equal,

$$Z_{\text{orb}}(\alpha'^{1/2}, \tau) = Z_{\text{tor}}(2\alpha'^{1/2}, \tau) , \tag{8.5.21}$$

as can be verified by theta function identities.

This equivalence holds only at these radii. For example, at generic R the toroidal theory has a $U(1) \times U(1)$ gauge symmetry while the orbifold has no gauge symmetry. Thus, the two moduli spaces are connected as shown in figure 8.2. This structure, the moduli space of vacua having distinct branches meeting at special points, will be a very common and important feature in *supersymmetric* theories. On each branch there is in general a distinct low energy effective field theory, here differing in the low energy gauge symmetry. In the neighborhood of the special point there are more light fields, different subsets of which become massive as we move away along the various branches.

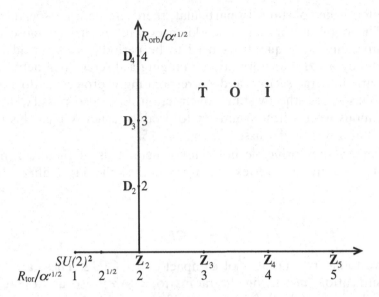

Fig. 8.2. Moduli space of $c = 1$ CFTs. The toroidal branch is horizontal and the orbifold branch is vertical. The special points obtained by twisting by a discrete subgroup of $SU(2)$ are indicated, including the three isolated points obtained from the tetrahedral, octahedral, and icosahedral groups. Other special radii, such as the torus at $R = (2\alpha')^{1/2}$, will arise later.

It is instructive to analyze the low energy physics near the meeting of the toroidal and orbifold branches. Describe this point in terms of twisting the $SU(2)$ theory by r', eq. (8.5.20). The massless scalars surviving the twist are $j^3\bar{j}^3$ and $j^i\bar{j}^j$ for $i, j \in \{1, 2\}$. Since these are all untwisted, the potential is the same as before twisting, namely $\det M$. Again the scalar field M_{ij} can be diagonalized, by a $U(1) \times U(1)$ rotation, and the potential is flat when exactly one diagonal element is nonzero. Now, however, the modulus M_{33} cannot be rotated into M_{11} or M_{22} so these directions are physically distinct. The modulus M_{33} leaves the $U(1) \times U(1)$ unbroken and corresponds to moving along the toroidal branch. The modulus M_{11} or M_{22} breaks the gauge symmetry and corresponds to moving along the orbifold branch. Since this modulus is charged, a reversal of its sign is a gauge transformation and so the line of physically distinct orbifolds ends at the point where the branches meet. We cannot simultaneously vary both moduli, because the linear combination is not a flat direction of the potential. That string theory on an orbifold can be equivalent to string theory on a circle is quite remarkable, one more indication that the string does not see spacetime geometry in the same way that we do.

Moving along the toroidal branch, there are extra massless states at all multiples of the $SU(2)$ radius, $R = k\alpha'^{1/2}$. These are the states with

$(n, w) = (\pm 2k, 0)$. We can think of these points as obtained, in the T-dual description $R = \alpha'^{1/2}/k$, by twisting the $SU(2) \times SU(2)$ theory by the $\mathbf{Z}_k$ symmetry

$$X^{25} \to X^{25} + \frac{2\pi\alpha'^{1/2}}{k} . \qquad (8.5.22)$$

This multiplies $j^1 + ij^2$ and $\bar{\jmath}^1 + i\bar{\jmath}^2$ by $\exp(2\pi i/k)$ and so the surviving massless scalars are

$$j^3\bar{\jmath}^3 , \quad j^1\bar{\jmath}^1 + j^2\bar{\jmath}^2 , \quad j^1\bar{\jmath}^2 - j^2\bar{\jmath}^1 . \qquad (8.5.23)$$

The first of these changes the radius of the torus and so is a flat direction as always. However, there are no other branches emerging from these points, because the conditions (8.3.23) on the potential for a flat direction are not met in the directions $M_{11} = M_{22}$ or $M_{12} = -M_{21}$.

The shift (8.5.22) generates a $\mathbf{Z}_k$ subgroup of $SU(2) \times SU(2)$. Let us consider twists by other subgroups. The dihedral group $\mathbf{D}_k$ generated by the shift (8.5.22) and the reflection $X^{25} \to -X^{25}$ produces the orbifold at radius $k\alpha'^{1/2}$. The other three discrete subgroups are the tetrahedral, octahedral, and icosahedral groups. These remove *all* moduli from the CFT, so the twisted CFTs are isolated points, not members of continuous families. Unlike the $\mathbf{Z}_k$ and $\mathbf{D}_k$ twists, which can be defined at all radii, these discrete groups involve elements of $SU(2)$ that exist only at the critical radius and so the radius cannot be varied after twisting. String theory with such a compact space would have only one scalar, the dilaton.

These are all of the known $c = 1$ CFTs, and so give all known 25-dimensional bosonic string backgrounds.

8.6 Open strings

The new feature in toroidal compactification of the open string is the possibility of a nontrivial *Wilson line,* a flat background for the gauge field. Studying first a $U(1)$ gauge theory with some charged fields, consider the constant background

$$A_{25}(x^M) = -\frac{\theta}{2\pi R} = -i\Lambda^{-1}\frac{\partial\Lambda}{\partial x^{25}} , \quad \Lambda(x^{25}) = \exp\left(-\frac{i\theta x^{25}}{2\pi R}\right) , \qquad (8.6.1)$$

where θ is a constant. Locally this is pure gauge. The field strength vanishes and the field equations are trivially satisfied. However, the gauge parameter Λ does not satisfy the spacetime periodicity, and so the background has a physical effect. The gauge-invariant quantity that measures this is the Wilson line

$$W_q = \exp\left(iq \oint dx^{25} A_{25}\right) = \exp(-iq\theta) . \qquad (8.6.2)$$

Consider first a point particle of charge q, with gauge-fixed world-line action

$$S = \int d\tau \left(\frac{1}{2} \dot{X}^M \dot{X}_M + \frac{m^2}{2} - iq A_M \dot{X}^M \right) . \tag{8.6.3}$$

The gauge action is simply $-iq \int dx^M A_M$, so paths that wind around the compact dimension pick up a phase just equal to the Wilson line W_q. The canonical momentum is

$$p_{25} = -\frac{\partial L}{\partial v^{25}} = v^{25} - \frac{q\theta}{2\pi R} , \tag{8.6.4}$$

where $v^{25} = i\dot{X}^{25}$ as in eq. (8.4.6). The wavefunctions must be periodic in the compact space, so $p_{25} = l/R$ for integer l and

$$v_{25} = \frac{2\pi l + q\theta}{2\pi R} . \tag{8.6.5}$$

The Hamiltonian, which annihilates physical states, is

$$H = \frac{1}{2}(p_\mu p^\mu + v_{25}^2 + m^2) , \tag{8.6.6}$$

so the masses $-p_\mu p^\mu$ are shifted due to the dependence of v_{25} on θ. The same spectrum is readily obtained in a field rather than particle description. Note that v_{25} is just the gauge-invariant momentum $-i\partial_{25} - q A_{25}$.

One can also make the gauge transformation Λ^{-1} to set A_{25} to zero. Charged fields in this gauge are no longer periodic, picking up a phase $\exp(iq\theta)$ under $x^{25} \to x^{25} + 2\pi R$. Physical (gauge-invariant) objects are still periodic. The canonical momentum is now shifted, and we have

$$v_{25} = p_{25} = \frac{2\pi l + q\theta}{2\pi R} \tag{8.6.7}$$

with the same result as before for the gauge-invariant momentum v_{25}.

Let us turn now to the string, and include $U(n)$ Chan–Paton factors. A general constant A_{25} can be diagonalized by a gauge transformation,

$$A_{25} = -\frac{1}{2\pi R} \text{diag}(\theta_1, \theta_2, \ldots, \theta_n) . \tag{8.6.8}$$

The gauge field is in the diagonal subgroup of $U(n)$, namely $U(1)^n$. The gauge field couples to the Chan–Paton factor of a general state as $[A_M, \lambda]$, so a string in Chan–Paton state $|ij\rangle$ has charge $+1$ under $U(1)_i$, -1 under $U(1)_j$, and is neutral under the others. Thus it has

$$v_{25} = \frac{2\pi l - \theta_j + \theta_i}{2\pi R} . \tag{8.6.9}$$

The open string spectrum is then

$$m^2 = \frac{(2\pi l - \theta_j + \theta_i)^2}{4\pi^2 R^2} + \frac{1}{\alpha'}(N - 1) . \tag{8.6.10}$$

Consider in particular the gauge bosons, where $l = 0$ and $N = 1$ and so

$$m^2 = \frac{(\theta_j - \theta_i)^2}{4\pi^2 R^2} \,. \tag{8.6.11}$$

In generic backgrounds, all the θs are distinct and the only massless vectors are the diagonal ones, $i = j$. The unbroken gauge group in this case is $U(1)^n$. If r of the θs are equal, the corresponding $r \times r$ matrix of vectors is massless, carrying the gauge symmetry $U(r)$. With the n θs equal in sets of r_i, the gauge symmetry is

$$U(r_1) \times \ldots \times U(r_s) \,, \quad \sum_{i=1}^{s} r_i = n \,. \tag{8.6.12}$$

As always, there is a low-energy interpretation. The gauge field A_{25} is a 25-dimensional scalar in the adjoint representation of $U(n)$, and its vacuum expectation value breaks the symmetry spontaneously to $U(r_1) \times \ldots \times U(r_s)$.

The reader can fill in the details of the effective low-energy action. It is important that with k compact dimensions, the potential contains a term

$$\mathrm{Tr}\left([A_m, A_n]^2\right) \tag{8.6.13}$$

from the field strength in the 26-dimensional Yang–Mills action. This forces the gauge fields in different directions to commute in a static solution, so that they are simultaneously diagonalizable. There are then kn moduli from the gauge field and k^2 from the metric and antisymmetric tensor.

T-duality

Now consider the $R \to 0$ limit of the open string spectrum. Open strings with Neumann boundary conditions have no quantum number comparable to w; they can always be unwound from the periodic dimension. So when $R \to 0$ the states with nonzero momentum go to infinite mass, but there is no new continuum of states. The behavior is as in field theory: the resulting states move in 25 spacetime dimensions.

A seeming paradox arises when one remembers that theories with open strings always have closed strings as well, so that in the $R \to 0$ limit the closed strings move in 26 spacetime dimensions but the open strings only in 25. One can reason out what is happening as follows. The interior of the open string is made of the same 'stuff' as the closed string, and so should still be vibrating in 26 dimensions. What distinguishes the open string is its endpoints, so it is these that must be restricted to a 25-dimensional hyperplane.

Indeed, recall that one describes the T-dual theory by using the new

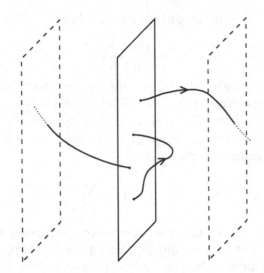

Fig. 8.3. Open strings with endpoints attached to a hyperplane. The dashed planes are periodically identified. Two strings are shown, with winding numbers zero and one.

embedding coordinate

$$X'^{25}(z, \bar{z}) = X_L^{25}(z) - X_R^{25}(\bar{z}) . \tag{8.6.14}$$

Then

$$\partial_n X^{25} = -i\partial_t X'^{25}, \tag{8.6.15}$$

with n the normal and t the tangent at the boundary. The Neumann condition on the original coordinate is a Dirichlet condition on the dual coordinate: the X'^{25} coordinate of each string endpoint is fixed.

Let us first consider compactification without Wilson lines. Then all endpoints are actually constrained to lie on *the same* hyperplane. To see this, integrate

$$X'^{25}(\pi) - X'^{25}(0) = \int_0^\pi d\sigma^1 \partial_1 X'^{25} = -i \int_0^\pi d\sigma^1 \partial_2 X^{25}$$

$$= -2\pi\alpha' v^{25} = -\frac{2\pi\alpha' l}{R} = -2\pi l R' . \tag{8.6.16}$$

The total change in X'^{25} between the two ends is an integral multiple of the periodicity $2\pi R'$ of the dual dimension, so the ends lie on the same hyperplane in the periodic T-dual space. For two *different* open strings, consider the connected world-sheet that results from graviton exchange between them. One can carry out the same argument (8.6.16) on a path connecting any two endpoints, so all endpoints of all strings lie on the

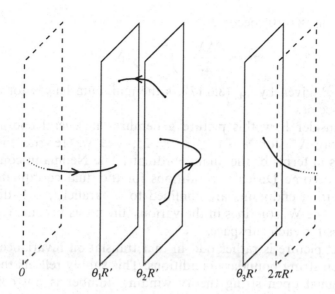

Fig. 8.4. $n = 3$ hyperplanes at different positions, with various strings attached. Ignoring the tachyon, the lightest strings with both ends on the same hyperplane are massless, while strings stretched between noncoincident hyperplanes are massive due to their tension. When two hyperplanes become coincident, the lightest strings stretched between them become massless.

same hyperplane, as in figure 8.3. The ends are still free to move in the other 24 spatial dimensions.

Now we consider the effect of Wilson lines. Due to the shift (8.6.9) in v_{25}, the difference (8.6.16) is replaced by

$$\Delta X'^{25} = X'^{25}(\pi) - X'^{25}(0) = -(2\pi l - \theta_j + \theta_i)R' . \qquad (8.6.17)$$

In other words, up to an arbitrary additive normalization, the endpoint in state i is at

$$X'^{25} = \theta_i R' = -2\pi\alpha' A_{25,ii}. \qquad (8.6.18)$$

Thus there are in general n hyperplanes at different positions as depicted in figure 8.4.

The mode expansion in the sector with Chan–Paton state $|ij\rangle$, for an open string winding l times around the compact dimension, is

$$X'^{25}(z,\bar{z}) = \theta_i R' - \frac{iR'}{2\pi}(2\pi l - \theta_j + \theta_i)\ln(z/\bar{z}) + i\left(\frac{\alpha'}{2}\right)^{1/2}\sum_{\substack{m=-\infty\\m\neq 0}}^{\infty}\frac{\alpha_m^{25}}{m}(z^{-m} - \bar{z}^{-m})$$

$$= \theta_i R' + \frac{\sigma^1}{\pi}\Delta X'^{25} - (2\alpha')^{1/2}\sum_{\substack{m=-\infty\\m\neq 0}}^{\infty}\frac{\alpha_m^{25}}{m}\exp(-m\sigma^2)\sin m\sigma^1.$$

$$(8.6.19)$$

The spectrum (8.6.10) becomes

$$m^2 = \left(\frac{\Delta X'^{25}}{2\pi\alpha'}\right)^2 + \frac{1}{\alpha'}(N-1) \,, \tag{8.6.20}$$

where $\Delta X'^{25}$, given by eq. (8.6.17), is the minimum length for a string in the given sector.

Now consider how this picture generalizes if several coordinates are periodic, say X^m for $p+1 \leq m \leq 25$, and we rewrite the periodic dimensions in terms of the dual coordinate. The Neumann conditions on X^m again become Dirichlet conditions for the dual coordinates X'^m, so the open string endpoints are confined to n parallel $(p+1)$-dimensional subspaces. The Wilson lines in the various directions become independent coordinates for each subspace.

The dual picture is rather odd in that translation invariance is broken by the open string boundary conditions. This simply reflects the fact that in the original open string theory winding number is not conserved (a closed winding string can break open and unwind) and winding number is T-dual to momentum. It is important to remember that T-duality is just a different description of the same theory, and further that when the compactification radii in the original theory are small the T-dual picture is the natural one to use. Figure 8.4 then shows how to interpret the resulting physics.

8.7   D-branes

We have just learned that when an open string theory is compactified on a small torus, the physics is described by compactification on a large torus but with the open string endpoints restricted to lie on subspaces. In fact, these subspaces are new dynamical objects in their own right.

Consider the spectrum of massless open strings for the generic configuration in which all θ_i are distinct, taking for simplicity the case in which only one coordinate is dualized. Ignoring the tachyon, the mass (8.6.20) vanishes only for $N = 1$ and $\Delta X'^{25} = 0$. That is, both endpoints are on the same hyperplane, with zero winding. We have therefore the massless states

$$\alpha_{-1}^{\mu}|k;ii\rangle \,, \quad \mathscr{V} = i\partial_t X^{\mu} \,, \tag{8.7.1a}$$

$$\alpha_{-1}^{25}|k;ii\rangle \,, \quad \mathscr{V} = i\partial_t X^{25} = \partial_n X'^{25} \,. \tag{8.7.1b}$$

These are of course the same massless states as in the original theory, T-duality simply giving a different picture of the same spectrum. For the generic Wilson line, the only massless open string states in the original theory would have been the n massless $U(1)$ vectors. In eq. (8.7.1) we have separated these according to whether the polarization is tangent or perpendicular to the hyperplane. The 25 states with tangent polarization comprise a gauge field living on the hyperplane. The state with perpendicular polarization was the gauge field in the compact direction of the original theory, and it has a simple and important interpretation in the T-dual theory: it is a collective coordinate for the shape of the hyperplane. This is already evident from the fact that a constant gauge background (8.6.18) corresponds to a uniform translation of the hyperplane. An x^μ-dependent background would by the same token go over to an x^μ-dependent translation, a curved hyperplane, and the quanta of the field A_{ii}^{25} correspond to small oscillations of the hyperplane.

This is the same phenomenon as with spacetime itself. We start with strings in a flat background and discover that a massless closed string state corresponds to fluctuations of the geometry. Here we found first a flat hyperplane, and then discovered that a certain open string state corresponds to fluctuations of its shape. We should not be surprised that the hyperplane has become dynamical. String theory contains gravity. A gravitational wave passing through the hyperplane would warp spacetime itself, so the hyperplane could hardly remain rigid.

Thus the hyperplane is indeed a dynamical object, a Dirichlet membrane, or *D-brane* for short. The p-dimensional D-brane, from dualizing $25 - p$ dimensions, is a Dp-brane. In this terminology, the original $U(n)$ open string theory contains n D25-branes. A D25-brane fills space, so the string endpoint can be anywhere: it just corresponds to an ordinary Chan–Paton factor.

Since T-duality interchanges Neumann and Dirichlet boundary conditions, a further T-duality in a direction tangent to a Dp-brane reduces it to a D($p-1$)-brane, while a T-duality in an orthogonal direction turns it into a D($p+1$)-brane. The case of a nontrivial angle will arise shortly.

We could have started by investigating Dirichlet boundary conditions in their own right. The route we have taken, discovering them via T-duality, is a good way to develop many of their properties. It also shows that one can reach a state containing D-branes from an 'ordinary' state by a continuous process. That is, take the $R \to 0$ limit of the original theory, leaving n D-branes in a noncompact space. In the superstring, we will use this to argue that D-branes are an essential part of the nonperturbative definition of the theory. For the bosonic string we will not try to make this argument because there is no strong argument that the theory has a nonperturbative definition. It is possible that bosonic string theory exists only as a mutilation of superstring theory.

It is interesting to look at the $U(n)$ symmetry breaking in the T-dual picture. When no D-branes coincide there is just one massless vector on each, giving gauge group $U(1)^n$ in all. If r D-branes coincide, there are new massless states because strings that are stretched between these branes can have vanishing length: the $\Delta X'^{25}$ in the mass formula (8.6.20) vanishes for $n = 0$ if both i and j are in this set of r. Thus, there are r^2 vectors, forming the adjoint of a $U(r)$ gauge group. This is just as found in the dual Wilson line picture. It is striking, however, that there will also be r^2 massless scalars from the components normal to the D-brane. We will discuss the significance of this below.

The D-brane action

The massless fields on the world-volume of a Dp-brane are a $U(1)$ vector plus $25 - p$ world-brane scalars describing the fluctuations. The low energy effective action for this system is as always a very useful thing to consider. We now take the dual radius R' to infinity, focusing on one D-brane in the resulting 26-dimensional spacetime. The world-brane fields are in interaction with the massless closed string fields, whose action has already been discussed in section 3.7. Introduce coordinates ξ^a, $a = 0, \ldots p$ on the brane. The fields on the brane are the embedding $X^\mu(\xi)$ and the gauge field $A_a(\xi)$. We claim that the action is

$$S_p = -T_p \int d^{p+1}\xi \, e^{-\Phi}[-\det(G_{ab} + B_{ab} + 2\pi\alpha' F_{ab})]^{1/2} \,, \qquad (8.7.2)$$

with T_p a constant to be determined later. Here

$$G_{ab}(\xi) = \frac{\partial X^\mu}{\partial \xi^a}\frac{\partial X^\nu}{\partial \xi^b}G_{\mu\nu}(X(\xi)) \,, \quad B_{ab}(\xi) = \frac{\partial X^\mu}{\partial \xi^a}\frac{\partial X^\nu}{\partial \xi^b}B_{\mu\nu}(X(\xi)) \qquad (8.7.3)$$

are the induced metric and antisymmetric tensor on the brane.

All features of the action (8.7.2) can be understood from general reasoning. Considering first just the spacetime metric and the embedding, the simplest and lowest-derivative coordinate-invariant action is the integral of $(-\det G_{ab})^{1/2}$, the world-volume. Note that this term has an implicit dependence on $X^\mu(\xi)$ through the induced fields (8.7.3). Expanded around a flat D-brane it gives the action for the fluctuations, just as the Nambu–Goto action describes the fluctuations of a string.

The dilaton dependence $e^{-\Phi} \propto g_c^{-1}$ arises because this is an open string tree-level action. The self-interactions of the open string fields, and their couplings to the closed string fields, arise first from the disk.

The dependence on F_{ab} can be understood by using T-duality. Consider a D-brane that is extended in the X^1- and X^2-directions, with the other dimensions unspecified, and let there be a constant gauge field F_{12} on it. Go to the gauge $A_2 = X^1 F_{12}$. Now take the T-dual along the 2-direction. The Neumann condition in this direction is replaced by a Dirichlet condition, so the D-brane loses a dimension. However, the T-duality relation (8.6.18) between the potential and coordinate implies that the D-brane is tilted in the (1–2)-plane,

$$X'^2 = -2\pi\alpha' X^1 F_{12} .$$ (8.7.4)

The tilt gives a geometric factor in the action,

$$\int dX^1 \left[1 + (\partial_1 X'^2)^2\right]^{1/2} = \int dX^1 \left[1 + (2\pi\alpha' F_{12})^2\right]^{1/2} .$$ (8.7.5)

For any D-brane, by boosting to align it with the coordinate axes and then rotating to bring F_{ab} to block-diagonal form, one can reduce the action to a product of factors (8.7.5) in each plane, equivalent to the F_{ab} term in the determinant (8.7.2). This determinant form for the gauge field action is known as the Born–Infeld action. It was originally proposed, unsuccessfully, as a solution to the short-distance divergences of quantum electrodynamics.

Finally, the dependence on B_{ab} is given by the following argument. The closed string field $B_{\mu\nu}$ and the open string field A^μ appear in the *string* world-sheet action as[3]

$$\frac{i}{4\pi\alpha'} \int_M d^2\sigma\, g^{1/2} \epsilon^{ab} \partial_a X^\mu \partial_b X^\nu B_{\mu\nu} + i \int_{\partial M} dX^\mu A_\mu .$$ (8.7.6)

Associated with each of these fields is a spacetime gauge invariance, which must be preserved for the consistency of the spacetime theory. The ordinary gauge transformation

$$\delta A_\mu = \partial_\mu \lambda$$ (8.7.7)

is an invariance of the action (8.7.6), the boundary term changing by the integral of a total derivative. The antisymmetric tensor variation

$$\delta B_{\mu\nu} = \partial_\mu \zeta_\nu - \partial_\nu \zeta_\mu$$ (8.7.8)

similarly changes the bulk action by a total derivative, but on a bounded world-sheet this gives rise to a surface term. This can only be canceled if the open string field A_μ also transforms under the tensor gauge symmetry,

$$\delta A_\mu = -\zeta_\mu / 2\pi\alpha' .$$ (8.7.9)

Now, however, only the combination

$$B_{\mu\nu} + 2\pi\alpha' F_{\mu\nu} \equiv 2\pi\alpha' \mathcal{F}_{\mu\nu}$$ (8.7.10)

---

[3] For the reader familiar with differential forms, this is

$$\frac{i}{2\pi\alpha'} \int_M B + i \int_{\partial M} A .$$

The reader can translate the subsequent equations similarly.

is invariant under both symmetries, and it is this combination that must appear in the action. Thus the form (8.7.2) of the action is fully determined.

As a check, it is natural that the combination $G_{\mu\nu} + B_{\mu\nu}$ should appear, since this same combination appears in the world-sheet action in conformal gauge. Incidentally, the action (8.7.2) can also be determined by taking the low energy limit of various open and open plus closed string amplitudes, but this is rather more laborious.

For n separated D-branes, the action is n copies of the action for a single D-brane. We have seen, however, that when the D-branes are coincident there are n^2 rather than n massless vectors and scalars on the brane, and we would like to write down the effective action governing these. The fields $X^{\mu}(\xi)$ and $A_a(\xi)$ will now be $n \times n$ matrices. For the gauge field the meaning is obvious — it becomes a non-Abelian $U(n)$ gauge field. For the collective coordinate X^{μ}, however, the meaning is mysterious: the collective coordinates for the embedding of n D-branes in spacetime are now enlarged to $n \times n$ matrices. This 'noncommutative geometry' has proven to play a key role in the dynamics of D-branes, and there are conjectures that it is an important hint about the nature of spacetime.

We can gain more insight by considering the effective action. There is now a nonderivative term in the action, a potential for the collective coordinates, which can be deduced by T-duality from a constant A_m field. For such a vector potential, the field strength is just $[A_m, A_n]$, which becomes $(2\pi\alpha')^{-2}[X_m, X_n]$ in the T-dual picture. Expanding in the field strength, the leading term in the action is roughly

$$V \propto \mathrm{Tr}\left([X_m, X_n][X^m, X^n]\right) . \tag{8.7.11}$$

This potential has the property that at the point $X^m = 0$ its second derivative vanishes, so that all kn^2 scalars are massless; again $k = 25 - p$ is the number of dualized dimensions. However, the space of flat dimensions is smaller. The potential is the sum of squares and so vanishes only if all of the $[X^m, X^n]$ vanish. We can then use the $U(n)$ gauge symmetry to go to a gauge in which the X^m are all diagonal. There are thus kn flat directions, just the number of diagonal elements. These kn diagonal elements can then be interpreted as the collective coordinates of the n Dp-branes. Thus the potential (8.7.11) correctly interpolates the physics of separated and coincident D-branes.

When the X^m do commute, the action should reduce to that of n separate D-branes, so we must have

$$S_p = -T_p \int d^{p+1}\xi \, \mathrm{Tr}\left\{ e^{-\Phi} [-\det(G_{ab} + B_{ab} + 2\pi\alpha' F_{ab})]^{1/2} \right.$$
$$\left. + O([X^m, X^n]^2) \right\} . \tag{8.7.12}$$

The determinant is on the world-volume indices ab, while the trace is over the n Chan–Paton indices. The trace is the appropriate $U(n)$ invariant which for diagonal matrices reduces to a sum over the separate D-branes. The full dependence on the commutator is more complicated than the simple potential (8.7.11), with additional couplings to other fields and corrections at higher orders in the commutator. However, the key property, the form of the flat directions, is unaffected. The full dependence can be determined by starting from the Born–Infeld action for the fully Neumann case and performing the T-duality. Incidentally, higher derivative terms involving commutators of field strengths cannot be determined by T-duality alone.

The D-brane tension

It is instructive to compute the constant T_p, and for the superstring the precise value will be significant. Before we carry out the explicit calculation, it is interesting to note that one gets a recursion relation for T_p from T-duality. Note that in a constant dilaton background, the tension of a Dp-brane is $T_p e^{-\Phi}$: this is minus the action per unit volume for a static Dp-brane. Consider now a Dp-brane in which the p directions tangent to the D-brane have been periodically identified. That is, the Dp-brane is *wrapped* on a p-torus in spacetime. Its mass is its tension times the volume of the torus,

$$T_p e^{-\Phi} \prod_{i=1}^{p} (2\pi R_i) , \tag{8.7.13}$$

Now take the T-dual on one of the periodic dimensions X^p. This does not change the mass, being just a new description of the same state. In terms of the dilaton (8.3.31) of the T-dual theory, the mass (8.7.13) is

$$2\pi \alpha'^{1/2} T_p e^{-\Phi'} \prod_{i=1}^{p-1} (2\pi R_i) . \tag{8.7.14}$$

In the dual theory, however, it is a D$(p-1)$-brane wrapped on a $(p-1)$-torus, so its mass is

$$T_{p-1} e^{-\Phi'} \prod_{i=1}^{p-1} (2\pi R_i) . \tag{8.7.15}$$

Equating the masses (8.7.14) and (8.7.15) gives

$$T_p = T_{p-1}/2\pi \alpha'^{1/2} , \tag{8.7.16}$$

which determines the T_p up to one overall normalization.

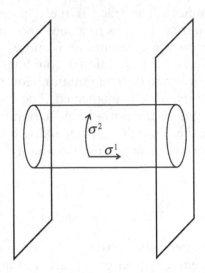

Fig. 8.5. Exchange of a closed string between two D-branes. Equivalently, a vacuum loop of an open string with one end on each D-brane.

To determine the actual value of the D-brane tension we need to calculate a string amplitude. For example, we could deduce it from the gravitational coupling to the D-brane, given by the disk with a graviton vertex operator. However, this involves the yet-unknown ratio g_c/g_o^2, the closed string coupling arising from the vertex operator and the open string coupling from the disk. Instead we can obtain the absolute normalization as follows. Consider two parallel Dp-branes at positions $X_1^m = 0$ and $X_2^m = y^m$. These two objects can feel each other's presence by exchanging closed strings as shown in figure 8.5. This string amplitude is an annulus, with no vertex operators, and so can be calculated by the same method as in the previous chapter. The poles from graviton and dilaton exchange then give the coupling T_p of closed string states to the D-brane.

Recall that in section 7.4 we calculated the annulus vacuum amplitude in terms of a loop of open strings, but then found closed string poles from the $t \to 0$ limit of the modular integration. Here it is just those poles that we are interested in, but once again the simplest way to calculate the amplitude is to think of it as an open string loop. In fact, the earlier result (7.4.1) carries over with just a few changes. The number of momentum integrations is reduced from 26 to $p + 1$ and similarly V_{26} becomes V_{p+1}; the weight h_i acquires an additional term $y^2/4\pi^2\alpha'$ from the tension of the stretched open string; the Chan–Paton weight n^2 becomes 2 (since the string can attach with either orientation). Thus we have

$$\mathcal{A} = iV_{p+1} \int_0^\infty \frac{dt}{t} (8\pi^2\alpha't)^{-(p+1)/2} \exp(-ty^2/2\pi\alpha')\eta(it)^{-24}$$

$$= \frac{iV_{p+1}}{(8\pi^2\alpha')^{(p+1)/2}} \int_0^\infty dt\, t^{(21-p)/2} \exp(-ty^2/2\pi\alpha')$$

$$\times [\exp(2\pi/t) + 24 + \ldots], \qquad (8.7.17)$$

where the asymptotics have been obtained just as in section 7.4. The first term is from tachyon exchange and so is again an uninteresting bosonic artifact. The integral over the second term gives

$$\mathcal{A} = iV_{p+1}\frac{24}{2^{12}}(4\pi^2\alpha')^{11-p}\pi^{(p-23)/2}\Gamma\left(\frac{23-p}{2}\right)|y|^{p-23}$$

$$= iV_{p+1}\frac{24\pi}{2^{10}}(4\pi^2\alpha')^{11-p}G_{25-p}(y), \qquad (8.7.18)$$

where $G_d(y)$ is the massless scalar Green's function in d dimensions, the inverse of $-\nabla^2$.

We now need to compare this with a field theory calculation of the same amplitude. Since the antisymmetric tensor does not couple to the D-branes, the relevant terms from the spacetime action (3.7.25) are

$$S = \frac{1}{2\kappa^2} \int d^{26}X\, (-\tilde{G})^{1/2}\left(\tilde{R} - \frac{1}{6}\nabla_\mu\tilde{\Phi}\tilde{\nabla}^\mu\tilde{\Phi}\right). \qquad (8.7.19)$$

Recall that the tilde denotes the Einstein metric; this is the convenient form to use because its action decouples from that of the dilaton. The tilded dilaton has been shifted so that its expectation value is zero. In terms of the same variables, the relevant terms from the D-brane action (8.7.2) are

$$S_p = -\tau_p \int d^{p+1}\xi\, \exp\left(\frac{p-11}{12}\tilde{\Phi}\right)(-\det \tilde{G}_{ab})^{1/2}. \qquad (8.7.20)$$

We have defined $\tau_p = T_p e^{-\Phi_0}$; this is the actual physical tension of the Dp-brane when the background value of the dilaton is Φ_0.

The field theory graph analogous to figure 8.5 is the exchange of a single graviton or dilaton between the D-branes. To obtain the propagator we expand the spacetime action to second order in $h_{\mu\nu} = \tilde{G}_{\mu\nu} - \eta_{\mu\nu}$ and to second order in $\tilde{\Phi}$. Also, we need to choose a gauge for the gravitational calculation. The simplest gauge for perturbative calculation is

$$F_\nu = \partial^{\hat{\mu}}h_{\mu\nu} - \frac{1}{2}\partial_\nu h^{\hat{\mu}}_{\ \mu} = 0, \qquad (8.7.21)$$

where a hat indicates that an index has been raised with the flat space metric $\eta^{\mu\nu}$. Expanding the action to second order and adding a gauge-fixing term $-F_\nu F^{\hat{\nu}}/4\kappa^2$, the spacetime action becomes

$$S = -\frac{1}{8\kappa^2} \int d^{26}X \left(\partial_\mu h_{\nu\lambda} \partial^{\hat{\mu}} h^{\hat{\nu}\hat{\lambda}} - \frac{1}{2}\partial_\mu h^{\hat{\nu}}_\nu \partial^{\hat{\mu}} h^{\hat{\lambda}}_\lambda + \frac{2}{3}\partial_\mu \tilde{\Phi}\partial^{\hat{\mu}}\tilde{\Phi} \right) . \qquad (8.7.22)$$

Inverting the kinetic term, one obtains the momentum space propagators

$$\langle \tilde{\Phi}\tilde{\Phi} \rangle = -\frac{(D-2)i\kappa^2}{4k^2} , \qquad (8.7.23a)$$

$$\langle h_{\mu\nu} h_{\sigma\rho} \rangle = -\frac{2i\kappa^2}{k^2}\left(\eta_{\mu\sigma}\eta_{\nu\rho} + \eta_{\mu\rho}\eta_{\nu\sigma} - \frac{2}{D-2}\eta_{\mu\nu}\eta_{\sigma\rho} \right) . \qquad (8.7.23b)$$

For later reference we have given these for general D. The D-brane action expanded around a flat configuration is

$$S_p = -\tau_p \int d^{p+1}\xi \left(\frac{p-11}{12}\tilde{\Phi} - \frac{1}{2}h_{aa} \right) . \qquad (8.7.24)$$

Note that $h_{\mu\nu}$ here is traced only over the directions tangent to the D-brane; we have taken ξ to be rectilinear coordinates with metric δ_{ab}.

We can now read off the Feynman graph,

$$\begin{aligned}
\mathcal{A} &= \frac{i\kappa^2\tau_p^2}{k_\perp^2}V_{p+1}\left\{ 6\left[\frac{p-11}{12}\right]^2 + \frac{1}{2}\left[2(p+1) - \frac{1}{12}(p+1)^2 \right] \right\} \\
&= \frac{6i\kappa^2\tau_p^2}{k_\perp^2}V_{p+1} .
\end{aligned} \qquad (8.7.25)$$

Comparison with the string result (8.7.18) yields

$$\tau_p^2 = \frac{\pi}{256\kappa^2}(4\pi^2\alpha')^{11-p} . \qquad (8.7.26)$$

It is a useful check that, once the dust has cleared, the recursion relation from T-duality is satisfied.

As one application, consider a state with n 25-branes, which is the same as an ordinary (fully Neumann) n-valued Chan–Paton factor. Expanding the 25-brane action (8.7.12) to second order in the gauge field gives

$$\frac{\tau_{25}}{4}(2\pi\alpha')^2\text{Tr}(F_{\mu\nu}F^{\mu\nu}) . \qquad (8.7.27)$$

This relates the open string gauge coupling to the gravitational coupling. Using eqs. (6.5.14), (6.5.16), (6.6.15), and (6.6.18), we can write it as a relation between the vertex operator normalizations g_c and g_o,

$$\frac{g_o^2}{g_c} = \frac{4\pi\alpha' g_o'^2}{\kappa} = 2^{18}\pi^{25/2}\alpha'^6 . \qquad (8.7.28)$$

This has the right form, the square of the open string coupling being proportional to the closed string coupling, but now with the numerical coefficient included. We have essentially obtained this via unitarity, factoring on the closed string poles of the annulus diagram. We could also have done this more directly, as in exercise 7.9, without the excursion through D-branes.

8.8   *T*-duality of unoriented theories

The $R \to 0$ limit in an unoriented string theory also leads to interesting new physics. Consider closed strings first. To form the unoriented theory we impose $\Omega = +1$ on the states. Following the discussion in section 8.5, we can think about this as *gauging* Ω. In particular, the transition functions used to build the world-sheet may now include orientation reversal, since it is a gauge symmetry, and this produces unoriented world-sheets.

The *T*-dual theory is obtained by using the coordinate $X'^m(z,\bar z) = X_L^m(z) - X_R^m(\bar z)$ in place of $X^m(z,\bar z) = X_L^m(z) + X_R^m(\bar z)$. In the original description we are gauging world-sheet parity Ω, which acts as

$$\Omega : \quad X_L^M(z) \leftrightarrow X_R^M(z) . \tag{8.8.1}$$

In terms of the *T*-dual coordinates, this is

$$\Omega : \quad X'^m(z,\bar z) \leftrightarrow -X'^m(\bar z,z) , \tag{8.8.2a}$$

$$X^\mu(z,\bar z) \leftrightarrow X^\mu(\bar z,z) , \tag{8.8.2b}$$

where again m indexes the coordinates on which the *T*-dual has been taken and μ those on which it has not. In the dual picture, the symmetry (8.8.2) is thus the *product* of a world-sheet parity transformation and a spacetime reflection. We have seen that gauging world-sheet parity alone produces the unoriented theory, while gauging the reflection alone produces an orbifold. The result of the combined projection is known as an *orientifold*.

Orientifolds are rather similar to orbifolds. Separate the string wave-function into its internal part and its dependence on the center of mass x^m, and take the internal wavefunction to be an eigenstate of Ω. The projection onto $\Omega = +1$ then determines the string wavefunction at $-x^m$ to be the same as that at x^m, up to a sign. For example, the various components of the metric and antisymmetric tensor satisfy

$$G_{\mu\nu}(x') = G_{\mu\nu}(x) , \quad B_{\mu\nu}(x') = -B_{\mu\nu}(x) , \tag{8.8.3a}$$

$$G_{\mu n}(x') = -G_{\mu n}(x) , \quad B_{\mu n}(x') = B_{\mu n}(x) , \tag{8.8.3b}$$

$$G_{mn}(x') = G_{mn}(x) , \quad B_{mn}(x') = -B_{mn}(x) , \tag{8.8.3c}$$

where $(x^\mu, x^m)' = (x^\mu, -x^m)$. There is a minus sign for each m,n and an additional minus sign for each B_{MN}; the orbifold is the same except that the latter sign would be absent. In other words, the *T*-dual spacetime is the torus T^k identified under a $\mathbf{Z}_2$ reflection in the k compact directions, just as in the orbifold construction. In the case of a single periodic dimension,

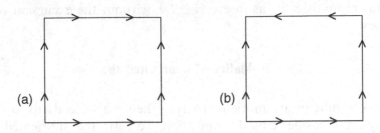

Fig. 8.6.   Identifications defining (a) the torus and (b) the Klein bottle.

for example, the dual spacetime is the line segment $0 \leq x^{25} \leq \pi R'$, with orientifold fixed planes at the ends.

Notice also that away from the orientifold fixed planes, the local physics is that of the *oriented* string theory. Unlike the original unoriented theory, where the projection removes half the states locally, here it relates the string wavefunction at any point to its value at the image point, as in eq. (8.8.3).

One difference between the orbifold and orientifold constructions is that in the latter there is no direct analog of the twisted states, because the Klein bottle does not have the modular transformation $\tau \to -1/\tau$. Consider figure 8.6, which shows the torus and Klein bottle. On the torus, the projection operator inserts a twist in the timelike direction of figure 8.6(a); rotating the figure by $90°$, this becomes a twist in the spacelike direction, implying twisted states in the spectrum. However, if the Klein bottle of figure 8.6(b) is rotated by $90°$, the directions of time on the two opposite edges do not match and there is no interpretation in terms of intermediate states in this channel. It should be noted then that orientifold planes cannot be dynamical. Unlike the case of D-branes, there are no string modes tied to the orientifold plane to represent fluctuations in its shape. Our heuristic argument that a gravitational wave forces a D-brane to oscillate does not apply to the orientifold plane. Essentially, the identifications (8.8.3) become boundary conditions at the fixed plane, such that the incident and reflected waves cancel. For the D-brane, the reflected wave is higher order in the string coupling.

We draw oriented strings with arrows to indicate the orientation. In the unoriented theory, we can either omit the arrow or take a linear combination of the two orientations. The latter picture is more consistent with the idea of gauging the discrete symmetry, and is more general: in an orientifold the parity operation is accompanied by a spacetime transformation, so we cannot just forget the arrow.

Although we have introduced the orientifold construction via T-duality, one can consider more general orientifolds that are not simply T-duals

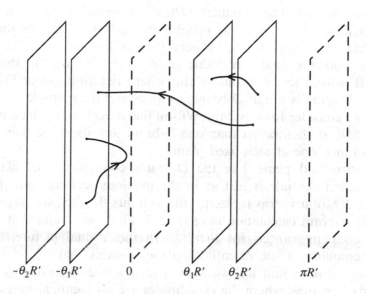

$$-\theta_2 R' \qquad -\theta_1 R' \qquad 0 \qquad \theta_1 R' \qquad \theta_2 R' \qquad \pi R'$$

Fig. 8.7. Orientifold planes at 0 and $\pi R'$, D-branes at $\theta_1 R'$ and $\theta_2 R'$, and D-brane images at $-\theta_1 R'$ and $-\theta_2 R'$. The twist operator Ω acts on any string by a combination of a spacetime reflection and reversing the orientation arrow, so that the string running from 2 to the image of 1 becomes a string running from 1 to the image of 2.

of toroidal compactifications, gauging a group that combines world-sheet parity with various spacetime symmetries.

Open strings

In the case of open strings, the situation is similar. Let us focus for convenience on a single compact dimension. Again there is one orientifold fixed plane at 0 and another at $\pi R'$. Introducing $SO(n)$ Chan–Paton factors (the symplectic case works in the same way), a general Wilson line can be brought to diagonal form, and for n even the eigenvalues are paired:

$$W = \mathrm{diag}\Big(e^{i\theta_1}, e^{-i\theta_1}, e^{i\theta_2}, e^{-i\theta_2}, \cdots, e^{i\theta_{n/2}}, e^{-i\theta_{n/2}}\Big) . \qquad (8.8.4)$$

Thus in the dual picture there are $\frac{1}{2}n$ D-branes on the line segment $0 \leq X'^{25} \leq \pi R'$, and $\frac{1}{2}n$ at their image points under the orientifold identification. Strings can stretch between D-branes and their images as shown in figure 8.7. The generic gauge group is $U(1)^{n/2}$. As in the oriented case, if r D-branes are coincident there is a $U(r)$ gauge group. However, if now the r D-branes in addition lie at one of the fixed planes, then strings stretching between one of these branes and one of the image branes also become massless and we have the right spectrum of additional states to fill out $SO(2r)$. The maximal $SO(n)$ is restored if all of the branes are coincident at a single orientifold plane. Note that this maximally symmetric case is asymmetric between the two fixed planes.

If n is odd, the final eigenvalue of W is ± 1, so that in the T-dual picture it is fixed to lie on one or the other orientifold plane. Having no image, it is actually a half D-brane, as measured for example by its mass. Also, if we consider for $n = 2$ the Wilson line $\text{diag}(1, -1)$ (which is in $O(2)$ but not $SO(2)$), then rather than one D-brane and its image there are two half D-branes, one at each fixed plane.

The orientifold plane, like the D-brane, couples to the dilaton and metric. The amplitude is just as in the previous section, with the Klein bottle and Möbius strip replacing the annulus. In fact, we have already done the relevant calculation in chapter 7. There we found that the total dilaton coupling cancels for $SO(2^{13})$. In the T-dual picture, the total dilaton coupling of the orientifold planes cancels that of 2^{12} D-branes (the images don't count!). If we T-dualize on $k = 25 - p$ dimensions there are 2^k fixed planes, where the coordinates are all combinations of 0 and $\pi R'_m$. Thus the effective action for a single fixed plane is

$$2^{12-k} T_p \int d^{p+1}\xi \, e^{-\Phi} (-\det G_{ab})^{1/2} \,, \tag{8.8.5}$$

the integral running over the fixed plane.

Although the orientifold has no direct analog of twisted sectors, it has been argued that there is a sense in which the twisted states are *open strings*. We prefer not to emphasize this, as the analogy is not complete: gauging Ω does not of itself introduce world-sheet boundaries. However, there is a definite physical sense in which it is natural to add open strings to an orientifold theory, to cancel the dilaton tadpole. This cancellation will play an important role in the superstring.

Exercises

8.1 (a) Derive the equal time commutator (8.2.21).
(b) Work out the details showing that vertex operators of the form (8.2.22) commute at equal time and finite separation.

8.2 For a single periodic dimension, calculate the scattering amplitude for four closed string tachyons with winding numbers $+1$, -1, 0, and 0, and vanishing compact momenta.

8.3 We have shown T-duality for spacetime-independent backgrounds. We would expect that a background with a varying metric $G_{25,25}(x^\mu)$ would be T-dual to one with the inverse metric $G_{25,25}(x^\mu)^{-1}$. In principle we can show this by expanding around the static configuration, but it is useful to show it directly as follows.

(a) Write the world-sheet action for a string moving in a general massless background, with only the requirement that the background be invariant under translations of X^{25}. Let the coordinate X^{25} have periodicity R.

(b) Show that replacing $\partial_a X^{25}$ with $\partial_a X^{25} + A_a$ makes the X^{25} translations a *local* symmetry on the world-sheet.

(c) Add a term $i\phi F_{ab}$ to the world-sheet action, with $\phi(\sigma)$ a new field and F_{ab} the field strength for A_a. The path integral over ϕ forces the field strength to vanish, so the gauge field is trivial: this effectively undoes step (b). Argue that in the gauge $\partial_a A^a = 0$ the resulting theory is just a rewriting of the original one.

(d) Show that you can reach the gauge $X^{25} = 0$, and then carry out the Gaussian path integration over A_a (treat it classically: ignore the determinant). Show that the resulting action has the expected form for the T-dual theory, if ϕ is identified with X'^{25} times an appropriate constant. To get the transformation (8.3.31) of the dilaton one would need to carry the A_a path integration to one loop. Also, to get the coordinate periodicity of X'^{25} requires a somewhat careful argument. (References: Buscher (1987, 1988), Roček & Verlinde (1992).)

8.4 Verify the equality of the string partition function for several compact dimensions as calculated directly and via the path integral; see the comment below eq. (8.4.12).

8.5 (a) Show that the spectrum (8.4.26) for a single compact dimension defines an even self-dual lattice.

(b) Do the same for the spectrum (8.4.7).

8.6 In the example of two periodic dimensions, section 8.4, show that the general kinetic term (8.4.35) reduces to the form (8.4.39). Verify the assertions about the action of the various dualities on τ and ρ in the special case that τ and ρ are both imaginary.

8.7 (a) Calculate the scattering amplitude for ripples on a Dp-brane. That is, consider the tree-level amplitude with four collective coordinate vertex operators. Note that the polarizations are orthogonal to all momenta, a major simplification.

(b) Examine the Regge and hard scattering limits.

(c) Take the low momentum limit and compare with what is obtained from the action (8.7.2).

8.8 Repeat the previous problem for two coincident Dp-branes. The collective coordinates now include Chan–Paton factors. Note in particular the change in part (c).

8.9 (a) Calculate the expectation value of n exponential vertex operators in the interior of the disk, with Neumann boundary conditions on $p + 1$ coordinates and Dirichlet conditions on $25 - p$.

(b) Extend the calculation to include operators $\partial_a X^M$ in the interior.

8.10 From the result of the previous exercise, calculate the amplitude for a Dp-brane to emit a graviton. Compare with the Dp-brane effective action and the result (8.7.28) for the couplings.

8.11 (a) For the boundary conditions of exercise 8.9, calculate the disk amplitude with two closed string tachyons: this corresponds to scattering a closed string from a D-brane.
(b) Examine the Regge and hard scattering limits.
(c) Show that the amplitude has poles from intermediate open strings. Give a physical interpretation.

8.12 (a) Do the calculation of the previous problem for the messier but more interesting case of massless closed strings.
(b) Show that there are poles in the graviton amplitude due to the open string collective coordinate, and compare with a calculation using the action (8.7.2).
(c) Do the same for the antisymmetric tensor and the open string photon.

9
Higher order amplitudes

In this chapter we discuss higher order amplitudes in string theory. The main goal of the first half of the chapter is an understanding of the finiteness and unitarity of string perturbation theory. After discussing the tree-level amplitude in some detail, we consider general amplitudes, using the idea of factorization to relate higher order amplitudes to lower order ones. In the second half, we briefly discuss some ideas that have been put forward to understand all-order and even nonperturbative string theory. The more powerful methods based on supersymmetry will not be developed until volume two. The present chapter is outside the main line of development in this book and can be skipped on a first reading.

9.1 General tree-level amplitudes

Chapter 6 was primarily concerned with the calculation of some specific tree-level amplitudes. It is useful to begin the present chapter with an analysis of the general tree amplitude, as it illustrates some patterns that will arise in the higher order amplitudes. We focus on the closed oriented string, and on the three- and four-point amplitudes, which illustrate the important features. The main issues are convergence and 'unitarity.' The last term is used somewhat loosely to cover both unitarity of the S-matrix, $SS^\dagger = 1$, and positivity of the Hilbert space. Of course we already know from chapter 4 that the BRST cohomology has a positive inner product, so what we need to show is that amplitudes respect the BRST invariance: that only BRST-invariant states are produced, and that the amplitudes for BRST-equivalent states are equal. We have given a general argument for the latter in section 5.4, but it is useful to illustrate it explicitly.

Before going on to a general analysis let us record some results that are useful in calculation. Suppose we have d infinite flat dimensions and the rest compact (or replaced by a compact unitary CFT), and the tachyon vertex operator is normalized

$$g_{c,d} :e^{ik\cdot X}: .\tag{9.1.1}$$

For example, with toroidal compactification of one dimension we would have $g_{c,25} = g_c(2\pi R)^{-1/2}$ as in the previous chapter. Then for the matter part of a general OCQ vertex operator, the correct normalization can be fixed by the most singular term in the OPE

$$\mathscr{V}_j(k;z,\bar{z})\overline{\mathscr{V}_{j'}(k;0,0)} = \frac{g_{c,d}^2}{z^2\bar{z}^2} +\cdots .\tag{9.1.2}$$

A bar denotes the Euclidean adjoint (6.7.28). The singularity is $|z|^{-4}$ because the vertex operator is of weight $(1,1)$. A repetition of the tachyon pole calculation (6.6.7) then determines the normalization

$$e^{-2\lambda}\left\langle \tilde{c}(\bar{z}_1)c(z_1)\,\tilde{c}(\bar{z}_2)c(z_2)\,\tilde{c}(\bar{z}_3)c(z_3)\ :e^{ik\cdot X}(z,\bar{z}): \right\rangle$$

$$= \frac{8\pi i}{\alpha' g_{c,d}^2}|z_{12}z_{13}z_{23}|^2(2\pi)^d\delta^d(k) .\tag{9.1.3}$$

So all normalization constants in the S-matrix are known in terms of $g_{c,d} = \kappa_d/2\pi$.

Three-point amplitudes

Let us first express the tree amplitude in several useful forms. The gauge-fixed expression (5.3.9) for the string S-matrix reduces, for the general tree-level, three-point amplitude, to a path integral with three fixed vertex operators:

$$S_{S_2}(1;2;3) = e^{-2\lambda}\left\langle \hat{\mathscr{V}}_1(\infty,\infty)\hat{\mathscr{V}}_2(1,1)\hat{\mathscr{V}}_3(0,0) \right\rangle_{S_2} ,\tag{9.1.4}$$

where each numeral 'n' is an abbreviation for string momentum and state, (k_n, j_n). A $\hat{}$ is used to designate a vertex operator in the 'fixed' picture — that is, not including the b-ghost insertions for the integral over position. As we have discussed in section 5.4, this is the form that comes naturally out of the state–operator mapping, and in the OCQ case takes the form

$$\hat{\mathscr{V}}_j = \tilde{c}c\mathscr{V}_j .\tag{9.1.5}$$

Using the results from section 6.7 for translating between path integrals on the sphere and matrix elements, this amplitude can also be represented as a matrix element or as an operator product coefficient:

$$S_{S_2}(1;2;3) = e^{-2\lambda}\langle\!\langle \hat{\mathscr{V}}_1|\hat{\mathscr{V}}_2(1,1)|\hat{\mathscr{V}}_3\rangle = e^{-2\lambda}c_{123} .\tag{9.1.6}$$

This relates the three-point amplitude in spacetime to a matrix element on the world-sheet, and further to the operator product coefficients which encode the local properties of the CFT.

The first issue is the decoupling of BRST-null states, so that BRST-equivalent states have equal amplitudes. Suppose that the state $|\hat{\mathscr{V}}_3\rangle$ in the three-point amplitude is null,

$$|\hat{\mathscr{V}}_3\rangle = Q_B|\chi\rangle \ . \tag{9.1.7}$$

Then

$$e^{2\lambda} S_{S_2}(1;2;3) = \langle\!\langle \hat{\mathscr{V}}_1| [\hat{\mathscr{V}}_2(1,1), Q_B] |\chi\rangle + \langle\!\langle \hat{\mathscr{V}}_1|Q_B \hat{\mathscr{V}}_2(1,1)|\chi\rangle$$
$$= 0 \ , \tag{9.1.8}$$

the terms vanishing respectively by the BRST invariance of states 2 and 1. Similarly, if $\hat{\mathscr{V}}_2 = \{Q_B, \chi\}$ or $|\hat{\mathscr{V}}_1\rangle = Q_B|\chi\rangle$ the amplitude also vanishes. We have used the Hilbert space formalism here, but it is worthwhile to phrase this same argument in path integral language. Suppose $\mathscr{V}_3$ is null,

$$\hat{\mathscr{V}}_3 = Q_B \cdot \mathscr{X} \ . \tag{9.1.9}$$

Expand the contour integral on the sphere away from $\mathscr{X}$ until it becomes a small loop around $\hat{\mathscr{V}}_1$ plus a small loop around $\hat{\mathscr{V}}_2$. These then vanish by BRST invariance, $Q_B \cdot \hat{\mathscr{V}}_1 = Q_B \cdot \hat{\mathscr{V}}_2 = 0$.

To this order then, amplitudes are well defined on the physical Hilbert space, the BRST cohomology. The decoupling of the null graviton, associated with spacetime coordinate invariance, is a special case of the decoupling of BRST-null states. It is then natural to wonder whether spacetime coordinate invariance is a piece of some much larger symmetry in string theory. We will discuss some ideas in this direction in section 9.6.

We now consider the unitarity of the S-matrix,

$$\sum_n S_{mn}S_{pn}^* \stackrel{?}{=} \delta_{mp} \ , \tag{9.1.10}$$

which is equivalent to the conservation of probability. It is useful to break up the S-matrix into a term in which nothing happens and a scattering term,

$$S_{mn} = \delta_{mn} + iT_{mn} \ . \tag{9.1.11}$$

For the scattering matrix T, unitarity implies

$$T_{mp} - T_{pm}^* = i\sum_n T_{mn} T_{pn}^* \ . \tag{9.1.12}$$

The T-matrix by definition involves a total of at least three particles, so for three-particle amplitudes on the sphere the right-hand side of eq. (9.1.12) vanishes. Unitarity then requires

$$T_{S_2}(1;2;3) = T_{S_2}(-1;-2;-3)^* \ , \tag{9.1.13}$$

where a minus sign is shorthand for replacing an incoming vertex operator by an outgoing vertex operator for the same state, or vice versa. The condition (9.1.13) is satisfied if the incoming and outgoing vertex operators are related

$$\hat{\mathscr{V}}_{-i} = -\overline{\hat{\mathscr{V}}_i} \, . \tag{9.1.14}$$

This is a natural definition: conjugation takes $e^{ik \cdot X}$ into $e^{-ik \cdot X}$, turning a vertex operator that injects momentum k^μ into the world-sheet into one that absorbs it, and similarly for other quantum numbers. The minus sign is included to undo the reversal of the order of $\tilde{c}c$ by the adjoint, so that $\hat{\mathscr{V}}_{-i}$ is obtained simply by conjugating all explicit factors of i in the vertex operators. Complex conjugation of the path integral then gives

$$\left\langle \hat{\mathscr{V}}'_1(\infty,\infty)\hat{\mathscr{V}}_2(1,1)\hat{\mathscr{V}}_3(0,0) \right\rangle_{S_2} = -\left\langle \hat{\mathscr{V}}'_{-1}(\infty,\infty)\hat{\mathscr{V}}_{-2}(1,1)\hat{\mathscr{V}}_{-3}(0,0) \right\rangle^*_{S_2} ,$$
$$\tag{9.1.15}$$

implying the unitarity relation (9.1.13). Naively, the path integral is manifestly real and so the relation (9.1.15) should hold without the minus sign. However, we must remember that the path integral has to be defined by specific analytic continuations, and the explicit result (6.2.6) acquired an overall factor of i in this way.

Four-point amplitudes and world-sheet duality

The connected four-point amplitude is

$$S_{S_2}(1;2;3;4) = e^{-2\lambda} \int_{\mathbf{C}} d^2z \left\langle \hat{\mathscr{V}}'_1(\infty,\infty) \, \hat{\mathscr{V}}_2(1,1) \, \hat{\mathscr{V}}_3(0,0) \, \mathscr{V}_4(z,\bar{z}) \right\rangle_{S_2} . \tag{9.1.16}$$

Writing the expectation value as a matrix element gives

$$\langle\!\langle \hat{\mathscr{V}}_1 | \mathrm{T}[\, \hat{\mathscr{V}}_2(1,1) \, \mathscr{V}_4(z,\bar{z})\,] | \hat{\mathscr{V}}_3 \rangle \, . \tag{9.1.17}$$

The ghost expectation value is just a constant for OCQ vertex operators, but to make the connection with the more general case it is useful to move the ghosts around a bit. Using a contour argument one can write the ghost insertion for the d^2z integration as

$$\mathscr{V}_4(z,\bar{z}) = (z\bar{z})^{-1}\{b_0, [\tilde{b}_0, \hat{\mathscr{V}}_4(z,\bar{z})]\} \, . \tag{9.1.18}$$

Using this and the condition (4.3.29) that b_0 and $\tilde{b}_0$ annihilate the physical states, the matrix element becomes

$$\theta(1 - |z|)\langle\!\langle \hat{\mathscr{V}}_1 | \hat{\mathscr{V}}_2(1,1) b_0 \tilde{b}_0 z^{L_0-1}\bar{z}^{\tilde{L}_0-1}\hat{\mathscr{V}}_4(1,1) | \hat{\mathscr{V}}_3 \rangle$$
$$+ \theta(|z| - 1)\langle\!\langle \hat{\mathscr{V}}_1 | \hat{\mathscr{V}}_4(1,1) b_0 \tilde{b}_0 z^{-L_0-1}\bar{z}^{-\tilde{L}_0-1}\hat{\mathscr{V}}_2(1,1) | \hat{\mathscr{V}}_3 \rangle \, . \tag{9.1.19}$$

Let us also write the matrix element after insertion of a complete set of intermediate states,

$$\theta(1 - |z|) \sum_{i,i'} z^{\alpha'(k_i^2 + m_i^2)/4 - 1} \bar{z}^{\alpha'(k_i^2 + \tilde{m}_i^2)/4 - 1}$$

$$\times \langle\!\langle \hat{\mathcal{V}}_1 | \hat{\mathcal{V}}_2(1,1) b_0 \tilde{b}_0 | i \rangle \mathcal{G}^{ii'} \langle\!\langle i' | \hat{\mathcal{V}}_4(1,1) | \hat{\mathcal{V}}_3 \rangle$$

$$+ \theta(|z| - 1) \sum_{i,i'} z^{-\alpha'(k_i^2 + m_i^2)/4 - 1} \bar{z}^{-\alpha'(k_i^2 + \tilde{m}_i^2)/4 - 1}$$

$$\times \langle\!\langle \hat{\mathcal{V}}_1 | \hat{\mathcal{V}}_4(1,1) b_0 \tilde{b}_0 | i \rangle \mathcal{G}^{ii'} \langle\!\langle i' | \hat{\mathcal{V}}_2(1,1) | \hat{\mathcal{V}}_3 \rangle . \quad (9.1.20)$$

In this last form the behavior of the integral at $z \to 0$ is evident, and it is the same as for the tachyon amplitudes studied in chapter 6. If $k_i^2 + m_i^2$ and $k_i^2 + \tilde{m}_i^2$ are positive for all i, that is,

$$k_i^2 = (k_1 + k_2)^2 > \frac{4}{\alpha'} , \quad (9.1.21)$$

then the integral converges. Elsewhere it is defined as in eq. (6.6.7),

$$\int_{|z|<1} d^2z \, z^{\alpha'(k^2 + m^2)/4 - 1} \bar{z}^{\alpha'(k^2 + \tilde{m}^2)/4 - 1} = \frac{8\pi}{\alpha'} \frac{\delta_{m^2, \tilde{m}^2}}{k^2 + m^2 - i\epsilon} . \quad (9.1.22)$$

One can think of this as analytic continuation from large positive values of m^2, keeping m^2 slightly below the real axis. Similarly, the integral as $z \to \infty$ converges for $(k_1 + k_4)^2 > 4/\alpha'$ and is defined elsewhere by analytic continuation. The form (9.1.20) is not useful for studying the $z \to 1$ limit. Instead use the symmetry of the amplitude to interchange vertex operators 2 and 3. The integral for $z \to 1$ is then defined by analytic continuation from the convergent region $(k_1 + k_3)^2 > 4/\alpha'$. The need to decompose the range of integration and to make possibly different analytic continuations for the different regions poses no difficulty.

After the expansion (9.1.20) is integrated as in (9.1.22), the amplitude is expressed as a series of poles in s and u. We know from chapter 6 that the amplitude also has poles in t, but these are not explicit: they come from the divergences of the infinite sums. Similarly the open string amplitude $I(s,t)$ has poles in s and t, but can be written as an infinite sum of terms having poles only in s, or only in t. This property is known as *duality*, or *world-sheet duality* to distinguish it from T- and other dualities. We can think of this as arising because the world-sheet can be cut open on different curves and so written as sums over intermediate states in different ways. Another example is the sum over open string states giving closed string poles on the cylinder.

Unitarity of the four-point amplitude

Now let us check that four-point amplitudes are equal for equivalent states. Let $|\hat{\mathcal{V}}_3\rangle = Q_B|\chi\rangle$ be null. Using the form (9.1.19), we can commute

Q_B to the left until it annihilates $\langle\langle \hat{\mathcal{V}}_1 |$. The BRST operator commutes with L_0, $\tilde{L}_0$ and with the vertex operators, but not with b_0 and $\tilde{b}_0$. The nonzero commutators involve

$$\left[b_0\tilde{b}_0 z^{\pm L_0-1}\bar{z}^{\pm\tilde{L}_0-1}, Q_B \right] = \pm b_0\partial_{\bar{z}}\left(z^{\pm L_0-1}\bar{z}^{\pm\tilde{L}_0} \right) \mp \tilde{b}_0\partial_z\left(z^{\pm L_0}\bar{z}^{\pm\tilde{L}_0-1} \right) .$$
$$(9.1.23)$$

These are total derivatives, illustrating the general principle (5.4.6) that the couplings of null states are proportional to total derivatives on moduli space. We have already seen an example of this for the massless vertex operators in section 6.6. As in that example, the integral vanishes by the canceled propagator argument. The integral at $z \to 0$ is defined by analytic continuation from large positive $(k_1 + k_2)^2$. The surface term vanishes identically there, and so by analyticity must vanish everywhere. The same argument applies to the other limits of moduli space.

For the four-particle amplitude on the sphere, the unitarity condition is

$$T_{S_2}(1;2;3;4) - T_{S_2}(-1;-2;-3;-4)^*$$

$$= i\sum_{j_5} \int \frac{d^{d-1}\mathbf{k}_5}{2E_5(2\pi)^{d-1}} T_{S_2}(1;2;5)T_{S_2}(-5;3;4)$$

$$+ 2 \text{ permutations}$$

$$= -ie^{-4\lambda} \sum_{j_5} \int \frac{d^{d-1}\mathbf{k}_5}{2E_5(2\pi)^{d-1}} \langle\langle \hat{\mathcal{V}}_1 | \hat{\mathcal{V}}_2(1,1) | \hat{\mathcal{V}}_5 \rangle\langle \hat{\mathcal{V}}_{-5} | \hat{\mathcal{V}}_3(1,1) | \hat{\mathcal{V}}_4 \rangle$$

$$+ 2 \text{ permutations} . \quad (9.1.24)$$

The intermediate particle is on-shell, the sum j_5 is over a complete set of physical states, and the right-hand side is summed over all ways to break 1234 into pairs: 12–34, 13–24, and 14–23.

Eq. (9.1.24) is what we wish to show. Using the reality condition (9.1.13) in the matrix element (9.1.20), the left-hand side of eq. (9.1.24) vanishes in the region where the z integral converges. The nonzero part comes from the $i\epsilon$ introduced in the analytic continuation,

$$\frac{1}{k^2 + m^2 - i\epsilon} - \frac{1}{k^2 + m^2 + i\epsilon} = 2\pi i\delta(k^2 + m^2) . \quad (9.1.25)$$

Thus,

$$T_{S_2}(1;2;3;4) - T_{S_2}(-1;-2;-3;-4)^*$$

$$= \frac{16\pi^2 i}{\alpha' e^{2\lambda}} \sum_{i,i'} \delta_{m_i^2,\tilde{m}_i^2}\delta(k_i^2 + m_i^2)\langle\langle \hat{\mathcal{V}}_1 | \hat{\mathcal{V}}_2(1,1)b_0\tilde{b}_0 | i \rangle \mathcal{G}^{ii'} \langle\langle i' | \hat{\mathcal{V}}_3(1,1) | \hat{\mathcal{V}}_4 \rangle$$

$$+ 2 \text{ permutations} . \quad (9.1.26)$$

The resemblance to the desired unitarity relation (9.1.24) is obvious, and a little bookkeeping and definition-chasing established the result. The index i in the sum (9.1.26), running over a complete set of string states, labels both internal state and momentum,

$$i \leftrightarrow (j, k) . \tag{9.1.27}$$

The delta functions and $b_0 \tilde{b}_0$ factor in eq. (9.1.26) project onto the space $\hat{\mathscr{H}}$ annihilated by L_0, $\tilde{L}_0$, b_0, and $\tilde{b}_0$, which was introduced in the discussion of the BRST cohomology. Thus,

$$\sum_{i,i'} \delta_{m_i^2, \tilde{m}_i^2} \delta(k_i^2 + m_i^2) b_0 \tilde{b}_0 |i\rangle \mathscr{G}^{ii'} \langle\!\langle i'|$$

$$= \delta_{L_0, \tilde{L}_0} \delta(4L_0/\alpha') b_0 \tilde{b}_0$$

$$= -\frac{i\alpha'}{16\pi^2 e^{2\lambda}} \int \frac{d^{d-1}\mathbf{k}}{2E(2\pi)^{d-1}} \sum_{\substack{j \in \hat{\mathscr{H}} \\ k^0 = \pm \omega_k}} |\hat{\mathscr{V}}_j(k)\rangle\langle\!\langle \overline{\hat{\mathscr{V}}}_j(k)| . \tag{9.1.28}$$

The normalization of the basis $|\mathscr{V}_j(k)\rangle$ follows from the OPE (9.1.2) and the expectation value (9.1.3), which imply

$$\langle\!\langle \overline{\hat{\mathscr{V}}}_j(k)|\tilde{c}_0 c_0|\hat{\mathscr{V}}_{j'}(k')\rangle = \frac{8\pi i e^{2\lambda}}{\alpha'} \delta_{jj'} (2\pi)^d \delta^d(k - k') . \tag{9.1.29}$$

Using the relation (9.1.28), the form and normalization of eq. (9.1.26) agree with the unitarity condition (9.1.24), leaving one important loose end. The sum in the unitarity relation runs only over the physical Hilbert space, the BRST cohomology, while the final sum in (9.1.28) runs over a complete basis for $\hat{\mathscr{H}}$. We now show that these sums are equal. Consider the following basis for $\hat{\mathscr{H}}$. First, take a complete set of 'unphysical states' $|a\rangle_U$. That is, no linear combination of these is annihilated by Q_B. Second, take a complete set of BRST-invariant states $|b\rangle_P$ which are independent in the sense that no linear combination of them is null. Finally, take the complete set of BRST-null states $|a\rangle_N = Q_B|a\rangle_U$ to complete the basis. Define the operators $\mathscr{P}$ and $\mathscr{U}$:

$$\mathscr{P}|b\rangle_P = |b\rangle_P , \quad \mathscr{P}|a\rangle_U = 0 , \quad \mathscr{P}|a\rangle_N = 0 ; \tag{9.1.30a}$$

$$\mathscr{U}|b\rangle_P = 0 , \quad \mathscr{U}|a\rangle_U = 0 , \quad \mathscr{U}|a\rangle_N = |a\rangle_U . \tag{9.1.30b}$$

Then

$$1 = \mathscr{P} + Q_B \mathscr{U} + \mathscr{U} Q_B . \tag{9.1.31}$$

Insert this decomposition into the factorization relation. The first term projects onto one state from each cohomology class, the physical spectrum appearing in the unitarity sum. The second and third terms each contain Q_B, and so vanish by the earlier argument that null states decouple. Thus unitarity is established.

It is worth noting that the argument just given is very similar to the argument that amplitudes in gauge theory are simultaneously Lorentz-invariant and unitary. The numerator of the covariant photon propagator

$$-i\frac{\eta^{\mu\nu}}{k^2} , \tag{9.1.32}$$

has a negative eigenvalue, corresponding to a negative norm intermediate state. Choose any null vector n^μ such that $n \cdot k = 1$, and decompose

$$\eta^{\mu\nu} = (\eta^{\mu\nu} - k^\mu n^\nu - n^\mu k^\nu) + k^\mu n^\nu + n^\mu k^\nu . \tag{9.1.33}$$

At $k^2 = 0$, the first term has $d - 2$ positive eigenvalues and no negative eigenvalues. If the photon is coupled to conserved currents at both ends, the last two terms vanish, so the unitarity relation is satisfied with just the $d - 2$ positive norm states expected for a photon in d dimensions. The decomposition (9.1.31) is the generalization of eq. (9.1.33), with $k^\mu \to |a\rangle_\mathrm{N}$ and $n^\mu \to |a\rangle_\mathrm{U}$.

To conclude, let us review how various general principles are illustrated by this example. BRST null states decouple, but only after integration on moduli space. The potential surface terms from the latter vanish by the canceled propagator argument. The ghost manipulations leading to eq. (9.1.19) have a natural interpretation in terms of the Riemann surface picture from chapter 5. Consider the sphere with vertex operators at $0, 1, z, \infty$. To be specific take $|z| > 1$ as in the second term of eq. (9.1.19). We can think of this sphere as being formed from two disks, with coordinates z_1 and z_2. Put vertex operators at $z_1 = 0, 1$ and at $z_2 = 0, 1$, and join the disks by the identification $z_1 = z/z_2$. For the modulus z, we have

$$\left.\frac{\partial z_1}{\partial z}\right|_{z_2} = \frac{z_1}{z} . \tag{9.1.34}$$

The ghost insertion (5.4.15) for dz is then b_0/z, and similarly for $d\bar{z}$, as illustrated by eq. (9.1.19).

9.2　Higher genus Riemann surfaces

There are many different representations of higher genus Riemann surfaces. The one that will be most useful for our purposes is based on the idea of building higher genus surfaces by sewing together lower genus surfaces. This will be described in the next section, but first for completeness we describe some other common constructions. The treatment is necessarily rather brief.

Schottky groups

Imagine removing two disks from a sphere and joining the edges of the resulting holes. This produces a handle, so the surface obtained in this way is a torus. Repeating g times, with a total of $2g$ holes, produces a genus g surface.

We can make this more precise as follows. Regarding the sphere as the complex z-plane plus point at infinity, consider the mapping

$$\gamma : \quad \frac{z' - U}{z' - V} = K \frac{z - U}{z - V} , \tag{9.2.1}$$

which depends on three complex parameters U, V, and K. This is a Möbius transformation, and in particular is holomorphic and one-to-one. The points U and V are fixed under γ. Taking $|K| > 1$, the surface expands away from U and contracts toward V. A fundamental region for γ is the annular region

$$|K|^{-1/2} < \left| \frac{z - U}{z - V} \right| < |K|^{1/2} . \tag{9.2.2}$$

Identifying points under γ glues together the inner and outer boundaries of the annulus (9.2.2), producing a torus. The group generated by g such elements (9.2.1) produces the genus g surface, with restrictions on the parameters so that the sewn circles do not overlap.

Every closed oriented Riemann surface can be obtained in this way. The range of the parameters is not simple, in part because some of the modular transformations act in a complicated way. Riemann surfaces with boundary and unoriented Riemann surfaces can be constructed by dividing by one or two reflections of a closed oriented surface, as we have illustrated in the lower genus examples.

Let us see that the Schottky construction produces the expected number of moduli. Each transformation (9.2.1) has three complex parameters for a total of $3g$. One can think of the six real parameters for each handle as being two for the position of each hole, one for the product of the radii of the holes (only the product, because if one hole is made smaller and the other larger we get the same surface), and one for the relative angle at which the holes are sewn. There is redundancy, because an overall Möbius transformation of the sphere gives an equivalent surface. Subtracting the six real parameters for this group leaves $6g - 6$, the correct number of moduli for genus $g \geq 2$ surfaces. For $g = 1$, two Möbius transformations leave the mapping invariant, so two real parameters remain.

The Schottky parameterization shows that the moduli space has natural complex coordinates. The path integral has certain holomorphicity properties with respect to these coordinates, though we will not emphasize this. Having constructed the general surface as a coset of the sphere, it is evident that Green's functions can be constructed by use of the method of images, summing over all images of a source. Functional determinants and other objects of interest can be expressed similarly.

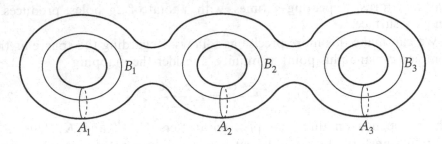

Fig. 9.1. Genus-three surface with A- and B-cycles marked.

Fuchsian groups

For this construction, consider first the world-sheet metric. It is a theorem that by a Weyl transformation one can always make the world-sheet curvature R a constant. At genus zero this gives the round sphere, at genus one the flat torus, and at higher genus a surface of constant negative curvature. The simply connected covering space for any of these will similarly have constant curvature. There are precisely three such simply connected spaces — the round sphere again, the flat plane (the covering space of the torus) and the upper half-plane with metric

$$ds^2 = \frac{dz\,d\bar{z}}{(\operatorname{Im} z)^2} \,. \tag{9.2.3}$$

The last is the covering space for all the genus $g \geq 2$ surfaces.

Correspondingly, all genus $g \geq 2$ surfaces can be obtained from the upper half-plane by dividing by the action of some discrete group. The elements of this group must be isometries of the metric (9.2.3); these isometries are just the Möbius transformations with real coefficients, $PSL(2,\mathbf{R})$. Figure 9.1 shows the genus-three surface with some closed curves drawn in, an A-cycle and B-cycle for each handle. These are nontrivial curves, in that they cannot be contracted to points. If we traverse one of these curves, on the covering space we come back not to the original point but to one of those identified with it. Thus to each of the cycles is associated an element of $PSL(2,\mathbf{R})$, also denoted A_i or B_i.

The A- and B-cycles are complete, in that every closed path is equivalent (homotopic) to a succession of A- and B-cycles. In fact they are slightly overcomplete: the path

$$(A_1 B_1^{-1} A_1^{-1} B_1)(A_2 B_2^{-1} A_2^{-1} B_2) \dots (A_g B_g^{-1} A_g^{-1} B_g) \tag{9.2.4}$$

can be contracted to a point, and the corresponding product must be the identity. Counting moduli, there are three real parameters in each $PSL(2,\mathbf{R})$ element, giving $6g$, minus three from the constraint (9.2.4), minus three more because overall conjugation by $PSL(2, \mathbf{R})$ gives an equivalent surface, for $6g - 6$ in all.

Although the Schottky and Fuchsian parameterizations are similar, they have rather different properties. In particular, the complex structure is much less evident in the latter, and the extension to the superstring correspondingly less simple.

The period matrix

The modulus τ of the torus has a natural $g \times g$ matrix generalization at genus g. To see this, we need first the *Abelian differentials* or holomorphic one-forms. These are world-sheet fields ω_a with the property that in complex coordinates

$$\partial_{\bar{z}} \omega_z = 0 , \quad \omega_{\bar{z}} = 0 . \tag{9.2.5}$$

There are precisely g such fields, and g more satisfying the conjugate relation. This follows from the Riemann–Roch theorem (5.3.22). The Abelian differentials (9.2.5) and their conjugates are just the kernel of the operator P_0^T, for which the Riemann–Roch theorem is

$$\dim \ker P_0^T - \dim \ker P_0 = -\chi = 2g - 2 . \tag{9.2.6}$$

The operator P_0 acts on scalar fields c and $\tilde{c}$ with $h = 0$, and so reduces to the ordinary gradient. This always has two zero modes ($c = $ constant or $\tilde{c} = $ constant), and so $\dim \ker P_0^T = 2g$.

The line integral of a one-form is a coordinate-invariant quantity. A natural basis for the Abelian differentials is then ω_{zi}, $i = 1, \ldots, g$, defined by

$$\oint_{A_i} dz \, \omega_{zj} = \delta_{ij} . \tag{9.2.7}$$

Then the *period matrix*

$$\tau_{ij} = \oint_{B_i} dz \, \omega_{zj} \tag{9.2.8}$$

characterizes the complex structure of the surface. At genus one, where ω_z is constant, this is the usual τ.

It can be shown that τ_{ij} is symmetric, so it has $\frac{1}{2} g(g + 1)$ complex parameters. For $g = 1, 2, 3$ this is equal to the number of complex moduli, but for $g \geq 4$ it is greater, so not all period matrices correspond to Riemann surfaces. The theta function (7.2.36) has a natural genus-g generalization, where τ becomes the period matrix and v, n, a, b become g-component vectors.

Hyperelliptic surfaces

Here is another description which is sometimes useful. Consider the space $S_2 \times S_2$, with coordinates z and y. Consider the two-dimensional submanifold given by the equation

$$y^2 = \prod_{i=1}^{2k} (z - Z_i) \qquad (9.2.9)$$

for complex parameters Z_i. Projecting down to the z-coordinate, there are two sheets joined across k branch cuts. This produces a surface with $g = k - 1$ handles. The branch points are singular when projected down to z, but the full surface (9.2.9) is nonsingular there: y is locally a good coordinate.

The surface depends on the $2k$ complex Z_i, but three of these are again redundant under a Möbius transformation, leaving $2k-3 = 2g-1$ complex parameters. For $g = 1, 2$ this is the same as the number of moduli, but for $g \geq 3$ the number of moduli is greater and only special surfaces have the form (9.2.9). These are known as *hyperelliptic surfaces*.

9.3　Sewing and cutting world-sheets

A useful concept for analyzing string amplitudes is the relation between higher and lower genus amplitudes. A key ingredient is the *plumbing fixture*, for building up higher genus surfaces from lower ones. Consider two world-sheets, M_1 with coordinate z_1 in some patch, and M_2 with coordinate z_2 in some patch. We can join these to form a new world-sheet M as follows. Let q be a complex parameter. Cut out from M_1 the disk $|z_1| < (1 - \epsilon)|q|^{1/2}$ for some small ϵ, and from M_2 the disk $|z_2| < (1 - \epsilon)|q|^{1/2}$. Now form a single surface by identifying pairs of points such that

$$z_1 z_2 = q . \qquad (9.3.1)$$

This is shown in figure 9.2. The annular region

$$(1 - \epsilon)^{-1}|q|^{1/2} > |z_1| > (1 - \epsilon)|q|^{1/2} \qquad (9.3.2)$$

is sewn as indicated. One notation for this construction is

$$M = M_1 \infty M_2(z_1, z_2, q) . \qquad (9.3.3)$$

The notation emphasizes that the resulting surface depends on the choice of local coordinates. With different local coordinates z_1' and z_2', one would be cutting out different circles and the joined surface will in general be different. The two patches might also be in different parts of a single

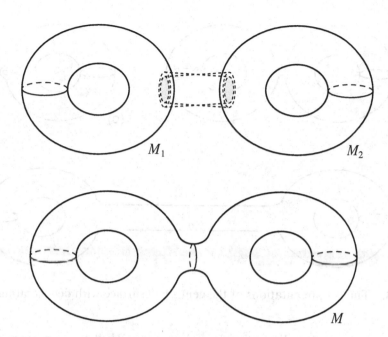

Fig. 9.2. Joining two world-sheets by removing a disk (shaded) from each and gluing the edges as indicated by the dashed lines.

surface M_0. The plumbing fixture construction then adds a handle to the surface. This is denoted

$$M = M_0 8(z_1, z_2, q) . \qquad (9.3.4)$$

The symbols '∞' and '8' are supposed to depict the two sewing processes. The parameter q is not actually needed — taking $z_1' = z_1 q^{-1/2}$, $z_2' = z_2 q^{-1/2}$, and $q' = 1$ gives an equivalent surface. It is useful to introduce q in order to discuss the behavior as a function of q with fixed coordinate patches.

We can also reverse this construction. Given any Riemann surface M, and any non-self-intersecting curve C on it, one can always find holomorphic coordinates $z_1 = z_2^{-1}$ in a neighborhood of C such that C is given by $|z_1| = |z_2| = |q|^{1/2}$. Cutting on C (undoing the identification $z_1 z_2 = q$) and gluing in the disks $|z_1| < |q|^{1/2}$ and $|z_2| < |q|^{1/2}$ produces M_1 and M_2 (or M_0) such that M would be reconstructed by the sewing operation — either (9.3.3) or (9.3.4) depending on whether the cut along C makes the surface disconnected.

For the genus-two example in figure 9.2, consider the case that $q \ll 1$. Figure 9.3(a) shows M in this case. We have smoothly interpolated the metric between $dz_1 d\bar{z}_1$ and $dz_2 d\bar{z}_2$ in the overlap region. The surface

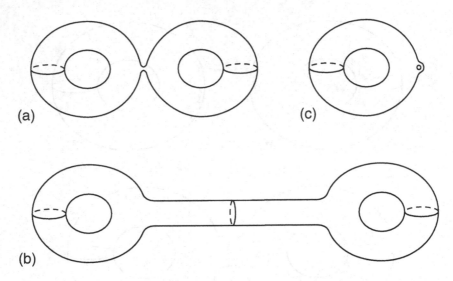

Fig. 9.3.   Three representations of the genus two surface with degenerating cycle.

pinches down to a radius of order $|q|^{1/2}$. By a Weyl transformation, one can find a metric in which the radius stays of order 1, as shown in figure 9.3(b). Since a Weyl transformation acts isotropically, the effect is that the *length* of the handle becomes large, of order $-\ln|q|$. The handle region $|z_1| < 1$, $|z_2| < 1$ is equivalent to the cylinder

$$\ln|q| < \operatorname{Im} w < 0 , \quad w \cong w + 2\pi , \tag{9.3.5}$$

where $z_1 = \exp(-iw)$ and $z_2 = q\exp(iw)$. Figure 9.3(c) shows one more representation of the same surface. The annular region $O(|q|) < |z| < O(1)$ surrounding the small handle is conformally equivalent to the long cylinder of figure 9.3(b). In terms of figure 9.3(a), one can think of this picture as a Weyl transformation where all of M_2 has been reduced.

In all the pictures, it is clear that the surface becomes special in the $q \to 0$ limit. The handle is said to *pinch* or *degenerate*. This is an example of a *boundary of moduli space*. The boundary of moduli space is important, because any divergences must come from there. In fact, we will see that every boundary is of this form, where one or more handles pinches. For the genus-two surface, there is one other way for a single handle to degenerate, if one of the A- or B-cycles pinches. This would correspond to building the genus-two surface from a single torus via the '8' operation (rather than from two tori via the '∞' operation) and taking $q \to 0$.

It is a useful fact that by repeated use of the cutting operation, every Riemann surface can be reduced to spheres with three punctures. A puncture is a marked point, with a specified complex coordinate near

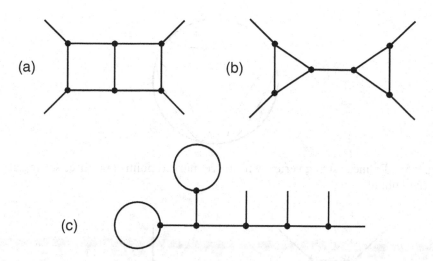

Fig. 9.4. Some sewing constructions of the genus-two surface with four operators. Each vertex is a sphere with three punctures, each internal line represents the plumbing fixture construction, and each external line represents a vertex operator.

the point; this is the data needed to sew in a plumbing fixture. In other words, every Riemann surface can be constructed from spheres with three punctures via sewing. For example, figure 9.4 shows three of the many ways to construct the genus-two surface with four vertex operators. There is one complex modulus q for each sewing operation, giving seven in all for each of the constructions in figure 9.4. These correspond to the $3g - 3 = 3$ complex moduli for the genus-two surface plus the positions of the four operators.

A graphical decomposition

The graphs in figure 9.4 are reminiscent of Feynman diagrams. We now describe a Feynman-diagram-like decomposition of moduli space. The vertices are rather more complicated than in figure 9.4, with any number of external lines and with internal corrections from string loops. The point of the construction is to make explicit the nature of the boundary of moduli space.

The statement is that the integral over moduli space is given by a sum of Feynman graphs. Each internal propagator in a graph represents the sewing operation, with the sewing parameter q integrated over some standard region such as $|q| < 1$. The n-point vertex represents a Riemann surface with n punctures. We will see that the surfaces in the vertices must be integrated over some *interior* region of moduli space, and also summed over genus. The sewing operation refers to a specific complex

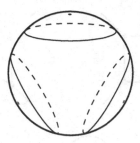

Fig. 9.5. Trilinear string vertex, with three marked points and three sewing circles on the sphere.

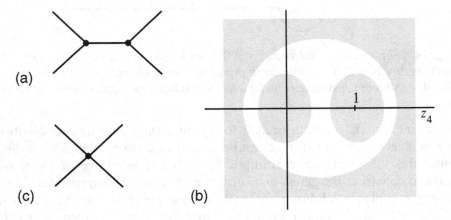

Fig. 9.6. (a) Constructing the sphere with four vertex operators. (b) The region of the z-plane covered by this graph. (c) The additional vertex needed to give the correct amplitude. The four-punctured sphere is integrated over the unshaded region.

coordinate for each point to be sewn, so the definition of the vertex must include such a choice at each point. The set of genus-g Riemann surfaces appearing in the n-point vertex is denoted $\mathcal{V}_{g,n}$. External lines represent vertex operators.

Let us illustrate this construction. Start with a three-point vertex $\mathcal{V}_{0,3}$, shown in figure 9.5 as a sphere with three marked points and three *sewing circles*. Sewing circles are unit circles in the complex coordinates associated with the respective points. Build a sphere with four points by taking two of these vertices, removing the interior of one sewing circle from each, and sewing them to the ends of a cylinder; this is equivalent to the plumbing fixture construction. This corresponds to the Feynman graph of figure 9.6(a); the length of the cylinder (with the circumference scaled to

2π) is $-\ln|q|$. Integrating over $|q| < 1$ and summing over the other two channels gives only part of the integration region for the sphere with four points, as shown in figure 9.6(b). The shaded regions are included but the unshaded interior region is missed.

Let us make this more explicit. For the trilinear vertex, put the points on the sphere at the standard $0, 1, \infty$ and choose coordinate frames $z^{(0)}$, $z^{(1)}$, and $z^{(\infty)}$ at each, so the three sewing circles are $|z^{(0,1,\infty)}| = 1$. The choice should be invariant (up to phase) under the six Möbius transformations that permute the vertex operators, so that the vertex is symmetric. There are many possible choices, for example

$$z^{(0)} = \frac{az}{z - 2} , \quad z^{(1)} = \frac{a(z - 1)}{z + 1} , \quad z^{(\infty)} = \frac{a}{2z - 1} , \tag{9.3.6}$$

for any constant a. For $a \geq 3$, the sewing circles do not overlap, guaranteeing that the plumbing fixture produces a good surface for all $q \leq 1$. As the reader can verify, sewing the $z^{(\infty)}$ and $z'^{(0)}$ patches from two spheres with parameter q produces a sphere with vertex operators at $0, 1, \frac{1}{2}(1 + a^2/q)$, and $\frac{1}{2}(1 - a^2/q)$. This is equivalent under a Möbius transformation to vertex operators at $0, 1, \infty$, and

$$z_4 = (q - a^2)^2/4a^2q . \tag{9.3.7}$$

As q runs over the unit disk, z_4 covers the outer shaded region of figure 9.6(b). Permuting the external vertex operators gives the other two regions. We could try to take a smaller, so the shaded regions grow, but they will eventually overlap and double-count some surfaces while other surfaces will still not be produced. They will never precisely fill the plane, so we need to add in an explicit four-point vertex $\mathscr{V}_{0,4}$ which is given by the sphere with four marked points, integrated over the unshaded part of the z-plane. This does not define $\mathscr{V}_{0,4}$ completely because there is also a choice of coordinate at each puncture for each value of z_4. It is always possible to make this choice in such a way as to connect smoothly at the boundary between the shaded and unshaded regions. Note that the Feynman graph of figure 9.6(a) includes all the asymptotic regions $z_4 \to 0, 1, \infty$, while the four-point vertex includes only nondegenerate surfaces. The behavior in the limits is then given explicitly by the $q \to 0$ limit of the plumbing fixture: this is why this construction is useful.

As another example, sewing two of the circles of a single vertex to the ends of a cylinder produces a torus with one vertex operator. This covers the limit $\text{Im}\,\tau \to \infty$ of the moduli space of the torus, but an interior region is missed and must be added by hand as a vertex region $\mathscr{V}_{1,1}$.

This construction can be extended inductively. The sewing operation always increases the parameter

$$r = 3g + n - 2 . \tag{9.3.8}$$

The vertex $\mathcal{V}_{0,3}$, the original sphere with three punctures, has $r = 1$. The 8 operation gives $r = r_0 + 1$, while the ∞ operation gives $r = r_1 + r_2$. Thus, $\mathcal{V}_{0,4}$ and $\mathcal{V}_{1,1}$ both have $r = 2$. It has been shown that one can construct $\mathcal{V}_{g,n}$ inductively in r, and so cover moduli space as claimed.

A more explicit construction is as follows (the proofs can be found in the references). On any Riemann surface there is a unique *minimal area metric*, defined as the metric of least area subject to the constraint that every nontrivial closed curve (those which cannot be continuously contracted to a point) has length at least 2π. Every point on the surface lies on at least one *saturating geodesic*, a nontrivial closed curve of length exactly 2π. Saturating geodesics that can be continuously deformed into one another (that is, are homotopically equivalent) cannot intersect. Each set of homotopically equivalent saturating geodesics forms a band (topologically an annulus) on the surface. The different bands cover the surface completely, and may overlap. The height h of such a band is defined as the minimum distance between its two bounding geodesics. The idea of a minimal area metric can be extended to surfaces with punctures, with the specification that a curve that circles a puncture is regarded as nontrivial. The minimal area metric near a puncture is of the form $dz d\bar{z}/z\bar{z}$, a semi-infinite cylinder of radius 2π. There are thus two kinds of bands: the finite ones in the interior of the surface, and the semi-infinite ones centered on the punctures, terminating in the interior of the surface.

We can now describe $\mathcal{V}_{g,n}$ in this construction. Take all Riemann surfaces with n punctures, with the restriction that the height of any internal band, in the minimal area metric, is less than 2π. Cut off the semi-infinite cylinders, leaving a stub of length π, and glue in the usual unit disk with vertex operator in the end of the stub. With these vertices, the Feynman graph sum covers moduli space exactly once. It can be shown that the sewing of these minimal area surfaces still produces a minimal area surface. Bands of height less than 2π are explicitly included in the vertices. Bands of height greater than 2π are produced by the sewing operation on the stubs, the resulting height being 2π from the stubs plus $-\ln|q|$ from the plumbing fixture; this runs from 2π to ∞ when q is integrated. The minimum length and maximum height restrictions guarantee that degenerating surfaces do not appear in the vertices, so they arise only from one or more qs going to zero, as claimed.

9.4 Sewing and cutting CFTs

We have described higher genus Riemann surfaces as built up by sewing together lower genus surfaces. Using the sewing properties of path integrals, the CFT on the higher genus surface can similarly be related to

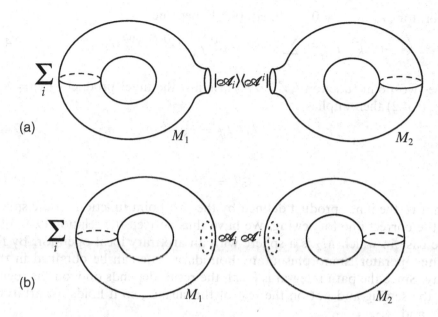

Fig. 9.7. (a) The path integral on M is cut open by inserting a complete set of boundary states. (b) Each state is replaced by a disk with local operator.

lower order amplitudes. The relation is shown pictorially in figure 9.7. For clarity we first discuss the case $q = 1$. The path integral on M is cut open on the circle $|z_1| = |z_2| = 1$, inserting a complete set of boundary states. Each boundary state is then replaced by a disk with the corresponding local operator, leaving the uncut world-sheets M_1 and M_2. The precise form of the result is

$$\left\langle \,\ldots_1\ldots_2\,\right\rangle_M = \sum_{ij} \left\langle\,\ldots_1\,\mathscr{A}_i^{(z_1)}\,\right\rangle_{M_1} \mathsf{G}^{ij} \left\langle\,\mathscr{A}_j^{(z_2)}\,\ldots_2\,\right\rangle_{M_2}. \qquad (9.4.1)$$

We are allowing arbitrary additional insertions $\ldots_1$ on M_1 and $\ldots_2$ on M_2; these must of course lie outside the circles $|z_1| = 1$ and $|z_2| = 1$ that are being cut. We have not specified any particular orthonormality for the basis; the appropriate inverse metric G^{ij} is to be determined. The insertions $\mathscr{A}_i$ and $\mathscr{A}_j$ that replace the handle are implicitly at the origin in the z_1 and z_2 frames respectively.

Now, let us consider the case in which M_2 is a sphere, with the sewing coordinate z_2 being u and with $\ldots_2$ being some local operator $\mathscr{A}_k$ at $z = 0$. We have the identifications $z_1 u = 1$ from the plumbing fixture and $zu = 1$ from the sphere, so we can just extend $z_1 = z$ through the whole sewn region. The sewn surface M is the same as M_1 with an additional operator $\mathscr{A}_k$ at $z_1 = 0$. Then eq. (9.4.1) becomes

$$\left\langle \cdots_1 \mathscr{A}_k^{(z_1)} \right\rangle_{M_1} = \sum_{ij} \left\langle \cdots_1 \mathscr{A}_i^{(z_1)} \right\rangle_{M_1} \mathsf{G}^{ij} \left\langle \mathscr{A}_j^{(u)} \mathscr{A}_k^{(z)} \right\rangle_{S_2} . \qquad (9.4.2)$$

The two-point function $\langle \mathscr{A}_j^{(u)} \mathscr{A}_k^{(z)} \rangle_{S_2}$ is just the inner product $\langle\!\langle j|k \rangle = \mathscr{G}_{jk}$. Eq. (9.4.2) then implies

$$\mathsf{G}^{ij} \mathscr{G}_{jk} = \delta^i_k \qquad (9.4.3)$$

or

$$\mathsf{G}^{ij} = \mathscr{G}^{ij} . \qquad (9.4.4)$$

That is, the inner product defined by the two-point function on the sphere is the correct one for sewing. We have thus derived the relation (9.4.1) for the case in which M_2 is a sphere with an arbitrary local operator. By the state–operator isomorphism, any boundary state can be obtained in this way. Since the path integral is local, the result depends only on the region of the sewing and not on the rest of the surface, so it holds for arbitrary M_2 and $\cdots_2$.

Sewing with parameter q is the same as sewing with $q = 1$ and local coordinates z_1 and $z'_2 = z_2/q$. We thus obtain eq. (9.4.1) with $\mathscr{A}_j$ in the primed frame, and so obtain an extra factor from the conformal transformation back to z_2,

$$\mathscr{A}_j^{(z'_2)} = q^{h_j} \bar{q}^{\bar{h}_j} \mathscr{A}_j^{(z_2)} . \qquad (9.4.5)$$

The result is

$$\left\langle \cdots_1 \cdots_2 \right\rangle_M = \sum_{ij} q^{h_j} \bar{q}^{\bar{h}_j} \left\langle \cdots_1 \mathscr{A}_i^{(z_1)} \right\rangle_{M_1} \mathscr{G}^{ij} \left\langle \mathscr{A}_j^{(z_2)} \cdots_2 \right\rangle_{M_2} , \qquad (9.4.6)$$

where we have taken a basis diagonal in the weights. For the '8' operation the path integrals on M and M_0 are related in the same way:

$$\langle \cdots \rangle_M = \sum_{ij} q^{h_j} \bar{q}^{\bar{h}_j} \mathscr{G}^{ij} \left\langle \cdots \mathscr{A}_i^{(z_1)} \mathscr{A}_j^{(z_2)} \right\rangle_{M_0} . \qquad (9.4.7)$$

It is interesting to consider again the boundary of moduli space, given by the $q \to 0$ limit of one or more handles. We see that the operators of lowest weight will control the behavior. The results (9.4.6) and (9.4.7) resemble the OPE, and in fact (9.4.6) reduces to the OPE when M_2 is a sphere and '$\cdots_2$' is a pair of vertex operators. This analogy is especially clear from figure 9.3(c), representing the limit as the insertion of a small handle. The sewing relation shows that not only a product of operators, but also a small handle, or any other disturbance that we might sew into a small circle on the world-sheet, is equivalent to a sum of local operators. The result (9.4.6) is an elegant generalization of the OPE. From the point of view of M_1, there is a tiny copy of M_2 glued in near $z_1 = 0$. This can be expanded in local operators $\mathscr{A}_i$, with coefficient functions $\langle \mathscr{A}^i \ldots_2 \rangle_{M_2}$. From the point of view of M_2, there is a tiny copy of M_1 glued in near $z_2 = 0$, which is expanded in local operators $\mathscr{A}^i$, with coefficient functions $\langle \ldots_1 \mathscr{A}_i \rangle_{M_1}$.

We have argued that every Riemann surface can be reduced to thrice-punctured spheres. The sewing construction thus relates all amplitudes to the three-point amplitudes on the sphere, which are essentially the operator product coefficients: *these are the data that define the CFT.* We will encounter many equivalences between CFTs — we have already seen T-duality, and bosonization (chapter 10) will be another notable example. The sewing construction provides the criteria for two theories to be the same: a one-to-one correspondence between their states (or local operators) and identical OPEs.[1] In fact, as we will show in chapter 15, the operator products of the primary fields alone determine those of all the others.

Let us explore this point of view further. One can cut a surface along many different sets of curves, so there must be consistency conditions to guarantee that the sewing relation always gives the same result for the expectation values on M. Consider first an infinitesimal change in the cutting curve C. A general infinitesimal change leaving M invariant can be obtained by an infinitesimal change in the coordinates used for the sewing,

$$z_1' = z_1 + \sum_{n=-\infty}^{\infty} \epsilon_n z_1^{n+1} , \tag{9.4.8a}$$

$$z_2' = z_2 - \sum_{n=-\infty}^{\infty} \epsilon_{-n} q^{-n} z_2^{n+1} . \tag{9.4.8b}$$

The difference of the sewn expressions is then proportional to

$$\mathscr{A}_i^{(z_1')} \mathscr{G}^{ij} \mathscr{A}_j^{(z_2')} - \mathscr{A}_i^{(z_1)} \mathscr{G}^{ij} \mathscr{A}_j^{(z_2)} = - \sum_{n=-\infty}^{\infty} \epsilon_n \Big[L_n \cdot \mathscr{A}_i^{(z_1)} \mathscr{G}^{ij} \mathscr{A}_j^{(z_2)}$$
$$- \mathscr{A}_i^{(z_1)} \mathscr{G}^{ij} L_{-n} \cdot \mathscr{A}_j^{(z_2)} \Big] + (\text{anti}) , \tag{9.4.9}$$

where we have used the conformal transformation (2.9.6), and 'anti' denotes the same form with $\epsilon_n \to \epsilon_n^*$ and $L_n \to \tilde{L}_n$. The vanishing of the quantity in braces can be shown by an argument parallel to that used to derive eq. (9.4.4). This is just the statement that one can pull the contour

---

[1] We should note that when there is more than one $(2,0)$ plus $(0,2)$ field that could play the role of T_{ab}, as in the free-field examples of section 2.5, it is necessary to specify which is the energy-momentum tensor. This choice determines the conformal transformations of operators, which enter implicitly into the sewing relation.

integral that defines the generator through the sewn handle; the change
from L_n to L_{-n} is from an inversion of coordinates. All other contour
integrals for conserved charges can be pulled through as well.

There are also sets of curves that are not related by a continuous
deformation. The four-point amplitude on the sphere can be written in
terms of three-point functions by cutting on curves circling different pairs
of operators. Equality between different sewings is the associativity of the
OPE, figure 2.8.

Another example is given by the torus: we can reduce the torus to
the sphere by cutting on the A-cycle, or the B-cycle, or any other non-
self-intersecting closed curve. In terms of the plane with identifications, a
closed curve C that begins at $(\sigma^1, \sigma^2) = (0,0)$ must end at an identified
point

$$2\pi(m,n) . \tag{9.4.10}$$

All curves with the same m and n values can be continuously deformed
into one another, so we need to consider the relation between sewing on
curves of different m and n. One can show that for a nontrivial non-self-
intersecting curve, m and n must be relatively prime, which implies that
there are integers p and q such that $mp - nq = 1$. There is then a modular
transformation

$$(\sigma^1, \sigma^2) = (\sigma'^1, \sigma'^2) \begin{bmatrix} m & n \\ q & p \end{bmatrix} , \tag{9.4.11}$$

such that the curve C comes back to

$$(\sigma'^1, \sigma'^2) = 2\pi(1,0) , \tag{9.4.12}$$

which is the A-cycle. So cutting on a general nontrivial non-self-intersecting
curve is equivalent to cutting on the A-cycle for a modular-equivalent
value of τ. The consistency of the sewing construction is thus equivalent
to modular invariance.

Every set of sewing curves may be transmuted into any other by a
series of two moves: the four-point associativity transformation on the
sphere and the modular transformation on the torus with a *single* vertex
operator. Figure 9.4 makes this plausible. For example, figures 9.4(a)
and 9.4(b) are related by one associativity condition. Repeated applications
of associativity can bring the sewing to the form of figure 9.4(c), where all
loops are expressed in terms of the one-loop, one-point function. So these
are the conditions for an arbitrary set of operator product coefficients
to define a sensible conformal field theory. The program of solving the
duality constraints is known as the *conformal bootstrap*. It has been carried
out only in cases where additional restrictions have been introduced, as
we will discuss in chapter 15.

All the considerations of this section extend to world-sheets with boundary, where in addition to degenerating handles there may be degenerating strips. The consistency conditions have been generalized to this case.

9.5 General amplitudes

We now have assembled all the ingredients that we need in order to understand the consistency of general string amplitudes.

Having given a graphical decomposition of moduli space, the sewing formalism allows us to give a diagrammatic expression for the general string amplitude. Consider first the propagator, from the plumbing fixture. The ghost insertion for the modulus q is

$$\oint \frac{dz}{2\pi i} b_{z_1 z_1} \frac{\partial z_1}{\partial q} \bigg|_{z_2} = \frac{b_0}{q}, \tag{9.5.1}$$

and similarly for $\bar{q}$ it is $\tilde{b}_0/\bar{q}$. Including these insertions, and using the sewing expression (9.4.6), the factors associated with the propagator are

$$\int_{|q|<1} \frac{d^2 q}{q\bar{q}} \, q^{\alpha'(k_i^2+m_i^2)/4} \bar{q}^{\alpha'(k_i^2+\tilde{m}_i^2)/4} \mathcal{G}^{ij} b_0 \tilde{b}_0 = \frac{8\pi \delta_{m_i^2, \tilde{m}_i^2} \mathcal{G}^{ij} b_0 \tilde{b}_0}{\alpha'(k_i^2 + m_i^2 - i\epsilon)}. \tag{9.5.2}$$

The n-string coupling is given by inserting vertex operators into the surfaces appearing in the vertices of the sewing construction,

$$g(i_1, \ldots, i_n) = \sum_g \int_{\mathcal{V}_{g,n}} d^m t \left\langle \prod_{k=1}^m B_k \prod_{l=1}^n \hat{\mathcal{V}}_{i_l} \right\rangle_g. \tag{9.5.3}$$

where $m = 6g + 2n - 6$ is the number of moduli. The Feynman graphs constructed from the couplings (9.5.3) and propagators (9.5.2) then reproduce the full string amplitude. The label i again stands for the state and momentum of each string. The tree amplitude in section 9.1 illustrates this structure.

The first issue is the finiteness of the amplitude. We assert that divergences can arise only from the boundary of moduli space, as in the various low order examples. At finite values of the moduli, the sewing formula gives a sum over an infinite number of string states, but this sum converges because it is ultimately just the expansion of a quantum state in a complete set. Actually, convergence is guaranteed only in a unitary CFT, but the nonunitary parts of the string CFT are quite tame. The X^0 CFT is defined by analytic continuation from a unitary CFT, while the ghost CFT just generates a Faddeev–Popov determinant and so is finite if we have made a good choice of gauge. Now, the boundary has been made explicit as the $q \to 0$ limit of the various propagators, and the integral (9.5.2) is one that we have encountered several times before. It is

defined via analytic continuation, giving an ordinary field-theoretic propagator. The only singularities are then the poles in propagators, which
represent production of on-shell particles in the intermediate states. In
particular they are long-, not short-, distance divergences.

The other issue is unitarity. We studied this for the four-point tree
amplitude in some detail. Having brought the general amplitude to a
Feynman-diagram form, rather than work out the combinatorics in detail
we will simply cite standard field theory results on the unitarity of the
diagrammatic expansion. Also, the demonstration at the end of section 9.1
that the sum over intermediate states reduces to a sum over physical states
generalizes as in field theory.

In summary, the issues involved in the perturbative finiteness and unitarity of string theory appear to be well understood. Observe once again
that we have not assumed 26 flat dimensions, but merely d flat dimensions
plus a compact positive norm CFT of central charge $26 - d$. Of course,
any discussion of the unitarity of the flat spacetime S-matrix for the
bosonic string is somewhat formal, since flat spacetime is not a solution
of the theory — first there is the tachyon instability, then the cosmological
constant induced by loops.

An alternative approach to the finiteness and unitarity of string theory,
based on the light-cone gauge, has also been extensively developed. This
gauge gives up covariance, but has certain advantages — it makes possible
a rather explicit description of moduli space, and the amplitudes can be
represented in a field-theoretic form with a Hamiltonian that is local in
the light-cone time x^+. Covariance has then been established by showing
that the perturbation theory is equivalent to the covariant perturbation
theory.

String divergences

Before moving on, let us expand somewhat further this point that string
divergences are long-distance effects. Consider the one-loop amplitude with
n vertex operators, taking the limit that m operators come together with τ
held fixed as in figure 9.8(a). The string loop of course includes gravity, and
figure 9.8(a) is reminiscent of the graviton loop of figure 1.1(c), which was
UV divergent when the interaction vertices came together in spacetime.
There is indeed a divergence in this string amplitude, but it is from long
spacetime distances, not short ones. Figure 9.8(b) shows a conformally
equivalent picture where the m vertex operators are connected to the rest
of the torus by a long cylinder. Cutting the cylinder open and inserting a
complete set of states, we obtain the usual expression (9.5.2), with $q \rightarrow 0$
being the limit where the vertex operators become coincident. So the only
singularities are the usual poles from propagation over long spacetime

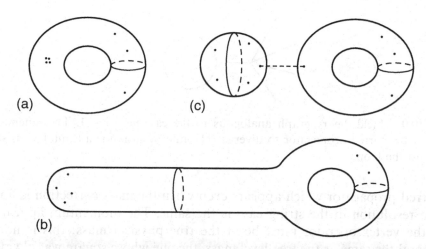

Fig. 9.8. (a) Torus with n vertex operators, in the limit that m of the operators come together. (b) Conformally equivalent picture: the m vertex operators at the end of a long cylinder. (c) Sewing construction, in terms of the product of a sphere with $m + 1$ vertex operators and a torus with $n - m + 1$.

distances. Conformal invariance once again changes short distances on the world-sheet into long distances in spacetime.

The cases $m = n$ and $m = n - 1$ present special problems. When $m = n$, meaning that all the vertex operators are coming together, momentum conservation forces k^μ to vanish. For massless states, the pole is then $1/0$ and the usual analytic continuation does not help because momentum conservation forces the amplitude to sit right on the massless pole. This is a tadpole, just as in section 7.4, and the resolution is the same, a correction to the background fields. The divergence here is proportional to the torus amplitude with one vertex operator.

In the other problematic case, $m = n - 1$, all the vertex operators but one, say $\mathscr{V}_n$, come together. Momentum conservation requires that the momentum flowing through the cylinder be $k^\mu = -k_n^\mu$. However, conformal invariance requires that $k_n^2 = -m_n^2$, so the intermediate momentum is fixed on top of a pole for those intermediate states with $m_j^2 = m_n^2$. Once again, there is an analogous divergence in field theory, shown in figure 9.9: the internal propagator is $1/0$. This divergence occurs whenever the mass of the internal line j is the same as that of the on-shell external particle n. The interpretation here is simple. The loop represents a radiative correction to the mass of the particle n. The S-matrix is defined with the external momentum taken on the *true* mass shell, including radiative corrections, and it is again free of this IR divergence. Note that the issue here is not some UV divergence internal to the loop, but the IR divergence on the

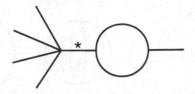

Fig. 9.9.   Field-theory graph analogous to the case $m = n - 1$. The kinematics force the starred propagator to diverge. The interpretation is a (finite) mass shift due to the loop.

starred propagator which appears even when the mass correction is finite. The resolution in the string case is the same. The momentum appearing in the vertex operator must be at the true physical mass; this is not in general the same as the tree-level mass, the one whose square was always a multiple of $1/\alpha'$. Note that for certain massless fields — in particular gauge bosons — gauge invariance guarantees that they will remain massless to all orders. Incidentally, for massive states the mass shift is usually complex, because these states are unstable to decay into lighter string states.

There is a puzzle here. We saw early on that consistency of string theory, the condition that the world-sheet theory be conformally invariant, required the beta functions (3.7.14) to vanish. These beta functions contain no information about higher genus surfaces — they correspond to the tree-level string field equations. What happens when we try to expand around a loop-corrected background? In fact things work out just right. The dangerous limit in figure 9.8 gives rise not only to a divergence, but also to a conformal anomaly. To see this heuristically, picture the limit as in figure 9.3(c), as a small handle glued into the sphere. The small handle can be represented by a series of local operators, and in particular from the massless level there is

$$\partial X^\mu \bar{\partial} X_\mu \frac{dq d\bar{q}}{q\bar{q}} \,. \tag{9.5.4}$$

Now, cut off the divergence in a coordinate-invariant way by requiring the size of the handle to be greater than a; that is, $|q|e^\omega > a$, where the metric on the sphere is $dz d\bar{z} e^{2\omega}$. The lower limit on the integral (9.5.4) gives

$$- 4\pi \partial X^\mu \bar{\partial} X_\mu \ln(a e^{-\omega}) \,. \tag{9.5.5}$$

Thus we have new dependence on the Weyl factor of the metric: this corrects the sigma-model Weyl anomaly $\beta^G_{\mu\nu}$, eq. (3.7.14). To obtain a Weyl-invariant theory, the background fields must satisfy the full *string-loop-corrected* Weyl invariance conditions (or, equivalently, BRST invariance). The Weyl anomaly from the shifted background cancels that from the small handle. This is the *Fischler–Susskind mechanism*. From figure 9.3(c), we can think of the divergence and the Weyl anomaly as world-sheet UV effects, not from the usual short-wavelength fluctuations of the fields, but from a small topological fluctuation. We should emphasize that on a *fixed* Riemann surface of any genus, the local structure is unaffected by the topology; it is the integration over Riemann surfaces that produces the divergence and Weyl anomaly.

This can also be regarded as a breakdown of the canceled propagator argument. Because the kinematics forces the amplitude to be at the pole, we cannot appeal to analytic continuation to eliminate possible surface terms under a BRST variation. In fact they appear, corresponding to the Weyl anomaly (9.5.5). In the loop-corrected background, the BRST anomaly from the shift of the background cancels that from the boundary of moduli space — the stringy technicalities all fit together to give a consistent spacetime picture.

To implement this in string perturbation theory is rather clumsy, but fortunately it is rarely necessary. First, in most supersymmetric theories the backgrounds are uncorrected. Second, for the analysis of long-distance physics in string theory it is usually most efficient to proceed in two steps, first using string theory to derive the form of the low energy effective action and then using the effective field theory at long distance. One then does not need to deal with the IR divergences at the string level. In the present case, the tadpoles imply a potential energy term

$$\delta S = -\sum_\chi \Lambda_\chi \int d^d x \, (-G)^{1/2} e^{-\chi\Phi} \tag{9.5.6}$$

in the effective action. Here Λ_χ is the tadpole from Euler number χ, and the dependence of the action on Φ follows from the fact that e^Φ is the string coupling. The separate tadpoles for the graviton and dilaton can be extracted from the string amplitude, with some effort.

The same idea extends to the vertex operator correction from the mass shift, and to other IR divergences such as those of exclusive amplitudes in theories with massless particles. These field-theory divergences must also appear in the corresponding string-theory amplitudes, since string theory reduces to field theory at long distance. Again we emphasize that while UV divergences in field theory generally signify a fundamental limitation on the range of validity of the theory, IR divergences usually mean that one is simply asking the wrong question (wrong background, wrong mass, exclusive versus inclusive cross-section, etc.) in what is otherwise a perfectly consistent theory. Any such divergences in string theory should then be dealt with precisely as in field theory.

Let us review what we have learned. Modulo problems particular to the bosonic string, the S-matrix is finite and unitary. The basic mechanism is modular invariance, which cuts off the UV divergences without spoiling the spacetime gauge invariance. Of course, knowing that string theory makes sense in weak coupling perturbation theory around flat spacetime is hardly enough. We would like to know that the theory makes sense beyond perturbation theory, and have some means of calculating nonperturbative corrections. Also we would like to know what string theory implies for the nature of spacetime at short distance. In the remainder of this chapter we briefly describe some of the attempts that have been made in these directions.

9.6   String field theory

It is interesting to compare the approach we have taken in string theory to the way quantum field theory is formulated. The world-sheet path integral is analogous to a sum over particle paths, often called the *first-quantized* description. In quantum field theory, on the other hand, we start with a path integral (or its Hamiltonian equivalent) over spacetime fields. One could instead develop quantum field perturbation theory in the first-quantized form; it is necessary to introduce explicit interaction vertices, spacetime points where n particle paths end. In string theory this second step is not necessary — there are no special points, and the sum over topologies produces the interactions. However, the result is only the perturbation expansion, and there are important phenomena that are not contained in this expansion. In the Standard Model, these include quark confinement and the finite-temperature baryon number violation of the weak interaction. There does not seem to be any useful sense in which these are contained in a resummation of the perturbation series; rather they require a more general formulation of the theory.

In the previous section we have expressed string amplitudes in a propagator and vertex form that looks like the perturbation theory for an infinite-component field theory. Could we instead derive this from a path integral over an infinite set of spacetime fields? And is there some principle, extending gauge invariance and general coordinate invariance, that determines the form of the theory? This is the subject of *string field theory*. We will be able to sketch here only a few of the ideas, ones that are particularly close to the approach that we have been taking. The reader should consult the literature for more details.

We will start with the guess that we should take the string wavefunctional $\Psi[X]$ and elevate it to a field operator, or equivalently to a path integral variable of integration — so-called second quantization. Now, $\Psi[X]$ depends on the curve $X^\mu(\sigma)$ traced out by the string in spacetime, and roughly speaking creates or destroys a string in this configuration. However, we need to be more precise. For example, should we require Ψ

to be invariant under reparameterization of the curve, and should it also depend on the Polyakov metric g_{ab}? Things seem to work much more nicely if one starts with the BRST-invariant form of the theory, including the Faddeev–Popov ghosts, and then second quantizes. From the commutation relations, b and c are conjugate to one another (as are $\tilde{b}$ and $\tilde{c}$), so if for example we treat c and $\tilde{c}$ as coordinates the wavefunctional is $\Psi[X, c, \tilde{c}]$. We then have a path integral running over all such functionals,

$$\int [d\Psi] \exp(iS[\Psi]) . \tag{9.6.1}$$

We will consider first the open string, describing one particular approach.

What symmetry principle will restrict the string action? From our study of the string vertex operators and null states, we might guess the symmetry

$$\delta \Psi = Q_B \Lambda \tag{9.6.2}$$

for any functional Λ. There is an extremely simple action with this invariance,

$$S_0 = \frac{1}{2} \langle\!\langle \Psi | Q_B | \Psi \rangle\!\rangle . \tag{9.6.3}$$

The equation of motion is just

$$Q_B \Psi = 0 . \tag{9.6.4}$$

One way to see that this is reasonable is by expanding the string field Ψ in terms of an infinite number of ordinary fields, one for each internal state of the string. Let $\Psi_i[X', c, \tilde{c}]$ be a complete set of wavefunctions for the internal modes — all but the center-of-mass x^μ. Then a general state can be expanded:

$$\Psi[X, c, \tilde{c}] = \sum_i \Phi_i(x) \Psi_i[X', c, \tilde{c}] , \tag{9.6.5}$$

the expansion coefficients being functions of the remaining variables x^μ. The string field path integral becomes an infinite product of path integrals over the component functions,

$$\int [d\Psi] \rightarrow \prod_i \int [d\Phi_i] . \tag{9.6.6}$$

As usual, we can start with the ground state functional Ψ_0 and build the remaining states with the raising operators. For example, keeping all

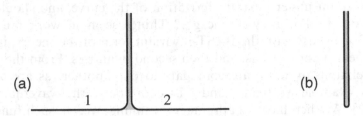

Fig. 9.10. (a) $\Psi_1 * \Psi_2$, convoluting the right half of Ψ_1 with the left half of Ψ_2. One can think of this as defined by a path integral over the shaded region, whose width is then taken to zero. (b) $\int \Psi$, convoluting the right and left halves of Ψ.

states at the first two levels,

$$\Psi = \Big[\varphi(x) + A_\mu(x)\alpha^\mu_{-1} + B(x)b_{-1} + C(x)c_{-1} + \varphi'(x)c_0$$
$$+ A'_\mu(x)\alpha^\mu_{-1}c_0 + B'(x)b_{-1}c_0 + C'(x)c_{-1}c_0 + \ldots\Big]\Psi_0 . \quad (9.6.7)$$

Here, φ is the tachyon field and A_μ the gauge field, while B and C are the Faddeev–Popov ghost fields associated with the spacetime gauge symmetry. The primed fields can all be gauged away or are auxiliary (that is, they satisfy algebraic rather than differential equations and can be eliminated). One can then check that the symmetry (9.6.2) includes the ordinary gauge transformation on A_μ, while the action is of Klein–Gordon form for φ and of Maxwell form for A_μ. The BRST string field formalism thus gives a very compact generalization of spacetime gauge invariance.

Interactions also work out rather nicely. Define a product operation $\Psi_1 * \Psi_2$ on string fields by the functional integral depicted in figure 9.10(a), forming the overlap of the right half of Ψ_1 with the left half of Ψ_2. Define also an operation $\int$, giving the overlap of the right and left halves of the single string wavefunctional as shown in figure 9.10(b). To make an interacting theory, add a nonlinear term to the gauge transformation,

$$\delta\Psi = Q_B\Lambda + g\Psi * \Lambda - g\Lambda * \Psi . \quad (9.6.8)$$

The $*$ product is associative, and one can show by a contour argument that Q_B acts as a derivation,

$$Q_B(\Psi_1 * \Psi_2) = (Q_B\Psi_1) * \Psi_2 + \Psi_1 * (Q_B\Psi_2) . \quad (9.6.9)$$

From these it follows that the symmetry (9.6.8) closes. Using the cyclic property,

$$\int \Psi_1 * \Psi_2 = \int \Psi_2 * \Psi_1 , \quad (9.6.10)$$

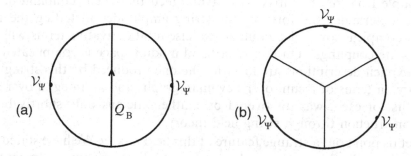

Fig. 9.11. CFT representation of the string field action: (a) the kinetic term; (b) the interaction. A $z^{2/3}$ conformal transformation is used to turn the usual semi-circle around each vertex operator into a wedge.

we can write the invariant action

$$S = \frac{1}{2} \int \Psi * Q_B \Psi + \frac{2g}{3} \int \Psi * \Psi * \Psi \, . \tag{9.6.11}$$

This is also usefully written in terms of CFT, replacing Ψ with a half-disk with corresponding vertex operator $\mathcal{V}_\Psi$,

$$S = \frac{1}{2} \langle \mathcal{V}_\Psi Q_B \cdot \mathcal{V}_\Psi \rangle_{D_2} + \frac{2g}{3} \langle \mathcal{V}_\Psi \mathcal{V}_\Psi \mathcal{V}_\Psi \rangle_{D_2} \, . \tag{9.6.12}$$

The vertex operator positions and conformal frames are not explicitly denoted in eq. (9.6.12) but are as shown in figure 9.11. A $z^{2/3}$ mapping is used to fit the three vertex operator half-disks into the disk. This action passes an important check. The path integral on the disk is nonzero only for insertions of total $Q^g = 3$. The vertex operators for the physical states all have ghost number 1, so both terms in the action (9.6.12) are nonvanishing. Since the action (9.6.11) is invariant, we might expect that for the gauge field it reduces to the non-Abelian Yang–Mills action, and this is the case.

The Siegel gauge is $b_0 \Psi = 0$. In the kinetic action, only the term $c_0 L_0$ from Q_B survives. Since in components $L_0 \propto p^2 + m^2$, this is analogous to the familiar covariant gauge for electrodynamics. The propagator

$$\int_0^\infty dt \, \exp(-t L_0) = L_0^{-1} \tag{9.6.13}$$

is simply the integral over strips of any length. The vertex is figure 9.11(b) with the wedges removed, joining the ends of three strips in a 'Y.' This has been shown to be equivalent to open string theory described as a sum over Riemann surfaces, including in particular a complete single cover of moduli space. On the one hand this is not surprising: the string field theory has the spacetime gauge invariance (9.6.8), so that null states will decouple from the S-matrix. We have argued that the decoupling of null states determines the form of the string amplitudes including the need for a complete cover of moduli space (else total derivative terms will give null state couplings). On the other hand moduli space is complicated and few explicit descriptions are known. The one produced by this string field theory, in terms of a sum over Feynman graphs and an integral over strip lengths for each, was discovered by mathematicians only shortly before its construction through string field theory.[2]

Let us point out a strange feature of this field theory. We have stated that the string field theory covers the entire moduli space, but we know from the annulus that this includes limits that produce *closed string* poles, even though there are no explicit closed string fields. The closed string poles come from strips going to zero, rather than infinite, length.[3] The precise statement is that the open string field path integral (9.6.1) generates the amplitudes of open *plus* closed string theory (except that surfaces with no boundaries at all, to which open strings cannot couple, are not generated). In this respect it is as though closed strings were in some imprecise sense 'bound states' of open strings.

For a theory with only closed strings we can try to repeat the above steps with a closed string field, though it is somewhat more complicated. Already for the free action (9.6.3) the ghost number does not work out: we need an extra c ghost to make the matrix element between physical states nonvanishing. There is a remedy, which is to require the string field Ψ and the gauge transformation Λ to be annihilated by $b_0 - \tilde{b}_0$ and by $L_0 - \tilde{L}_0$. The action

$$S_0' = \frac{1}{2} \langle\!\langle \Psi | (c_0 - \tilde{c}_0) Q_{\mathrm{B}} | \Psi \rangle \tag{9.6.14}$$

is then invariant under $\delta\Psi = Q_{\mathrm{B}}\Lambda$. Going on to the interacting theory there is no associative product analogous to $*$, and correspondingly no simple cover of moduli space. As in the previous section we need to add in higher vertices $\mathscr{V}_{g,n}$, with more powers of the field and also with internal handles, in order to cover moduli space. The resulting action, like the open string action, can be obtained from a symmetry principle using what is known as the *Batalin–Vilkovisky formalism*. Gauge fixing then leads to the perturbation theory described in the previous section.

---

[2] Another physically motivated description of moduli space comes from the light-cone gauge, where the moduli are the interaction times and the momenta p^+ of the intermediate states, and also a twist angle for each intermediate closed string.

[3] For this reason, this form of string field theory would not be suitable for our purposes of the previous section. We would need to add stubs to the vertex to eliminate the unexpected divergences, and then to add in closed string propagators and vertices to cover the omitted region of moduli space.

The derivation of string theory from spacetime gauge symmetry is extremely elegant, particularly for the open string. The inclusion of closed strings in the open string field theory is intriguing. However, it remains to be seen whether string field theory is a useful tool beyond perturbation theory. It has not played a role in the recent nonperturbative progress based on spacetime supersymmetry, which will be described in chapter 14. It is quite possible that strings, or string fields, are simply not the correct degrees of freedom for a nonperturbative formulation of 'string' theory.

9.7 Large order behavior

There is an interesting result about the large order behavior of string perturbation theory. Let us start with field theory. Consider the integral

$$\int_{-\infty}^{\infty} dy \, \exp\left[-\frac{y^2}{2!} - \lambda\frac{y^3}{3!}\right] = \lambda^{-1} \int_{-\infty}^{\infty} dz \, \exp\left[-\frac{1}{\lambda^2}\left(\frac{z^2}{2!} + \frac{z^3}{3!}\right)\right], \quad (9.7.1)$$

where $z = y\lambda$. This integral diverges for $\lambda \neq 0$, but for now we can ignore that and calculate it as a formal expansion in powers of λ,

$$\sum_{n=0}^{\infty} \frac{(-\lambda)^n}{6^n n!} \int_{-\infty}^{\infty} dy \, y^{3n} \exp(-y^2/2) = (2\pi)^{1/2} \sum_{k=0}^{\infty} \lambda^{2k} C_{2k} \,, \quad (9.7.2)$$

with

$$C_{2k} = \frac{2^k \, \Gamma(3k + 1/2)}{\pi^{1/2} \, 3^{2k} (2k)!} \,, \quad (9.7.3)$$

This integral is the zero-dimensional limit of a scalar field theory, with a mass term and cubic coupling. The expansion in λ can be represented in terms of Feynman graphs, with propagator 1 and cubic vertex $-\lambda$. The constant C_{2k} then simply counts the total number of vacuum graphs with $2k$ cubic vertices, or equivalently $k + 1$ loops, weighted by the appropriate symmetry factor. For example, the reader can check that there are two vacuum graphs with two vertices, with symmetry groups of order 8 and 12, and indeed $C_2 = \frac{5}{24} = \frac{1}{8} + \frac{1}{12}$. To get the large order behavior, we use Stirling's approximation $n! \approx n^{n+1/2}e^{-n}$. Then

$$C_{2k} \approx k^k \approx k! \,, \quad (9.7.4)$$

where the approximation holds up to factors of the form k^a and b^k.

One can also include a quartic term $\lambda^2 y^4/4!$, which makes the integral converge. The power of λ is determined by requiring that all terms in the action for z be proportional to λ^{-2}; equivalently, all $k + 1$ loop vacuum graphs are weighted by λ^{2k}. This additional term does not substantially change the result; the estimate (9.7.4) still holds to the given accuracy.

When considering field theory, the propagator and vertices become more complicated. However, the dominant behavior at large orders usually comes from simple counting of graphs — the propagators and vertices only contribute factors of order b^k, unless there is unusual IR or UV behavior. The series is of the form

$$\sum_{k=0}^{\infty} \lambda^{2k} k! f_{2k} \,, \tag{9.7.5}$$

where $f_{2k} \approx k^a b^k$ varies more slowly than $k!$. This diverges unless f_{2k} has unexpected cancellations. For small λ the terms become smaller for a while and then grow as the $k!$ dominates. The smallest terms in the sum are of order

$$\exp[-O(1/\lambda^2)] \,. \tag{9.7.6}$$

This is then an indication of the accuracy of perturbation theory as an asymptotic series. Indeed, effects like instantons and dynamical symmetry breaking, which vanish to all orders of perturbation theory, are typically of this order. For example, the large order behavior of the series (9.7.2) is related to the 'instanton' (stationary point) at $z = -2$, with action $2/3\lambda^2$.

In string theory, the Feynman graph decomposition of moduli space makes it possible to establish the same bounds. It has been shown explicitly that the series is of the form (9.7.5) with f_{2k} positive and bounded below by b^k. Thus the series is divergent, and is not even Borel summable. Borel summation rewrites the series (9.7.5) as the Laplace transform of a more convergent series

$$\int_0^{\infty} dt \, \exp(-t) \sum_{k=0}^{\infty} (t\lambda^2)^k f_{2k} \,. \tag{9.7.7}$$

However, the positivity of the f_{2k} implies that this new sum, even if it converges, has poles on the positive t-axis, so that the integral is ambiguous. Thus, we should expect that string theories will have interesting nonperturbative phenomena. This is not surprising because even in the low energy limit they reduce to nontrivial field theories.

The estimate (9.7.4) for closed string theory is a lower bound, which can be obtained even keeping only the tree-level cubic vertex $\mathscr{V}_{0,3}$. Because the closed string vertex receives corrections from Riemann surfaces of all genus, one might wonder whether perturbation theory grows even more rapidly. One way to see that in fact it does is to recall that the open string field theory contains all open plus closed string Riemann surfaces with at least one boundary. Let us consider the surfaces with one boundary, in the region where the boundary has shrunk to a point. This limit of open string moduli space is equivalent to the moduli space for *closed* surfaces with one vertex operator. One can show that this region is not vastly smaller than the full open string moduli space, so the same estimate holds, the perturbation series going as

$$\sum_{k=0}^{\infty} g_o^{2k} O(k!) = \sum_{k=0}^{\infty} g_c^k O(k!) . \tag{9.7.8}$$

The genus-g term, proportional to g_c^{2g}, has a coefficient of order $(2g)!$. This is larger than the $g!$ characteristic of field theory, which was also the lower bound obtained from $\mathcal{V}_{0,3}$ alone. Evidently the largest part of the large order amplitudes is contained within the vertices $\mathcal{V}_{g,n}$ and comes from the interior of moduli space. One then expects nonperturbative effects of order

$$\exp[-O(1/g_c)] , \tag{9.7.9}$$

much larger at weak coupling than the $\exp[-O(1/g_c^2)]$ characteristic of field theory. This large order behavior was first observed in some simplified models to be discussed in section 9.9. As we will see in chapter 13, these effects are now known to be related to D-branes, at least in some supersymmetric string theories.

9.8 High energy and high temperature

In order to obtain a better understanding of string dynamics, one might look at the behavior in different limits. In this section we consider various high energy limits of string theory.

Hard scattering

Consider again the hard scattering behavior, which we obtained from the Veneziano and Virasoro–Shapiro amplitudes by Stirling's approximation. Go back to the integral form (6.6.5),

$$\frac{8\pi i g_c^2}{\alpha'} (2\pi)^{26} \delta^{26}(\textstyle\sum_i k_i) \int_C d^2 z_4 \, |z_4|^{-\alpha' u/2 - 4} |1 - z_4|^{-\alpha' t/2 - 4} . \tag{9.8.1}$$

In the hard scattering region $s \to \infty$ with t/s fixed, the exponents are large and the integral is dominated by the stationary points of the exponent.[4] The saddle point, for $s, t, u \gg 1$, is given by

$$\frac{\partial}{\partial z_4} \left(u \ln |z_4|^2 + t \ln |1 - z_4|^2 \right) = \frac{u}{z_4} + \frac{t}{z_4 - 1} = 0 , \tag{9.8.2}$$

---

[4] As usual, the integral diverges at $z_4 \to \infty$ and is defined by analytic continuation. The saddle point approximation can be justified for the continuation.

with solution $z_4 = -u/s$; recall that $s + t + u = O(1)$. The saddle point approximation then gives an amplitude

$$S \approx \exp[-\alpha'(s \ln s + t \ln t + u \ln u)/2] , \qquad (9.8.3)$$

as given in eq. (6.6.13).

Let us also back up one step, from the integral over z_4 to the path integral over X^μ. This is dominated by the saddle point

$$X^\mu_{cl}(\sigma) = i \sum_i k^\mu_i G'(\sigma, \sigma_i) , \qquad (9.8.4)$$

where the σ_i are the vertex operator positions. Recall that the Green's function is proportional to α'. At high energy the size of the scattering region grows linearly with energy, $\delta X \approx \alpha' k$. In field theory, higher energy hard scattering probes shorter distances, $\delta X \approx 1/k$ by the uncertainty principle. In string theory this is not the case beyond the string scale. It appears to be impossible in string perturbation theory to probe distances below this scale, giving rise to the idea that it is a limiting distance in spacetime. The effective lower limit on compactification $R \approx \alpha'^{1/2}$ also supports this. However, we will see in chapter 13 that recent advances in nonperturbative string theory reveal the existence of shorter length scales.

At general orders in the string loop expansion, the dominant term in the amplitude will be

$$\exp\left[-\sum_{i<j} k_i \cdot k_j G'(\sigma_i, \sigma_j)\right] . \qquad (9.8.5)$$

We need to find the saddle points of the exponent, with respect to both the vertex operator positions and the moduli of the Riemann surface. This is like an electrostatics problem: the exponent is the energy of n point charges k^μ_i on a Riemann surface, where μ labels d different electrostatic fields and $\mu = 0$ has the opposite sign in the energy. The background charge density vanishes by momentum conservation, and the renormalized self-energy is negligible because k^2_i remains on-shell and small. Remarkably, one can find what is probably the dominant saddle point at each genus. Consider the Riemann surface

$$z^N_2 = \frac{(z_1 - a_1)(z_1 - a_2)}{(z_1 - a_3)(z_1 - a_4)} . \qquad (9.8.6)$$

Like the hypergeometric curve (9.2.9), it is a subspace of $S_2 \times S_2$. It can be regarded as N sheets glued together at branch cuts, and has genus $N - 1$. For the four-point amplitude, if the charges are placed at the branch cuts a_i then the amplitude is stationary with respect to their positions — the surface is 'vertical' at these points, $\partial z_2/\partial z_1 = \infty$, and the symmetry $z_2 \to \exp(2\pi i/N)z_2$ requires the vertical component of the gradient to vanish. It remains to extremize with respect to the positions a_i, but this is easy: the electric field is the same as on the sphere with charges at $z = a_i$, except that it is divided between N sheets. Thus the energy is $N \cdot N^{-2} = N^{-1}$ times that on the sphere. Three of the positions can be fixed by Möbius invariance, for example to $0, 1, \infty$, and the fourth is given by the same condition (9.8.2) as on the sphere. Thus at genus g,

$$S_g \approx \exp[-\alpha'(s \ln s + t \ln t + u \ln u)/2(g+1)] . \qquad (9.8.7)$$

With additional effort one can include the Gaussian fluctuations around the saddle, which determine the coefficient of S_g. For the N-sheeted saddle points, X^μ_{cl} is simply divided by $N = g + 1$.

There is a check that this result makes sense. In the small-angle region $-t/s \ll 1$ it becomes

$$\exp[\alpha' t \ln s/2(g+1)] . \qquad (9.8.8)$$

Now, $t^{1/2}$ is the total momentum transferred from the incoming particle 1 to the final particle 3. At small angles it is plausible that the genus-g scattering process can be decomposed into $g + 1$ successive scatterings. Each separate scattering has the same center of mass s, while the t_i are constrained so that the total momentum transfer is $t^{1/2}$. The individual scatterings go as $\exp(\alpha' t_i \ln s/2)$. It is then easy to check that it is most efficient to take the momentum transfers to be equal, $t_i^{1/2} = t^{1/2}/(g+1)$, leading to the result (9.8.8) for the product of the amplitudes. This is different from particle scattering, where it is most efficient to transfer momentum in a small number of hard scatterings.

For excited string states, the exponential part of the vertex operator is the same as for the tachyon. The same saddle point then dominates, with X^μ_{cl} inserted in the vertex operators. Since this same saddle point dominates for all states and, up to the factor $g + 1$, all orders, there are linear relations between the amplitudes for different states that hold for all genus and therefore plausibly for the exact amplitudes. In quantum field theory, high energy scattering often reveals symmetries that are not obvious at low energy. It has been suggested that the relations that emerge here are of the same sort.

It is not clear to what extent it makes sense to sum the series (9.8.7). It is interesting to contrast the result here with that in the previous section. In both cases it is the interior of moduli space that is producing the interesting physics (in contrast to the field-theoretic behavior found near the boundary). However, the large order behavior came from the whole of moduli space, dominated in particular by its large volume. The high energy behavior at fixed genus is dominated by a single point.

Regge scattering

The analysis of the Regge limit can also be carried to higher orders. We will not do this, but will develop an interesting interpretation of the tree-level Regge physics.

Regge scattering is less familiar to most physicists than hard scattering. It is a gentle probe of a highly boosted system. When a system is boosted, time dilation slows down its internal processes. One might then expect to see more of the virtual degrees of freedom as one goes to higher energies.

Let us in particular consider the quantum fluctuations of a single string, in particular the root mean size of the ground state. The mode expansion gives

$$\langle 0|[X^1(\sigma) - x^1]^2|0\rangle = \sum_{n=1}^{\infty} \frac{\alpha'}{2n^2} \langle 0|(\alpha_n\alpha_{-n} + \tilde{\alpha}_n\tilde{\alpha}_{-n})|0\rangle$$

$$= \alpha' \sum_{n=1}^{\infty} \frac{1}{n} . \tag{9.8.9}$$

This is divergent but has no direct physical significance. A measurement on a time-scale $\delta\tau$ is sensitive only up to modes of frequency less than $\delta\tau^{-1}$, essentially $n\alpha'^{-1/2} < \delta\tau^{-1}$. The logarithmic divergence becomes $\alpha' \ln(\alpha'^{1/2}/\delta\tau)$, and the size is the square root of this.

This growth in the effective size is directly visible in the Regge amplitude (6.6.12) at small t,

$$s^{2+\alpha't/2} \frac{\Gamma(-\alpha't/4 - 1)}{\Gamma(\alpha't/4 + 2)} \sim \frac{s^2}{t} \exp\left[-\frac{q^2\alpha'}{2} \ln(\alpha's)\right], \tag{9.8.10}$$

where $q^2 = -t$ is the momentum transfer. This is the gravitational amplitude s^2/t modified by a form-factor whose Fourier transform corresponds to an object of size $(\alpha' \ln \alpha's)^{1/2}$. From the point of view of one string the other has a boost of order $\alpha's$, so the time resolution of its internal processes is inverse to this, in agreement with the earlier estimate for the size.

The square root of a logarithm is a very slow function and ordinarily would be of little importance, but it has been argued that this behavior is very significant for strings near black holes. In chapter 14 we will revisit, from a different point of view, the idea that strings reveal their true nature when very highly boosted.

High temperature

A similar limit with interesting stringy behavior is high temperature. From the generic behavior (7.2.30) of the partition function at high weight h, and the relation $m^2 = 4(h-1)/\alpha'$, the density of string states $n(m)$ satisfies

$$\int_0^\infty dm\, n(m) \exp(-\alpha' \pi m^2 \ell) \approx \exp(4\pi/\ell) , \tag{9.8.11}$$

implying $n(m) \approx \exp(4\pi m \alpha'^{1/2})$. The thermal partition function of a single string is then

$$\int_0^\infty dm\, \exp(4\pi m \alpha'^{1/2}) \exp(-m/T) . \tag{9.8.12}$$

Because the density of states grows exponentially, the partition function (9.8.12) diverges for temperatures greater than the *Hagedorn temperature*

$$T_H = \frac{1}{4\pi \alpha'^{1/2}} . \tag{9.8.13}$$

One can understand the Hagedorn divergence as follows. The energy of a long string is proportional to its length. So also is its entropy, being extensive. At low temperature the energy dominates, but at high temperature the entropy dominates and it is favorable to produce strings of arbitrary length. In an ensemble of such strings the density will become large, and the interactions will be important near the transition.

This divergence suggests a phase transition. Indeed, in the strong interaction there is a similar growth in the density of states of highly excited hadrons, and there is a phase transition from a low temperature hadron phase to a high temperature quark–gluon phase. It is interesting to wonder whether the Hagedorn divergence indicates some analogy to the quark–gluon structure in the fundamental string, but there is no clear answer.

To conclude, let us discuss the partition function of free strings in somewhat more detail. The free energy density of a real scalar field of mass m in dimension d is

$$F(T, m^2) = T \int \frac{d^{d-1}\mathbf{k}}{(2\pi)^{d-1}} \ln[1 - \exp(-\omega_k/T)]$$

$$= -\int_0^\infty \frac{dt}{t} (2\pi t)^{-d/2} \sum_{r=1}^\infty \exp\left(-\frac{m^2 t}{2} - \frac{r^2}{2T^2 t}\right) . \tag{9.8.14}$$

In the second form, one can readily sum over the string spectrum. Following the derivation of the vacuum amplitudes (7.3.12) and (7.3.6), the free energy density in 26 flat dimensions is

$$F(T) = -\int_R \frac{d\tau d\bar{\tau}}{2\tau_2} (4\pi^2 \alpha' \tau_2)^{-13} |\eta(\tau)|^{-48} \sum_{r=1}^\infty \exp\left(-\frac{r^2}{4\pi T^2 \alpha' \tau_2}\right) . \tag{9.8.15}$$

The integral runs over the full region

$$R: \quad \tau_1 \leq \frac{1}{2}, \quad |\tau_2| > 0. \tag{9.8.16}$$

The derivation from a modular-invariant path integral is given as exercise 9.7. There is a possible divergence as $\tau_2 \to 0$; the integrand is largest when the real part of τ is zero, so let $\tau = i\tau_2$. From the modular transformation (7.2.44), the asymptotic behavior of the eta function in this limit is

$$\eta(i\tau_2) = \eta(i/\tau_2)\tau_2^{-1/2}$$
$$\approx \exp(-\pi/12\tau_2) \text{ as } \tau_2 \to 0. \tag{9.8.17}$$

The τ integral thus converges for $T < T_H$ and diverges for $T > T_H$.

The following curious observation has been made. In field theory the partition function is given by a Euclidean path integral, with time having period $1/T$. The same is true in string theory (exercise 9.7). T-duality should then relate the high and low temperature behaviors, and in fact the Hagedorn temperature is half the self-dual value. Define

$$\tilde{F}(T) = F(T) + \rho_0, \tag{9.8.18}$$

the thermal part (9.8.15) of the free energy plus the vacuum part (7.3.6). The logarithm of the path integral is $-\tilde{F}(T)/T$, so T-duality implies

$$\frac{1}{T}\tilde{F}(T) = \frac{T}{4T_H^2}\tilde{F}(4T_H^2/T). \tag{9.8.19}$$

This relates the partition function below the Hagedorn temperature to that above it. The Hagedorn divergence for $T > T_H$ is a reflection of the tachyon divergence in the low temperature theory, but a similar formal result holds for the supersymmetric string, which has no tachyonic divergence. If we take this relation seriously, we obtain

$$\tilde{F}(T) \to \frac{T^2}{4T_H^2}\rho_0 \tag{9.8.20}$$

as $T \to \infty$. If this is a measure of the number of high energy degrees of freedom, they are very few. A single quantum field in d spacetime dimensions has a free energy of order T^d, so it is as though the high energy limit of string theory looks like a field theory in two *spacetime* dimensions. This is entirely unlike the exponentially growing spectrum below the Hagedorn transition, and may be a clue to the nature of the phase transition.

9.9 Low dimensions and noncritical strings

Field theories often simplify as the number of spacetime dimensions is reduced, and these low-dimensional theories have been useful models for understanding dynamics. We might try to do the same in string theory, but string theory requires a critical dimension. We could compactify, but this does not reduce the number of degrees of freedom: the string still oscillates in the compact directions. We have, however, encountered a consistent background in which the string actually does move in fewer dimensions. This is the linear dilaton background (3.7.18),

$$G_{\mu\nu}(x) = \eta_{\mu\nu}\,, \quad B_{\mu\nu}(x) = 0\,, \quad \Phi(x) = V_\mu x^\mu\,, \qquad (9.9.1a)$$

$$V_\mu = \delta_\mu^{\ 1}\left(\frac{26-D}{6\alpha'}\right)^{1/2}. \qquad (9.9.1b)$$

For $D < 26$ the dilaton gradient is spacelike and we have taken it to lie in the 1-direction.

The effective string coupling e^Φ is a function of x^1, diverging at large x^1. This makes it hard to use string perturbation theory, but fortunately there is a slightly different background for which perturbation theory is good. Inserting the background (9.9.1) into the tachyon action (6.6.16) leads to the linearized tachyon field equation,

$$-\partial^\mu\partial_\mu T(x) + 2V^\mu\partial_\mu T(x) - \frac{4}{\alpha'}T(x) = 0\,. \qquad (9.9.2)$$

This is the condition for the exponential vertex operator to be Weyl invariant, with the linear dilaton energy-momentum tensor (2.5.1). The solutions are

$$T(x) = \exp(q\cdot x)\,, \quad (q-V)^2 = \frac{2-D}{6\alpha'}\,. \qquad (9.9.3)$$

If we look for solutions which, like the dilaton, depend only on x^1, then

$$q_1 = \alpha_\pm = \left(\frac{26-D}{6\alpha'}\right)^{1/2} \pm \left(\frac{2-D}{6\alpha'}\right)^{1/2}. \qquad (9.9.4)$$

For $D > 2$ this is complex, so the tachyon oscillates, but for $D \leq 2$ one gets a real exponential (times a linear function at the critical value $D = 2$). Of course, the tachyon field equation will have nonlinear corrections. It can be shown that this does not affect the qualitative form of the background, except that for rather subtle reasons only the solution α_- leads to a nonsingular background.

For $D \leq 2$ the term 'tachyon' for the string ground state is actually a misnomer, because this field is stable (exercise 9.8). Also for $D \leq 2$, the real exponential $T(x) = T_0 \exp(\alpha_- x^1)$ prevents propagation to the region

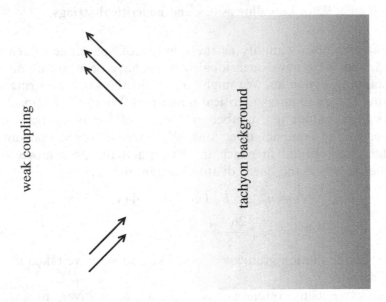

Fig. 9.12. Spacetime with dilaton and tachyon backgrounds. The interaction strength grows toward the right but is cut off by an effective potential from the tachyon background.

of large positive x^1 where the coupling diverges. To see this, consider the world-sheet action

$$S_\sigma = \frac{1}{4\pi\alpha'} \int_M d^2\sigma \, g^{1/2} \Big[g^{ab}\eta_{\mu\nu}\partial_a X^\mu \partial_b X^\nu + \alpha' R V_1 X^1 + T_0 \exp(\alpha_- X^1) \Big] \,. \tag{9.9.5}$$

At large x^1 the exponential tachyon dominates the linear dilaton and the path integral is suppressed, corresponding to an effective potential for the string due to the tachyon background. Physically one has an asymptotic region $x^1 \to -\infty$, where the coupling goes to zero. Strings propagating toward positive x^1 interact with one another and reflect off the potential and back to the asymptotic region. The maximum effective coupling is set by the energy and the strength T_0 of the potential. This picture, shown in figure 9.12, is a bit exotic, but in the interest of finding a solvable model it is worth pursuing. A model with rather similar physics, the static model, played a role in unraveling the nature of quantum field theory.

For $D > 2$ the physical effect of the tachyon background is less clear, though it has been argued that it still acts as a barrier. However, the focus has been on $D \leq 2$ for another reason, namely that these theories are actually solvable. Let us first observe that when we quantize the string, the gauge symmetries take away two families of oscillators, so that in $D \leq 2$ none will be left — there is only the center-of-mass motion, a single scalar.[5] Nevertheless, to solve the string theory directly is still quite hard. First one has to evaluate the world-sheet path integral with action (9.9.5). This is no longer a free theory, because of the exponential tachyon field. It is known as the *Liouville field theory*. It is classically integrable, and with substantial effort some quantum amplitudes have been determined. However, to solve the string theory one would still need to integrate the result at each genus over the moduli space of Riemann surfaces; we have seen that this is a complicated space. Remarkably, the full amplitude has been determined, to all orders, by very different means. To describe this, we must first introduce another idea.

Noncritical strings

Consider a statistical mechanical system defined by a sum over two-dimensional surfaces in a d-dimensional Euclidean space. Let us also suppose that the surfaces have some internal degree of freedom that can be interpreted as a measure of length. The natural fields to describe the system are the embedding $X^\mu(\sigma)$ and a metric $g_{ab}(\sigma)$. However, there is no particular reason for the length to drop out of the problem — that is, for the theory to be Weyl-invariant. This is therefore called the *noncritical string*. The lowest-dimensional action (or Boltzmann weight, since we are thinking of the path integral in the context of statistical mechanics) for this system would then be the Polyakov action with an additional world-sheet cosmological constant term, which in the critical string is forbidden by Weyl invariance,

$$S = \frac{1}{4\pi\alpha'} \int_M d^2\sigma\, g^{1/2} \left(g^{ab} \partial_a X^\mu \partial_b X_\mu + \mu \right) . \qquad (9.9.6)$$

Without Weyl invariance, we can use diff invariance to fix the metric partially, but one degree of freedom remains:

$$g_{ab}(\sigma) = \exp(2\phi(\sigma))\, \hat{g}_{ab}(\sigma) , \qquad (9.9.7)$$

where $\hat{g}_{ab}$ is a gauge choice and the scale ϕ of the metric is still path-integrated. The Faddeev–Popov determinant for this gauge-fixing is the same as in eq. (3.4.19), the Weyl-fixing contribution to the latter being trivial. Inserting the metric (9.9.7), for which the curvature is

$$g^{1/2}R = \hat{g}^{1/2} \left(\hat{R} - 2\hat{\nabla}^2\phi \right) , \qquad (9.9.8)$$

and dropping a field-independent term, the action plus Faddeev–Popov

---

[5] At special momenta one finds additional physical states. These can be interpreted as unbroken symmetries of string field theory, and likely play a role in the solvability of the theory.

determinant becomes

$$S = \frac{1}{4\pi\alpha'} \int_M d^2\sigma \, \hat{g}^{1/2} \left[\hat{g}^{ab} \partial_a X^\mu \partial_b X_\mu + \mu \exp 2\phi + \frac{13\alpha'}{3} \left(\hat{g}^{ab} \partial_a \phi \partial_b \phi + \hat{R}\phi \right) \right].$$
(9.9.9)

Note that the Faddeev–Popov determinant has provided a kinetic term for ϕ. Comparing this with the dilaton–tachyon action (9.9.5) we see a marked resemblance. If we let $d = D - 1$ but interpret the new field ϕ as x^1, and if we rename $g_{ab} \to \hat{g}_{ab}$ in the action (9.9.5) (since it can be fully gauge-fixed), then all the same terms appear. Some coefficients are a bit different — there is no rescaling of ϕ that makes them all agree — but in fact we claim that these are the same field theory. The point is that the critical string path integral (9.9.5) is defined using $\hat{g}_{ab}$ as the metric, and the noncritical string (9.9.9) using $\exp(2\phi)\hat{g}_{ab}$. A careful treatment of the Weyl transformation of the measure then converts one to the other. One can understand this quite easily by noting that only the full metric (9.9.7) appears in the definition of the noncritical string, so nothing can change under

$$\hat{g}_{ab}(\sigma) \to \exp(2\omega(\sigma)) \, \hat{g}_{ab}(\sigma) \,, \tag{9.9.10a}$$
$$\phi(\sigma) \to \phi(\sigma) - \omega(\sigma) \,. \tag{9.9.10b}$$

The noncritical string thus acquires a Weyl invariance, and this is enough to fix the various coefficients, other than the free parameter μ.

It may seem odd that Weyl-invariant and noninvariant theories can be equivalent. The point is that in field theory we can always add extra 'fake' gauge symmetries if we also add extra fields that we just use to fix the gauge. A gauge symmetry after all is just a redundancy of description, and we can always adopt a more redundant description. For the symmetry (9.9.10) it is clear that we can just cancel the Weyl symmetry against the field ϕ, but for general string backgrounds (for example flat space, where the embedding coordinates are Weyl-invariant) this is not possible and the Weyl invariance is 'real.'

In any case we are now nearing the punch line. One can approximate the noncritical string by a sum of triangulated surfaces, the number of edges between two points being a measure of length. These triangulated surfaces resemble large Feynman diagrams, and in fact one can find a quantum system, the *matrix model,* for which this is the diagram expansion. The fields of the matrix model are $N \times N$ matrices. Taking $N \to \infty$ and adjusting other couplings, there is (at least for $D \le 2$) a critical point dominated by large smooth surfaces, so we recover the noncritical string. Moreover, the matrix model is solvable (again for $D \le 2$) by diagonalizing the matrices. In the end one has a rather simple system (noninteracting fermions moving in the space of eigenvalues) which by retracing the above steps allows the calculation of string amplitudes to all genus.

Because these methods are so far afield from the rest of the book, we will have to leave the details to the references. In any case they have not yet been useful at $D > 2$ or in supersymmetric theories. (However, we will see in chapter 14 that a somewhat different theory of large-N matrices has recently been proposed as a nonperturbative string theory.) We should mention, though, at least one important lesson from these models. This is the $(2g)!$ behavior of perturbation theory, which was seen first in the explicit matrix model solution and then argued to hold in all string theories. In the matrix model the corresponding $\exp[O(-1/g_c)]$ effect involves tunneling of the fermions in the space of eigenvalues. It is not clear whether this can be related to the $\exp[O(-1/g_c)]$ effects in higher-dimensional string theories.

Exercises

9.1 (a) Use a conformal transformation to show that the OPE (9.1.2) and expectation value (9.1.3) imply the inner product (9.1.29).
(b) By taking matrix elements in the basis $|\hat{\mathscr{V}}_j(k)\rangle$, verify the second equality in equation (9.1.28).

9.2 (a) Write the four-point amplitude on the torus in operator form, parallel to eq. (9.1.17) for the sphere, but now involving a trace.
(b) Show that if one of the vertex operators is null, $\hat{\mathscr{V}}_1 = \{Q^B, \mathscr{X}\}$, then the trace is equal to total derivatives with respect to the vertex operator positions plus total derivatives with respect to τ and $\bar{\tau}$.

9.3 (a) For the torus with four vertex operators, use the result of part (a) of the previous problem to study the unitarity in the limit that w_1, w_2, and $w_3 - w_4$ are fixed and w_3 and $2\pi\tau - w_3$ approach $i\infty$. This is two handles degenerating, and unitarity relates the amplitude to the product of two tree-level four-point amplitudes.
(b) Repeat in particular the demonstration that the sum over the full Hilbert space $\mathscr{H}$ and that over the physical Hilbert space are equal. Show that this requires cancellation between intermediate states of different ghost numbers. This cancellation is a stringy generalization of the cancellation between Faddeev–Popov ghosts and longitudinal gauge bosons in gauge theory.

9.4 Verify the assertion in eq. (9.3.7) about the region covered by the Feynman graph of figure 9.6(a), and find the other two regions covered by this graph.

9.5 (a) For the terms (9.6.7) in the expansion of the string field, write out explicitly the gauge transformation (9.6.2) and the action (9.6.3).

(b) Eliminate auxiliary fields and fields that can be completely gauged away, and show that the resulting gauge transformation and action for A_μ have the expected form.

9.6 For $f_{2k} = k^a b^k$, show that the smallest term in the series (9.7.5) is of the order of $\exp(-C/g^2)$. Work to sufficient accuracy to find C. For the series (9.7.2), compare with the instanton action.

9.7 In quantum theory, the Euclidean path integral with Euclidean time having period T^{-1} gives the thermal partition function $\exp[-\tilde{F}(T)/T]$. Show that for the string path integral this agrees with the result (9.8.15). The one subtlety is an interplay between the region of integration and the sum over windings (Reference: McClain & Roth, (1987).)

9.8 For the tachyon wave equation (9.9.2) in the linear dilaton background (9.9.1), show that any *finite energy* initial configuration has stable (oscillatory) evolution provided that $D \leq 2$.

9.9 Study the physical state conditions (either in OCQ or BRST form) for the graviton level of the closed string in the $D = 2$ linear dilaton background. Let the momenta take complex values. You should find that the physical spectrum is trivial at generic momenta but that at two momenta there are nontrivial physical states.

9.10 Derive eq. 9.8.14. Hint: in the second line introduce an integral of $\mathbf{k}$ such that m^2 becomes $m^2 + \mathbf{k}^2$. Then integrate over t (the result is in terms of elementary functions. Finally, the sum on r is the series expansion of the log in the first line.

Appendix A
A short course on path integrals

Path integrals are a powerful means of representing quantum theories, especially on surfaces of nontrivial topology. The introduction we present here is similar to that which the reader will find in any modern field theory text. We include it in order to emphasize certain ideas that we will need, such as the relation between the path integral and Hilbert space formalisms and the use of operator equations inside the path integral.

A.1 Bosonic fields

Consider first a quantum mechanics problem, one degree of freedom with Hamiltonian $\hat{H}(\hat{p}, \hat{q})$, where

$$[\hat{q}, \hat{p}] = i\hbar \ . \tag{A.1.1}$$

Throughout the appendix operators are indicated by hats. A basic quantity of interest is the amplitude to evolve from one q eigenstate to another in a time T,

$$\langle q_f | \exp(-i\hat{H}T/\hbar)|q_i\rangle \ . \tag{A.1.2}$$

In field theory it is generally convenient to use the Heisenberg representation, where operators have the time dependence

$$\hat{A}(t) = \exp(i\hat{H}t/\hbar)\, \hat{A} \, \exp(-i\hat{H}t/\hbar) \ . \tag{A.1.3}$$

The state $|q, t\rangle$ is an eigenstate of $\hat{q}(t)$,

$$\hat{q}(t)|q, t\rangle = |q, t\rangle q \ . \tag{A.1.4}$$

In terms of the $t = 0$ Schrödinger eigenstates this is

$$|q, t\rangle = \exp(i\hat{H}t/\hbar)|q\rangle \ . \tag{A.1.5}$$

In this notation the transition amplitude (A.1.2) is $\langle q_f, T|q_i, 0\rangle$.

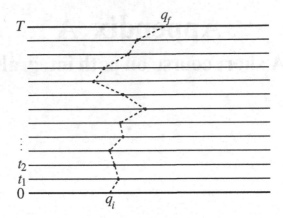

Fig. A.1. Transition amplitude broken down into time steps. The dashed line suggests a piecewise linear path, which is one of many ways to define path integration.

By inserting a complete set of states, we can write the transition amplitude as a coherent sum over all states q through which the system might pass at some intermediate time t:

$$\langle q_f, T | q_i, 0 \rangle = \int dq \, \langle q_f, T | q, t \rangle \langle q, t | q_i, 0 \rangle \; . \tag{A.1.6}$$

Divide the time interval further into N steps as in figure A.1,

$$t_m = m\epsilon \; , \quad \epsilon = T/N \; , \tag{A.1.7}$$

and corresponding to each intermediate time insert a complete set of states

$$\langle q_f, T | q_i, 0 \rangle = \int dq_{N-1} \ldots dq_1 \prod_{m=0}^{N-1} \langle q_{m+1}, t_{m+1} | q_m, t_m \rangle \; . \tag{A.1.8}$$

Translate back to Schrödinger formalism and introduce an integral over intermediate momenta,

$$\langle q_{m+1}, t_{m+1} | q_m, t_m \rangle = \langle q_{m+1} | \exp(-i\hat{H}\epsilon/\hbar) | q_m \rangle$$
$$= \int dp_m \, \langle q_{m+1} | p_m \rangle \langle p_m | \exp(-i\hat{H}\epsilon/\hbar) | q_m \rangle \; . \tag{A.1.9}$$

By commuting, we can always write $\hat{H}$ with all $\hat{p}$ to the left and all $\hat{q}$ to the right, so that

$$\langle p_m | \hat{H}(\hat{p}, \hat{q}) | q_m \rangle = H(p_m, q_m) \langle p_m | q_m \rangle \; . \tag{A.1.10}$$

Use this to evaluate the matrix element in eq. (A.1.9). From expanding the exponential there will be terms of order ϵ^2 and higher that have $\hat{q}$ to the left of $\hat{p}$; drop these. Then to this order $\langle q_{m+1}, t_{m+1} | q_m, t_m \rangle$ becomes

$$\int dp_m \exp[-iH(p_m, q_m)\epsilon/\hbar]\langle q_{m+1}|p_m\rangle\langle p_m|q_m\rangle$$

$$= \int \frac{dp_m}{2\pi\hbar} \exp\left\{-\frac{i}{\hbar}\left[H(p_m, q_m)\epsilon - p_m(q_{m+1} - q_m)\right] + O(\epsilon^2)\right\}. \quad (A.1.11)$$

Thus

$$\langle q_f, T | q_i, 0 \rangle \approx \int \frac{dp_{N-1}}{2\pi\hbar} dq_{N-1} \cdots \frac{dp_1}{2\pi\hbar} dq_1 \frac{dp_0}{2\pi\hbar}$$

$$\times \exp\left\{-\frac{i}{\hbar}\sum_{m=0}^{N-1}\left[H(p_m, q_m)\epsilon - p_m(q_{m+1} - q_m)\right]\right\}$$

$$\rightarrow \int [dpdq] \exp\left\{\frac{i}{\hbar}\int_0^T dt\,[p\dot{q} - H(p,q)]\right\}. \quad (A.1.12)$$

In the last line we have taken a formal $\epsilon \rightarrow 0$ limit, so the integral runs over all paths $p(t), q(t)$ with given $q(T)$ and $q(0)$. This is the *Hamiltonian path integral* (or *functional integral*), the sum over phase space paths weighted by $\exp(iS/\hbar)$ with S the action in Hamiltonian form.

Let us integrate over $p(t)$, making first the approximation that the integral is dominated by the point of stationary phase,

$$0 = \frac{\partial}{\partial p}[p\dot{q} - H(p,q)] = \dot{q} - \frac{\partial}{\partial p}H(p,q). \quad (A.1.13)$$

Solving for p in terms of q and $\dot{q}$ and recalling that $L = p\dot{q} - H$, the stationary phase approximation gives

$$\langle q_f, T | q_i, 0 \rangle = \int [dq] \exp\left[\frac{i}{\hbar}\int_0^T dt\,L(q,\dot{q})\right]. \quad (A.1.14)$$

Often the momentum integral is Gaussian, $\hat{H} = \frac{1}{2}\hat{p}^2 + V(\hat{q})$, so that the stationary phase approximation is exact up to a normalization that can be absorbed in the measure $[dq]$. In fact, we can do this generally: because the relation (A.1.13) between p and $\dot{q}$ is local, the corrections to the stationary phase approximation are local and so can be absorbed into the definition of $[dq]$ without spoiling the essential property that $[dq]$ is a product over independent integrals at each intermediate time. The result (A.1.14) is the *Lagrangian path integral*, the integral over all paths in configuration space weighted by $\exp(iS/\hbar)$, where now S is the Lagrangian action.

We have kept $\hbar$ so as to discuss the classical limit. As $\hbar \rightarrow 0$, the integral is dominated by the paths of stationary phase, $\delta S/\delta q(t) = 0$, which is indeed the classical equation of motion. Henceforth, $\hbar = 1$.

The derivation above clearly generalizes to a multicomponent q^i. One can always write a quantum field theory in terms of an infinite number of qs, either by making the spatial coordinate $\mathbf{x}$ discrete and taking a limit, or by expanding the x-dependence in terms of Fourier modes. Thus the above derivation of the path integral representation applies to bosonic field theory as well.

The derivation of the path integral formula treated the continuum limit $\epsilon \to 0$ rather heuristically. Renormalization theory is the study of this limiting process, and is rather well understood. The limit exists provided the Lagrangian satisfies certain conditions (renormalizability) and provided the coefficients in the Lagrangian (bare couplings) are taken to vary appropriately as the limit is taken. The result is independent of various choices that must be made, including the operator ordering in the Hamiltonian, the precise definition of the measure, and the differences between regulators (the limit of piecewise linear paths suggested by the figure versus a frequency cutoff on the Fourier modes, for example).[1] Being local effects these can all be absorbed into the bare couplings. Actually, we will evaluate explicitly only a few Gaussian path integrals. We use the path integral primarily to derive some general results, such as operator equations and sewing relations.

Relation to the Hilbert space formalism

Starting from the path integral one may cut it open to recover the Hilbert space of the theory. In equation (A.1.6), if one writes each of the amplitudes as a path integral one finds

$$\int [dq]_{q_i,0}^{q_f,T} \exp\left(i \int_0^T dt L\right)$$

$$= \int dq \int [dq]_{q,t}^{q_f,T} \exp\left(i \int_t^T dt L\right) \int [dq]_{q_i,0}^{q,t} \exp\left(i \int_0^t dt L\right) . \quad (A.1.15)$$

That is, the path integral on $[0, T]$ breaks up into separate path integrals on $[0, t]$ and $[t, T]$, plus an ordinary integral over $q(t)$. The boundary conditions on paths are denoted by super/subscripts. Note the restriction that $T \geq t \geq 0$. One can also replace the intermediate integral with a sum over any complete set of states. This cutting (or, inversely, sewing) principle is particularly useful on the surfaces of nontrivial topology that arise in string theory. There one may cut the surface open along many different closed curves; different choices give different Hilbert space representations of a given amplitude. This is known as *world-sheet duality*.

Consider now a path integral with the additional insertion of a factor

---

[1] As another example, one can freely rescale fields. The Jacobian is formally a product of separate Jacobians at each point; its logarithm is therefore a sum of a local expression at each point. Regulated, this is just a shift of the bare couplings.

of $q(t)$, where $0 < t < T$. Then, using the cutting relation (A.1.15),

$$\int [dq]_{q_i,0}^{q_f,T} \exp(iS)q(t) = \int dq \, \langle q_f, T | q, t \rangle \, q \, \langle q, t | q_i, 0 \rangle$$

$$= \int dq \, \langle q_f, T | \hat{q}(t) | q, t \rangle \, \langle q, t | q_i, 0 \rangle$$

$$= \langle q_f, T | \hat{q}(t) | q_i, 0 \rangle \,. \qquad \text{(A.1.16)}$$

Thus, $q(t)$ in the functional integral translates into $\hat{q}(t)$ in the matrix element. For a product of two insertions $q(t)q(t')$ (which is equal to $q(t')q(t)$, since these are simply variables of integration), where both t and t' are in the range $[0, T]$, one finds by the same method that

$$\int [dq]_{q_i,0}^{q_f,T} \exp(iS)q(t)q(t') = \langle q_f, T | \mathrm{T}[\hat{q}(t)\hat{q}(t')] | q_i, 0 \rangle. \qquad \text{(A.1.17)}$$

Here, T denotes the time-ordered product

$$\mathrm{T}[\hat{A}(t)\hat{B}(t')] = \theta(t - t')\hat{A}(t)\hat{B}(t') + \theta(t' - t)\hat{B}(t')\hat{A}(t). \qquad \text{(A.1.18)}$$

Because of the way the path integral is built out of successive infinitesimal time slices, two or more insertions in the path integral will always correspond to the time-ordered product of operators in the matrix element.

The equation of motion is $\delta S / \delta q(t) = 0$. Let us see how this works inside the path integral. Consider this equation in a path integral with other insertions $\mathscr{F}$,

$$\int [dq]_{q_i,0}^{q_f,T} \exp(iS)\frac{\delta S}{\delta q(t)}\mathscr{F} = -i \int [dq]_{q_i,0}^{q_f,T} \left[\frac{\delta}{\delta q(t)} \exp(iS) \right] \mathscr{F}$$

$$= i \int [dq]_{q_i,0}^{q_f,T} \exp(iS)\frac{\delta \mathscr{F}}{\delta q(t)} \,. \qquad \text{(A.1.19)}$$

We have integrated by parts in the second line.[2] If none of the fields in $\mathscr{F}$ are at time t, the right-hand side vanishes and the equation of motion holds in the path integral. We therefore say that the equation of motion holds as an *operator equation*. This is a technique that we will use extensively: an operator equation is one which holds when inserted into any path integral that has no other fields coincident with those involved in the equation. Translating to Hilbert space language, we can use the other insertions $\mathscr{F}$ to set up arbitrary initial and final states, so this corresponds

---

[2] The functional derivative is defined as

$$\frac{\delta}{\delta q(t)} q(t') = \delta(t - t') \,.$$

It can be regarded as the limit of the ordinary derivative

$$\frac{1}{\epsilon}\frac{\partial}{\partial q_m} \,, \quad m = t/\epsilon \,.$$

In particular, one can integrate a functional derivative by parts, as for an ordinary derivative.

to the ordinary notion that an operator equation is one that should hold for general matrix elements.

When additional insertions are present at t, operator equations inside the path integral hold only up to additional delta-function *source terms*. For example, let $\mathscr{F} = q(t')\ldots$, where t' is near t and the other fields in '…' are far away. Then eq. (A.1.19) becomes

$$\left\langle q(t')\frac{\delta S}{\delta q(t)} \ldots \right\rangle = i\langle \delta(t-t') \ldots \rangle \,. \qquad (A.1.20)$$

This is an example of the *Schwinger–Dyson equation,* the equation of motion for the expectation value of a product of fields. The same result written as an operator equation is

$$q(t')\frac{\delta S}{\delta q(t)} = i\delta(t-t') \,. \qquad (A.1.21)$$

As an explicit example, take the harmonic oscillator,

$$S = \frac{1}{2}\int dt\,(\dot{q}^2 - \omega^2 q^2)\,, \qquad (A.1.22a)$$

$$\frac{\delta S}{\delta q(t)} = -\ddot{q}(t) - \omega^2 q(t)\,. \qquad (A.1.22b)$$

Then the operator equations we have derived are

$$\left(\frac{\partial^2}{\partial t^2} + \omega^2\right)q(t) = 0 \qquad (A.1.23)$$

and

$$\left(\frac{\partial^2}{\partial t^2} + \omega^2\right)q(t)q(t') = -i\delta(t-t') \,. \qquad (A.1.24)$$

In Hilbert space language these become

$$\left(\frac{\partial^2}{\partial t^2} + \omega^2\right)\hat{q}(t) = 0 \,,$$

$$\left(\frac{\partial^2}{\partial t^2} + \omega^2\right)\mathrm{T}[\hat{q}(t)\hat{q}(t')] = -i\delta(t-t') \,. \qquad (A.1.25a)$$

The first is the Heisenberg equation for the field operator, while the delta function in the second comes from the time derivative acting on the step functions in the time-ordered product,

$$\left(\frac{\partial^2}{\partial t^2} + \omega^2\right)\mathrm{T}\,[\hat{q}(t)\hat{q}(t')] = \mathrm{T}\left\{\left[\ddot{\hat{q}}(t) + \omega^2\hat{q}(t)\right]\hat{q}(t')\right\} + \delta(t-t')[\dot{\hat{q}}(t),\hat{q}(t)]$$

$$= 0 \,-\, i\delta(t-t') \,. \qquad (A.1.26)$$

Euclidean path integrals

In Schrödinger language, the time-ordered product (A.1.17) is

$$\langle q_f | \exp[-i\hat{H}(T - t_>)]\, \hat{q}\, \exp[-i\hat{H}(t_> - t_<)]\, \hat{q}\, \exp[-i\hat{H}t_<]|q_i\rangle , \quad \text{(A.1.27)}$$

where $t_>$ is the greater of t and t', and $t_<$ is the lesser. By the nature of the path integral, which builds amplitudes from successive time steps, the time differences multiplying H are always positive in amplitudes such as (A.1.27): one only encounters $\exp(-iH\Delta t)$ for positive Δt. Now continue analytically by uniformly rotating all times in the complex plane, $t \to e^{-i\theta}t$ for any phase $0 \leq \theta \leq \pi$. The real part of the exponent is $-E \sin\theta \Delta t$, where E is the energy of the intermediate state. Because the Hamiltonian is bounded below for a stable system, the sum over intermediate states is safely convergent and no singularities are encountered — the continued amplitude is well defined.

Taking the times to be purely imaginary, $t = -iu$ for real u, defines the *Euclidean* amplitudes, as opposed to the *Minkowski* amplitudes studied above:

$$\langle q_f, u | \mathrm{T}\left[\prod_a \hat{q}(u_a)\right] |q_i, 0\rangle_{\mathrm{E}} \quad \text{(A.1.28)}$$

is defined to be the continuation of

$$\langle q_f, t | \mathrm{T}\left[\prod_a \hat{q}(t_a)\right] |q_i, 0\rangle \quad \text{(A.1.29)}$$

to $t = -iu$. One can repeat the derivation of the path integral representation for the Euclidean case, but the result is extremely simple: just replace t in the action with $-iu$,

$$\langle q_f, U | \mathrm{T}\left[\prod_a \hat{q}(u_a)\right] |q_i, 0\rangle_{\mathrm{E}} = \int [dq]_{q_i,0}^{q_f,U} \exp\left[\int_0^U du\, L(q, i\partial_u q)\right] \prod_a q(u_a) .$$
$$\text{(A.1.30)}$$

The weight in the Euclidean path integral is conventionally defined to be $\exp(-S_{\mathrm{E}})$, so the Euclidean action contains a minus sign,

$$S_{\mathrm{E}} = \int du\, L_{\mathrm{E}}(q, \partial_u q) = -\int du\, L(q, i\partial_u q) . \quad \text{(A.1.31)}$$

Euclidean path integrals are usually the more well-defined, because the exponential of the action is damped rather than oscillatory. The Minkowski matrix elements can then be obtained by analytic continuation. In the text our main concern is the Euclidean path integral, which we use to represent the string S-matrix.

One is often interested in traces, such as

$$\mathrm{Tr}\exp(-\hat{H}U) = \sum_i \exp(-E_i U) , \quad \text{(A.1.32)}$$

where the sum runs over all eigenvectors of $\hat{H}$ and the E_i are the corresponding eigenvalues. In the q basis the trace becomes

$$\int dq \, \langle q| \exp(-\hat{H}U)|q\rangle = \int dq \, \langle q, U|q, 0\rangle_{\mathrm{E}}$$

$$= \int dq \int [dq]_{q,0}^{q,U} \exp(-S_{\mathrm{E}})$$

$$= \int [dq]_{\mathrm{P}} \exp(-S_{\mathrm{E}}) \, . \qquad \text{(A.1.33)}$$

In the last line, $[dq]_{\mathrm{P}}$ denotes the integral over all *periodic* paths on $[0, U]$. An interesting variation on this is the integral over all *antiperiodic* paths:

$$\int [dq]_{\mathrm{A}} \exp(-S_{\mathrm{E}}) = \int dq \int [dq]_{-q,0}^{q,U} \exp(-S_{\mathrm{E}})$$

$$= \int dq \, \langle q, U| - q, 0\rangle_{\mathrm{E}}$$

$$= \int dq \, \langle q, U|\hat{R}|q, 0\rangle_{\mathrm{E}}$$

$$= \mathrm{Tr}\left[\exp(-\hat{H}U)\hat{R}\right] \, . \qquad \text{(A.1.34)}$$

Here $\hat{R}$ is the reflection operator, taking q to $-q$. We see that *twisting* the boundary conditions — for example, replacing periodic with antiperiodic — is equivalent to inserting an operator into the trace. Since $\hat{R}^2 = 1$, its eigenvalues are ± 1. If we take the average of the periodic and antiperiodic path integrals, this inserts $\frac{1}{2}(1 + \hat{R})$, which projects onto the $\hat{R} = +1$ space:

$$\frac{1}{2}(\text{periodic} + \text{antiperiodic}) = \sum_{R_i=+1} \exp(-E_i U) \, . \qquad \text{(A.1.35)}$$

In the same way, $\frac{1}{2}(\text{periodic} - \text{antiperiodic})$ gives the trace on the $\hat{R} = -1$ space.

A minor but sometimes confusing point: if $\hat{A}$ is a Hermitean operator, the Euclidean Heisenberg operator $\hat{A}(u) = \exp(\hat{H}u)\hat{A}\exp(-\hat{H}u)$ satisfies

$$\hat{A}(u)^\dagger = \hat{A}(-u) \qquad \text{(A.1.36)}$$

and so is not in general Hermitean. For this reason it is useful to define a *Euclidean adjoint*,

$$\overline{\hat{A}(u)} = \hat{A}(-u)^\dagger \, . \qquad \text{(A.1.37)}$$

A Hermitean operator in the Schrödinger picture becomes Euclidean self-adjoint in the Euclidean Heisenberg picture.

Diagrams and determinants

The diagrammatic expansion is easily obtained from the path integral. This subject is well treated in modern field theory texts. We will only derive here a few results used in the text. For notational clarity we switch to field theory, replacing $q(t)$ with a real scalar field $\phi(t, \mathbf{x})$. Begin with the free action

$$\frac{1}{2} \int d^d x \, \phi(x) \Delta \phi(x) \,, \tag{A.1.38}$$

where Δ is some differential operator. Consider the path integral with a source $J(x)$, taking the Euclidean case to be specific,

$$
\begin{aligned}
Z[J] &= \left\langle \exp\left[i \int d^d x \, J(x) \phi(x) \right] \right\rangle \\
&= \int [d\phi] \, \exp\left[-\frac{1}{2} \int d^d x \, \phi(x) \Delta \phi(x) + i \int d^d x \, J(x) \phi(x) \right] . \tag{A.1.39}
\end{aligned}
$$

This is the generating functional for Euclidean Green's functions. Taking functional derivatives with respect to $J(x)$ pulls down factors of $i\phi(x)$ into the integral. Shifting ϕ to $\phi + i\Delta^{-1} J$ completes the square, leaving

$$
\begin{aligned}
Z[J] &= Z[0] \exp\left[-\frac{1}{2} \int d^d x \, d^d y \, J(x) \Delta^{-1}(x, y) J(y) \right] \\
&= Z[0] \exp\left[\frac{1}{2} \int d^d x \, d^d y \, \Delta^{-1}(x, y) \frac{\delta^2}{\delta \phi(x) \delta \phi(y)} \right] \\
&\qquad\qquad \times \exp\left[i \int d^d x \, J(x) \phi(x) \right] \bigg|_{\phi=0} . \tag{A.1.40}
\end{aligned}
$$

In the second line, each ϕ derivative becomes iJ. Any functional $\mathscr{F}$ of ϕ can be written

$$\mathscr{F}[\phi] = \int [dJ] \, \exp\left[i \int d^d x \, J(x) \phi(x) \right] f[J] \,, \tag{A.1.41}$$

where $f[J]$ is the (functional) Fourier transform of $\mathscr{F}$. By linearity one obtains

$$\langle \mathscr{F} \rangle = Z[0] \exp\left[\frac{1}{2} \int d^d x \, d^d y \, \Delta^{-1}(x, y) \frac{\delta^2}{\delta \phi(x) \delta \phi(y)} \right] \mathscr{F}[\phi] \bigg|_{\phi=0} . \tag{A.1.42}$$

Expanding the exponential generates the diagrammatic expansion, all ways of contracting pairs of fields in $\mathscr{F}$. Also, when $\mathscr{F}$ is a product of factors, $\mathscr{F}_1 \mathscr{F}_2 \dots$, we can separate

$$\frac{\delta}{\delta \phi(x)} = \frac{\delta}{\delta \phi_1(x)} + \frac{\delta}{\delta \phi_2(x)} + \dots \,, \tag{A.1.43}$$

where $\delta/\delta\phi_i$ acts only on $\mathcal{F}_i$. The *self-contractions* are then the terms

$$\frac{\delta^2}{\delta\phi_i(x)\delta\phi_i(y)} \quad \text{(no sum on } i\text{)} . \tag{A.1.44}$$

Consider now the overall factor $Z[0]$. We assume that Δ is self-adjoint and so we expand $\phi(x)$ in a complete set of eigenfunctions $\Phi_i(x)$:

$$\phi(x) = \sum_i \phi_i \Phi_i(x) , \tag{A.1.45}$$

where

$$\Delta\Phi_i(x) = \lambda_i\Phi_i(x) , \tag{A.1.46a}$$

$$\int d^dx \, \Phi_i(x)\Phi_j(x) = \delta_{ij} . \tag{A.1.46b}$$

Define the path integral measure by

$$[d\phi] = \prod_i d\phi_i . \tag{A.1.47}$$

Then

$$\langle 1 \rangle = \prod_i \left[\int d\phi_i \, \exp\left(-\frac{1}{2}\phi_i^2\lambda_i\right) \right]$$

$$= \prod_i \left(\frac{2\pi}{\lambda_i}\right)^{1/2}$$

$$= \left(\det \frac{\Delta}{2\pi} \right)^{-1/2} . \tag{A.1.48}$$

The product of eigenvalues must be suitably regulated and renormalized. The value of $\langle 1 \rangle$ is usually given simply as $(\det\Delta)^{-1/2}$, because a rescaling of Δ is equivalent to a rescaling of ϕ, the Jacobian for which can be absorbed into the bare couplings. (In free field theory, the only 'coupling' that would be renormalized would be a c-number constant in the Lagrangian density; we will see this appearing in the example.)

All of the above discussion extends readily to multicomponent q^i or ϕ^i. One case of interest is a complex scalar field. One can break this up into two real scalars, $\phi = 2^{-1/2}(\phi^1 + i\phi^2)$, and the measure is $[d\phi \, d\phi^*] \equiv [d\phi^1 \, d\phi^2]$. For complex fields, (A.1.48) becomes

$$\int [d\phi \, d\phi^*] \exp\left(-\int d^dx \, \phi^*\Delta\phi\right) = (\det\Delta)^{-1} . \tag{A.1.49}$$

One similarly finds for two real fields

$$\int [d\phi^1 \, d\phi^2] \exp\left(i\int d^dx \, \phi^1\Delta\phi^2\right) = (\det\Delta)^{-1} . \tag{A.1.50}$$

This is a functional version of the delta-function integral

$$\int_{-\infty}^{\infty} dx \int_{-\infty}^{\infty} dy \, \exp(i\lambda xy) = 2\pi \int_{-\infty}^{\infty} dx \, \delta(\lambda x) = \frac{2\pi}{\lambda} \, . \qquad (A.1.51)$$

An example

As an example, we study the Euclidean transition amplitude for the harmonic oscillator,

$$\langle q_f, U | q_i, 0 \rangle_{\mathrm{E}} = \int [dq]_{q_i,0}^{q_f,U} \, \exp(-S_{\mathrm{E}}) \, , \qquad (A.1.52a)$$

$$S_{\mathrm{E}} = \frac{1}{2} \int_0^U du \left[(\partial_u q)^2 + \omega^2 q^2 \right] + S_{\mathrm{ct}} \, , \qquad (A.1.52b)$$

where S_{ct} is a cutoff-dependent piece to be determined later. To evaluate this, separate out the 'classical' part of q,

$$q(u) = q_{\mathrm{cl}}(u) + q'(u) \, , \qquad (A.1.53)$$

where

$$-\ddot{q}_{\mathrm{cl}}(u) + \omega^2 q_{\mathrm{cl}}(u) = 0 \, , \qquad (A.1.54a)$$

$$q_{\mathrm{cl}}(0) = q_i \, , \quad q_{\mathrm{cl}}(U) = q_f \, . \qquad (A.1.54b)$$

The boundary conditions on q' are then

$$q'(0) = q'(U) = 0 \qquad (A.1.55)$$

and the action separates

$$S_{\mathrm{E}} = S_{\mathrm{cl}}(q_i, q_f) + S' + S_{\mathrm{ct}} \, . \qquad (A.1.56)$$

Here

$$S_{\mathrm{cl}}(q_i, q_f) = \frac{1}{2} \int_0^U du \left[(\partial_u q_{\mathrm{cl}})^2 + \omega^2 q_{\mathrm{cl}}^2 \right]$$

$$= \omega \frac{(q_i^2 + q_f^2) \cosh \omega U - 2 q_i q_f}{2 \sinh \omega U} \, , \qquad (A.1.57a)$$

$$S' = \frac{1}{2} \int_0^U du \left[(\partial_u q')^2 + \omega^2 q'^2 \right] \, . \qquad (A.1.57b)$$

The amplitude becomes

$$\langle q_f, U | q_i, 0 \rangle = \exp[-S_{\mathrm{cl}}(q_i, q_f) - S_{\mathrm{ct}}] \int [dq']_{0,0}^{0,U} \, \exp(-S') \, . \qquad (A.1.58)$$

Expand in a complete set of eigenfunctions of $\Delta = -\partial_u^2 + \omega^2$ subject to the boundary condition (A.1.55),

$$f_j(u) = \left(\frac{2}{U}\right)^{1/2} \sin\frac{j\pi u}{U}, \quad \lambda_j = \frac{j^2\pi^2}{U^2} + \omega^2, \qquad \text{(A.1.59a)}$$

$$\det\frac{\Delta}{2\pi} = \prod_{j=1}^{\infty} \frac{j^2\pi^2 + \omega^2 U^2}{2\pi U^2}. \qquad \text{(A.1.59b)}$$

The infinite product diverges. The *Pauli–Villars regulator* is a simple way to define it. Divide by the amplitude for a *regulator* oscillator of very high frequency Ω,

$$\det\frac{\Delta}{2\pi} \xrightarrow{\text{regulate}} \prod_{j=1}^{\infty} \frac{j^2\pi^2 + \omega^2 U^2}{j^2\pi^2 + \Omega^2 U^2}. \qquad \text{(A.1.60)}$$

For j small, the terms in (A.1.60) are the same as those in (A.1.59b) up to a multiplicative constant, while for j large the product converges. The value of the product (A.1.60) is

$$\frac{\Omega\sinh\omega U}{\omega\sinh\Omega U}. \qquad \text{(A.1.61)}$$

This follows from a standard infinite product representation for the sinh. One can guess the result (A.1.61) by noting that it has the right poles and zeros in U, and the right value as U goes to zero.

Keeping only terms that are nonvanishing as $\Omega \to \infty$,

$$\langle q_f, U | q_i, 0 \rangle \to \left(\frac{\omega}{2\sinh\omega U}\right)^{1/2} \exp\left[-S_{\text{cl}}(q_i, q_f) + \frac{1}{2}(\Omega U - \ln\Omega) - S_{\text{ct}}\right]. \qquad \text{(A.1.62)}$$

To get a finite answer as $\Omega \to \infty$, we need first to include a term $\frac{1}{2}\Omega$ in the Lagrangian L_{ct}, canceling the linear divergence in (A.1.62). That is, there is a linearly divergent bare coupling for the operator '1'. It may seem strange that we need to renormalize in a quantum mechanics problem, but the power counting is completely uniform with quantum field theory. The logarithmic divergence is a wavefunction renormalization. To determine the finite normalization, compare the path integral expression (A.1.62) to the canonical result as $U \to \infty$. The $U \to \infty$ amplitude is easily evaluated by inserting a complete set of states, since only the ground state $|0\rangle$ contributes in the limit:

$$\langle q_f, U | q_i, 0 \rangle \to \langle q_f | 0 \rangle \langle 0 | q_i \rangle \exp(-E_0 U)$$

$$= \left(\frac{\omega}{\pi}\right)^{1/2} \exp\left[-\frac{\omega}{2}(q_f^2 + q_i^2 + U)\right]. \qquad \text{(A.1.63)}$$

On the other hand, the path integral expression (A.1.62) becomes

$$\langle q_f, U | q_i, 0 \rangle \to \omega^{1/2} \exp\left[-\frac{\omega}{2}(q_f^2 + q_i^2 + U) + \frac{1}{2}(\Omega U - \ln\Omega) - S_{\text{ct}}\right]. \qquad \text{(A.1.64)}$$

Thus

$$S_{\text{ct}} = \frac{1}{2}\left[\int_0^U du\,\Omega\right] - \frac{1}{2}\ln\frac{\Omega}{\pi} \tag{A.1.65}$$

and

$$\langle q_f, U|q_i, 0\rangle = \left(\frac{\omega}{2\pi\sinh\omega U}\right)^{1/2}\exp(-S_{\text{cl}}), \tag{A.1.66}$$

which is the correct result.

A.2 Fermionic fields

Consider the basic fermionic quantum system, two states $|\uparrow\rangle$ and $|\downarrow\rangle$ with raising operator $\hat{\chi}$ and lowering operator $\hat{\varphi}$:

$$\hat{\varphi}|\downarrow\rangle = 0, \quad \hat{\chi}|\downarrow\rangle = |\uparrow\rangle, \tag{A.2.1a}$$

$$\hat{\varphi}|\uparrow\rangle = |\downarrow\rangle, \quad \hat{\chi}|\uparrow\rangle = 0, \tag{A.2.1b}$$

$$\{\hat{\varphi}, \hat{\chi}\} = 1, \quad \hat{\varphi}^2 = \hat{\chi}^2 = 0. \tag{A.2.1c}$$

Amplitudes will be expressed as path integrals over the classical analogs of $\hat{\chi}$ and $\hat{\varphi}$. These are *anticommuting c-numbers* or *Grassmann variables*, elements χ_m and ψ_m of a Grassmann algebra

$$\{\chi_m, \psi_n\} = \{\chi_m, \chi_n\} = \{\psi_m, \psi_n\} = 0. \tag{A.2.2}$$

There is a natural notion of integration over Grassmann variables, the Berezin integral. Notice that if θ is a Grassmann variable then $\theta^2 = 0$, and so the most general function of θ is of the form $a + \theta b$. Define

$$\int d\theta = 0, \quad \int d\theta\,\theta = -\int \theta\,d\theta = 1. \tag{A.2.3}$$

Integrals of more general expressions are determined by requiring the integral to be linear and by defining $d\theta$ to anticommute with all Grassmann variables. The significance of the definition (A.2.3) is that the integral of a total derivative is zero:

$$\int d\theta\,\frac{d}{d\theta}(a + \theta b) = \int d\theta\,b = 0. \tag{A.2.4}$$

For multiple Grassmann variables the integral is

$$\int d\theta_1 \ldots d\theta_n\,f(\theta) = c, \tag{A.2.5}$$

where c is the coefficient of the highest term in the Taylor expansion of f,

$$f(\theta) = \ldots + \theta_n\theta_{n-1}\ldots\theta_1 c \tag{A.2.6}$$

Looking back on the discussion of the bosonic path integral, one sees that all of the properties of ordinary integrals that were used are also true for the Berezin integral: linearity, invariance (except for a sign) under change of order of integration, and integration by parts.[3]

In order to follow the bosonic discussion as closely as possible, it is useful to define states that are formally eigenstates of $\hat{\psi}$:

$$|\psi\rangle = |\downarrow\rangle + |\uparrow\rangle\psi \, , \tag{A.2.7a}$$

$$\hat{\psi}|\psi\rangle = |\psi\rangle\psi \, . \tag{A.2.7b}$$

Define $\langle\psi|$ to satisfy

$$\langle\psi|\psi'\rangle = \psi - \psi' \, . \tag{A.2.8}$$

The right-hand side is the Grassmann version of the Dirac delta function. It has the same properties as the ordinary delta function,

$$(\psi - \psi')f(\psi) = (\psi - \psi')f(\psi') \tag{A.2.9}$$

and

$$\int d\psi \, (\psi - \psi')f(\psi) = f(\psi') \tag{A.2.10}$$

for arbitrary $f(\psi)$. The state $\langle\psi|$ is a left eigenstate with opposite sign,

$$\langle\psi|\hat{\psi} = -\psi\langle\psi| \, , \tag{A.2.11}$$

as one verifies by taking the inner product of both sides with a general state $|\psi'\rangle$. Note also that

$$\int |\psi\rangle \, d\psi \, \langle\psi| = 1 \, . \tag{A.2.12}$$

One can verify this by taking the matrix element between arbitrary states $\langle\psi'|$ and $|\psi''\rangle$ and evaluating the Grassmann integral.

The delta function has an integral representation

$$- \int d\chi \, \exp[\chi(\psi' - \psi)] = \psi - \psi' = \langle\psi|\psi'\rangle \, . \tag{A.2.13}$$

By expanding the inner product (A.2.8) in powers of ψ' one obtains

$$\langle\psi|\downarrow\rangle = \psi \, , \quad \langle\psi|\uparrow\rangle = -1 \, , \tag{A.2.14}$$

and therefore

$$\langle\psi|\hat{\chi}|\psi'\rangle = \langle\psi|\uparrow\rangle$$
$$= - \int d\chi \, \chi \exp[\chi(\psi' - \psi)] \, . \tag{A.2.15}$$

---

[3] It is not useful to think of $\int d\theta$ as an integral in any ordinary sense: it is simply a linear mapping from functions of θ to the complex numbers. Since it shares so many of the properties of ordinary integrals, the integral notation is a useful mnemonic.

Using the further property that $|\psi\rangle$ is an eigenstate of $\hat{\psi}$, one obtains

$$-\langle\psi|f(\hat{\chi},\hat{\psi})|\psi'\rangle = \int d\chi\, f(\chi,\psi')\exp[\chi(\psi'-\psi)]\,. \qquad (A.2.16)$$

This holds for any function f with the prescription that it is written with the $\hat{\psi}$ on the right.

We can now emulate the bosonic discussion, using the completeness relation (A.2.12) to join together successive time steps, with the result

$$\langle\psi_f, T|\psi_i, 0\rangle \approx -\int d\chi_0 d\psi_1 d\chi_1 \ldots d\psi_{N-1} d\chi_{N-1}$$

$$\times \exp\sum_{m=0}^{N-1}\left[-iH(\chi_m,\psi_m)\epsilon - \chi_m(\psi_{m+1}-\psi_m)\right]$$

$$\rightarrow \int[d\chi d\psi]\exp\left\{\int_0^T dt\left[-\chi\dot{\psi} - iH(\chi,\psi)\right]\right\}\,. \qquad (A.2.17)$$

The inconvenient sign in the integral representation has been offset by reversing the order of Grassmann integrations. A typical fermionic Hamiltonian is $\hat{H} = m\hat{\chi}\hat{\psi}$. The corresponding Lagrangian is $L = i\chi\dot{\psi} - H$. For such first order Lagrangians the Hamiltonian and Lagrangian formulations coincide, and the result (A.2.17) is also the Lagrangian path integral.

There are two useful inner products for this system,

$$(A): \quad \langle\uparrow|\uparrow\rangle = \langle\downarrow|\downarrow\rangle = 1\,, \quad \langle\uparrow|\downarrow\rangle = \langle\downarrow|\uparrow\rangle = 0\,, \qquad (A.2.18a)$$

$$(B): \quad \langle\uparrow|\downarrow\rangle = \langle\downarrow|\uparrow\rangle = 1\,, \quad \langle\uparrow|\uparrow\rangle = \langle\downarrow|\downarrow\rangle = 0\,. \qquad (A.2.18b)$$

Inner product (A) is positive definite and gives $\hat{\psi}^\dagger = \hat{\chi}$. It is the inner product for an ordinary physical oscillator. Inner product (B) is not positive, and it makes $\hat{\psi}$ and $\hat{\chi}$ Hermitian. It is the appropriate inner product for the Faddeev–Popov ghost system, as discussed in chapter 4. The derivation of the path integral was independent of the choice of inner product, but the relations (A.2.14) determine the expansion of $\langle\psi|$ in basis states in terms of the inner product.

All of the discussion of the bosonic case applies to the fermionic path integral as well. The derivation is readily extended to multiple Fermi oscillators and to field theory. Renormalization theory works in the same way as for the bosonic path integral. Fermionic path integrals again satisfy cutting relations, which simply amount to taking one of the iterated integrals (A.2.17) and saving it for the end. Insertion of $\psi(t)$ and $\chi(t)$ in the path integral again gives the corresponding operator in the matrix element, and multiple insertions are again time-ordered. The path integral insertion

$$\psi(t)\psi(t') = -\psi(t')\psi(t) \qquad (A.2.19)$$

becomes the operator

$$T[\hat{\psi}(t)\hat{\psi}(t')] \equiv \theta(t - t')\hat{\psi}(t)\hat{\psi}(t') - \theta(t' - t)\hat{\psi}(t')\hat{\psi}(t) . \qquad (A.2.20)$$

Note that a minus sign appears in the definition of time-ordering. The equations of motion and the Euclidean continuation work as in the bosonic case, but there is one important change. Consider a general operator $\hat{A}$,

$$\hat{A}|i\rangle = |j\rangle A_{ji}, \qquad (A.2.21)$$

where i, j run over $\downarrow$ and $\uparrow$. Then by expanding $\hat{A}|\psi\rangle$ in terms of the basis states one finds

$$\int d\psi \, \langle\psi|\hat{A}|\psi\rangle = A_{\downarrow\downarrow} - A_{\uparrow\uparrow}$$

$$= \mathrm{Tr}\left[(-1)^{\hat{F}}\hat{A}\right] , \qquad (A.2.22)$$

where the fermion number $\hat{F}$ has been defined to be even for $|\downarrow\rangle$ and odd for $|\uparrow\rangle$. In particular,

$$\int d\psi \, \langle\psi, U|\psi, 0\rangle = \mathrm{Tr}\left[(-1)^{\hat{F}} \exp(-\hat{H}U)\right] . \qquad (A.2.23)$$

In contrast to the bosonic results (A.1.33) and (A.1.34), the *periodic* path integral gives the trace weighted by $(-1)^{\hat{F}}$ and the *antiperiodic* path integral gives the simple trace.

Feynman diagrams are as in the bosonic case, but the determinants work out differently. Consider

$$\langle 1 \rangle = \int [d\psi \, d\chi] \, \exp\left(\int d^d x \, \chi\Delta\psi\right) , \qquad (A.2.24)$$

where we have again switched to the notation of field theory. Expand ψ in eigenfunctions of Δ, and χ in eigenfunctions of Δ^T,

$$\psi(x) = \sum_i \psi_i\Psi_i(x) , \quad \chi(x) = \sum_i \chi_i\Upsilon_i(x) , \qquad (A.2.25a)$$

$$\Delta\Psi_i(x) = \lambda_i\Psi_i(x) , \quad \Delta^T\Upsilon_i(x) = \lambda_i\Upsilon_i(x) , \qquad (A.2.25b)$$

$$\int d^d x \, \Upsilon_i(x)\Psi_j(x) = \delta_{ij} . \qquad (A.2.25c)$$

Define $[d\psi d\chi] = \prod_i d\psi_i d\chi_i$. Then

$$\langle 1 \rangle = \prod_i \left[\int d\psi_i d\chi_i \, \exp(\lambda_i\chi_i\psi_i)\right]$$

$$= \prod_i \lambda_i$$

$$= \det\Delta . \qquad (A.2.26)$$

Similarly, one finds for a single field ψ that

$$\int [d\psi] \exp\left(\int d^d x\, \psi \Delta \psi\right) = (\det \Delta)^{1/2}, \qquad \text{(A.2.27)}$$

where Δ must be antisymmetric. These fermionic path integrals are just the inverse of the corresponding bosonic integrals.

For finite-dimensional integrals we have to be careful about 2πs,

$$\int_{-\infty}^{\infty} dx \int_{-\infty}^{\infty} dy\, \exp(2\pi i \lambda x y) = \frac{1}{\lambda} = \left[\int d\psi \int d\chi\, \exp(\lambda \chi \psi)\right]^{-1}. \qquad \text{(A.2.28)}$$

Curiously, we can be more cavalier for functional determinants, where these can be eliminated by rescaling the fields. Bosonic path integrals actually give the absolute values of determinants, while fermionic path integrals give the determinant with a sign that depends on the precise order of Grassmann factors in the measure. We will fix the sign by hand.[4]

If Δ has zero eigenvalues, the path integral vanishes in the fermionic case and diverges in the bosonic case. When this occurs there is always a good reason, as we will see in practice, and the integrals of interest will have appropriate insertions to give a finite result.

A useful result about Berezin integrals: from the definition it follows that if $\theta' = \theta a$ then $d\theta' = a^{-1} d\theta$. That is, the Jacobian is the inverse of the bosonic case. More generally, with bosonic variables x_i and Grassmann variables θ_j, under a general change of variables to $x_i'(x, \theta)$, $\theta_j'(x, \theta)$ the infinitesimal transformation of the measure has an extra minus sign,

$$\delta(d^m x\, d^n \theta) = d^m x\, d^n \theta \left(\sum_i \frac{\partial}{\partial x_i} \delta x_i - \sum_j \frac{\partial}{\partial \theta_j} \delta \theta_j\right). \qquad \text{(A.2.29)}$$

Exercises

A.1 (a) Evaluate the path integral (A.1.33) for the harmonic oscillator by expanding $[dq]_P$ in a set of periodic eigenfunctions (not by inserting the result already obtained for the transition function into the left-hand side). Compare with an explicit evaluation of the sum over states (A.1.32).
(b) Do the same for the antiperiodic path integral (A.1.34).

A.2 Consider two harmonic oscillators of the same frequency, and form the complex coordinate $q = 2^{-1/2}(q_1 + iq_2)$. Evaluate the following path

---

[4] A fine point: when the world-sheet theory is parity-asymmetric, even the phase of the path integral is ambiguous. The point is that Δ is not Hermitian in this case, because the Euclidean adjoint includes a time-reversal (which is the same as a parity flip plus Euclidean rotation). One must instead take the determinant of the Hermitian operator $\Delta^\dagger \Delta$, which determines that of Δ only up to a phase. We will encounter this in section 10.7, for example, where we will have to check that there is a consistent definition of the phase.

integral on $[0, U]$ by using a suitable mode expansion:

$$\int [dq\, dq^*]_\theta \, \exp(-S_{\mathrm{E}}) \, ,$$

where the integral is to run over all paths such that $q(U) = q(0)e^{i\theta}$. Again compare with the answer obtained by summing over states: it is useful here to form complex combinations of oscillators, $a_1 \pm ia_2$. This is a more exotic version of exercise A.1, and arises in various string theories such as orbifolds.

A.3 Consider a free scalar field X in two dimensions, with action

$$S = \frac{1}{4\pi\alpha'} \int d^2\sigma \left(\partial_1 X \partial_1 X + \partial_2 X \partial_2 X + m^2 X^2 \right) .$$

Evaluate the path integral with both coordinates periodic,

$$0 \le \sigma_1 \le 2\pi \,, \quad 0 \le \sigma_2 \le T \,.$$

[Expanding the spatial dependence in Fourier modes, the answer takes the form of an infinite product of harmonic oscillator traces.]

A.4 Consider a free particle on a circle,

$$H = -\frac{1}{2} \frac{\partial^2}{\partial \phi^2} \, .$$

By means of path integrals, evaluate $\langle \phi_f, T | \phi_i, 0 \rangle$. Be careful: there are topologically inequivalent paths, distinguished by the number of times they wind around the circle.

A.5 The fermionic version of exercise A.1: evaluate the fermionic path integral (A.2.23) and the corresponding antiperiodic path integral by using a complete set of periodic or antiperiodic functions, and compare with the sum over states.

References

The literature on the subjects covered in these two volumes easily exceeds 5000 papers, and probably approaches twice that number. To represent accurately the contributions of all who have worked in these many areas is an undertaking beyond the scope of this book. Rather, I have tried to assemble a list that is short enough to be useful in developing specific subjects and in giving points of entry to the literature. The resulting list includes many review articles, plus papers from the original literature where I felt they were needed to supplement the treatment in the text.

Papers listed as 'e-print: hep-th/yymmnnn' are available electronically at the Los Alamos physics e-print archive,

$$\text{http://xxx.lanl.gov/abs/hep-th/yymmnnn} \, .$$

For a list of corrections and other useful information see

$$\text{http://www.itp.ucsb.edu/\textasciitilde joep/bigbook.html} \, .$$

General references

The reader may find the two volume set by Green, Schwarz, & Witten (1987) (henceforth GSW) a useful supplement. The mix of covariant and light-cone methods used in that book, and the advanced topics emphasized, are somewhat complementary to the approach taken here. The book by Polyakov (1987) emphasizes conformal field theory and the path integral approach to strings, but the focus is less on unified string theories and more on string theories of the strong interaction and on the statistical mechanics of random surfaces. Some other lectures on string theory are Peskin (1987), Lüst & Theisen (1989), Alvarez-Gaumé & Vazquez-Mozo (1992), D'Hoker (1993), Ooguri & Yin (1997), and Kiritsis (1997).

A brief history

There have been three periods of intense activity in the history of string theory, covering roughly the years 1968–73, 1984–9, and 1995 to the present. The first of these began with the Veneziano amplitude as a phenomenological proposal for the S-matrix of the strong interaction. This led to *dual resonance models,* where the Veneziano formula was generalized and systematized. Further discoveries were the connection of dual models to strings, the critical dimension (26), superconformal invariance, and the existence of strings with critical dimension 10. This first period of activity was ended by the discovery of QCD as the correct theory of the strong interaction, but those who had worked on string theory were left with the belief that they had discovered a rich physical theory that should have some application in nature.

During the succeeding 'quiet' period, a small group of theorists developed string theory as a unified theory of gravity and gauge interactions. String theories with spacetime supersymmetry were discovered during this period, and the argument made that superstrings were a finite theory of quantum gravity. Three discoveries in 1984 established that string theory was also a strong candidate for the unification of the Standard Model. These discoveries were the existence of anomaly-free chiral theories in ten dimensions, the heterotic string, and Calabi–Yau compactifaction. The ensuing period of activity was largely devoted to working out their consequences. The main difficulty that remained, and still remains, was the enormous number of potential vacua and an insufficient understanding of the dynamics that chooses among them.

The latest period of activity is marked by new dynamical tools. Again, many important results date from the preceding quiet period, but their importance and their relation to one another took time to emerge. The key steps were the systematic analysis of the consequences of supersymmetry in gauge theories and then in string theory. This led to many new discoveries: weak–strong duality in string theory, the unification of all string theories as limits of a single theory, a limit in which a new, eleventh, dimension appears, the significance of D-branes, and new understanding of the quantum mechanics of black holes.

The developments from 1984 on are covered in the literature and review articles below and in volume two, but the earlier period is somewhat neglected. Many of the key papers up to 1985 can be found in Schwarz (1985), which also includes summaries of the history. Chapters 1 and 2 of the reference section of GSW give a more detailed history of the dual model period, with extensive references. The review by Scherk (1975) also covers this period.

Chapter 1

There are now many treatments of the Standard Model; Quigg (1983) gives a survey and Cheng & Li (1991) a more detailed treatment, to name just two. Weinberg (1979) discusses the UV problem of quantum gravity in field theory. References for the various unifying ideas will be given in the appropriate chapters.

The classical motion of the string and light-cone quantization are discussed in Goddard, Goldstone, Rebbi, & Thorn (1973) and in the reviews by Rebbi (1974) and Scherk (1975). The first two of these include a careful treatment of Lorentz invariance. The heuristic argument that we have given is based on Brink & Nielsen (1973). Weinberg (1964) discusses the couplings of massless vector and tensor particles.

Chapter 2

In addition to the general references, Friedan, Martinec, & Shenker (1986) covers many of the subjects in this chapter. Belavin, Polyakov, & Zamolodchikov (1984) is a classic reference on CFT, but directed at the advanced topics to be treated in chapter 15. Ginsparg (1990) is a review of CFT; Gatto (1973) and Fradkin & Palchik (1978) treat conformal invariance in four-dimensional quantum field theory; although this is much smaller than the two-dimensional symmetry, many of the same ideas apply.

Chapter 3

Additional references on the Polyakov path integral are Polyakov (1981), Friedan (1982), and Alvarez (1983). The Faddeev–Popov procedure is treated in all modern field theory textbooks. The basic steps follow the original reference Faddeev & Popov (1967). The book by Birrell & Davies (1982) discusses Weyl anomalies in various dimensions (though it uses the common terminology *conformal anomaly* instead). Gravitational anomalies are discussed in Alvarez-Gaumé & Witten (1983).

An LSZ-like derivation of the vertex operators in light-cone gauge is in Mandelstam (1973); see also chapter 11 of GSW (1987). For calculation of beta functions in general nonlinear sigma models see Alvarez-Gaumé, Freedman, & Mukhi (1981) and Friedan (1985). A general reference on strings in background fields is Callan, Friedan, Martinec, & Perry (1985); reviews include Callan & Thorlacius (1989) and Tseytlin (1989). The linear dilaton theory is discussed as a CFT by Chodos & Thorn (1974) and as a string background by Myers (1987).

Chapter 4

Classic proofs of the OCQ no-ghost theorem are Brower (1972) and Goddard & Thorn (1972). The basic reference on quantization with constraints is Dirac (1950); a review is Henneaux (1985). The BRST symmetry in gauge theory is discussed in modern field theory texts. The structure of its cohomology is discussed in Kugo & Ojima (1979). The BRST operator for field theory is derived in Kato & Ogawa (1983); our proof of the no-ghost theorem is loosely patterned on the one in that paper. Other proofs in the BRST context are Freeman & Olive (1986), Frenkel, Garland, & Zuckerman (1986), and Thorn (1987).

Chapter 5

D'Hoker & Phong (1988) is an extensive review of the subjects in this and subsequent chapters. The measure on moduli space is obtained in Moore & Nelson (1986) and D'Hoker & Phong (1986); see also Alvarez (1983). The representation of the measure in terms of the transition functions is discussed in Martinec (1987) and Alvarez-Gaumé, Gomez, Moore, & Vafa (1988). General properties of the amplitude are also discussed in Friedan, Martinec, & Shenker (1986) and Mansfield (1987).

Chapter 6

Some of the physics of the tree-level amplitudes is discussed in Veneziano (1974). The appearance of gauge and gravitational interactions in the long-distance limit is developed in Neveu & Scherk (1972), Yoneya (1974), and Scherk & Schwarz (1974). Constraints on open string gauge groups are discussed in Marcus & Sagnotti (1982) and Schwarz (1982). For more on operator methods see GSW and the dual model references Schwarz (1973) and Mandelstam (1974). The relation of closed to open string tree-level amplitudes is given in Kawai, Lewellen, & Tye (1986).

Chapter 7

The two volumes by Mumford (1983) are a beautiful introduction to theta functions; the one-loop functions are discussed in the first half of volume one. Constraints from modular invariance on general conformal field theories are in Cardy (1986). The zero-point energy in quantum field theory is discussed in Coleman & Weinberg (1973). The cosmological constant problem is reviewed in Weinberg (1989).

Cancellation of divergences for open superstring gauge group $SO(32)$ is discussed in Green & Schwarz (1985). The analysis using the boundary state formalism is in Callan, Lovelace, Nappi, & Yost (1987) and Polchinski & Cai (1988).

Chapter 8

Kaluza–Klein theory is reviewed in Duff, Nilsson, & Pope (1986) and in the reprint volume Appelquist, Chodos, & Freund (1987). Giveon, Porrati, & Rabinovici (1994) review toroidal compactification and T-duality. Polchinski (1997) reviews T-duality of open and unoriented strings, and D-branes. Further references on toroidal compactification are Narain (1986) and Narain, Sarmadi, & Witten (1987).

Dixon, Harvey, Vafa, & Witten (1985) is an introduction to orbifolds. Twisted state vertex amplitudes are obtained in Dixon, Friedan, Martinec & Shenker (1987) and Hamidi & Vafa (1987). More extensive orbifold references are given with chapter 16. The space of $c = 1$ CFTs is described in Dijkgraaf, Verlinde, & Verlinde (1988) and Ginsparg (1988) and in the review Ginsparg (1990).

Chapter 9

Unitarity of tree-level amplitudes is discussed in the dual model reviews Schwarz (1973) and Mandelstam (1974) (these discussions are in OCQ, whereas section 9.1 gives the BRST version of the argument). One-loop amplitudes were originally constructed by imposing unitarity on the tree-level amplitudes; see Scherk (1975) for a review.

The moduli space of higher genus Riemann surfaces is reviewed in Martinec (1987) and D'Hoker & Phong (1988). Some more details of the Schottky construction can be found in Mandelstam (1974) and references therein. Sewing of CFTs has been discussed by many authors. The treatment here is based on Vafa (1987), Polchinski (1988), Sonoda (1988a,b,1989) and LeClair, Peskin, & Preitschopf (1989a,b). The extension to the open string is in Lewellen (1992). The graphical decomposition of moduli space and the construction of minimal area metrics are reviewed in Zwiebach (1993). A discussion of the combinatorics of unitarity in quantum field perturbation theory can be found in 't Hooft & Veltman (1973). The Fischler–Susskind mechanism is reviewed in Polchinski (1988). For a different approach to string unitarity, based on showing the equivalence between the light-cone and Polyakov amplitudes, see D'Hoker & Giddings (1987). The light-cone formalism is discussed in Mandelstam (1973) and GSW, chapter 11.

The form of open string field theory that we have described is from Witten (1986). String field theory is reviewed by Siegel (1988), Thorn (1989), and Zwiebach (1993). Borel summability of string perturbation theory is discussed in Gross & Periwal (1988). The 2g! behavior is discussed in Shenker (1991). High energy hard scattering is examined in Gross & Mende (1988) and Gross (1988). Another approach to high energy scattering is Amati, Ciafaloni, & Veneziano (1989). The spreading of the boosted string is described by Susskind (1993). The Hagedorn transition has been a source of much speculation since the original work of Hagedorn (1965). The discussion here is based in part on Atick & Witten (1988).

The review Polchinski (1996) covers many of the subjects in this chapter, and in particular gives an introduction to matrix models. Other matrix model reviews are Kazakov & Migdal (1988), Klebanov (1992), Martinec (1992), and an extensive review by Ginsparg & Moore (1993).

References

Alvarez, O. (1983). Theory of strings with boundaries. *Nuclear Physics,* **B216**, 125.

Alvarez-Gaumé, L., Freedman, D. Z., & Mukhi, S. (1981). The background field method and the ultraviolet structure of the supersymmetric nonlinear sigma model. *Annals of Physics,* **134**, 85.

Alvarez-Gaumé, L., Gomez, C., Moore, G., & Vafa, C. (1988). Strings in the operator formalism. *Nuclear Physics,* **B303**, 455.

Alvarez-Gaumé, L., & Vazquez-Mozo, M. A. (1992). Topics in string theory and quantum gravity. In *Les Houches Summer School on Gravitation and Quantizations,* eds. J. Zinn-Justin & B. Julia, pp. 481–636. Amsterdam: North-Holland. E-print hep-th/9212006.

Alvarez-Gaumé, L., & Witten, E. (1983). Gravitational anomalies. *Nuclear Physics,* **B234**, 269.

Amati, D., Ciafaloni, M., & Veneziano, G. (1989). Can spacetime be probed below the string size? *Physics Letters,* **B216**, 41.

Appelquist, T., Chodos, A., & Freund, P. G. O. (1987). *Modern Kaluza–Klein Theories.* Menlo Park: Addison Wesley.

Atick, J. J., & Witten, E. (1988). The Hagedorn transition and the number of degrees of freedom of string theory. *Nuclear Physics,* **B310**, 291.

Belavin, A. A., Polyakov, A. M., & Zamolodchikov, A. B. (1984). Infinite conformal symmetry in two-dimensional quantum field theory. *Nuclear Physics,* **B241**, 333.

Birrell, N. D., & Davies, P. C. W. (1982). *Quantum Fields in Curved Space.* Cambridge: Cambridge University Press.

Brink, L., & Nielsen, H. B. (1973). A simple physical interpretation of the critical dimension of spacetime in dual models. *Physics Letters,* **B45**, 332.

Brower, R. C. (1972). Spectrum-generating algebra and no-ghost theorem for the dual model. *Physical Review,* **D6**, 1655.

Buscher, T. (1987). A symmetry of the string background field equations. *Physics Letters,* **B194**, 59.

Buscher, T. (1988). Path integral derivation of quantum duality in nonlinear sigma models. *Physics Letters,* **B201**, 466.

Callan, C. G., Friedan, D., Martinec, E. J., & Perry, M. J. (1985). Strings in background fields. *Nuclear Physics,* **B262**, 593.

Callan, C. G., Lovelace, C., Nappi, C. R., & Yost, S. A. (1987). Adding holes and crosscaps to the superstring. *Nuclear Physics,* **B293**, 83.

Callan, C. G., & Thorlacius, L. (1989). Sigma models and string theory. In *Particles, Strings, and Supernovae, TASI 1988,* eds. A. Jevicki & C.-I. Tan, pp. 795–878. Singapore: World Scientific.

Cardy, J. L. (1986). Operator content of two-dimensional conformally invariant theories. *Nuclear Physics,* **B270**, 186.

Cheng, T.-P., & Li, L.-F. (1991). *Gauge Theory of Elementary Particle Physics.* Oxford: Oxford University Press.

Chodos, A., & Thorn, C. B. (1974). Making the massless string massive. *Nuclear Physics,* **B72**, 509.

Coleman, S., & Weinberg, E. (1973). Radiative corrections as the origin of spontaneous symmetry breaking. *Physical Review,* **D7**, 1888.

D'Hoker, E. (1993). TASI Lectures on critical string theory. In *Recent Directions in Particle Theory, TASI 1992,* eds. J. Harvey & J. Polchinski, pp. 1–98. Singapore: World Scientific.

D'Hoker, E., & Giddings, S. B. (1987). Unitarity of the closed bosonic Polyakov string. *Nuclear Physics,* **B291**, 90.

D'Hoker, E., & Phong, D. H. (1986). Multiloop amplitudes for the bosonic Polyakov string. *Nuclear Physics,* **B269**, 205.

D'Hoker, E., & Phong, D. H. (1988). The geometry of string perturbation theory. *Reviews of Modern Physics,* **60**, 917.

Dijkgraaf, R., Verlinde, E., & Verlinde, H. (1988). $c = 1$ conformal field theories on Riemann surfaces. *Communications in Mathematical Physics,* **115**, 649.

Dirac, P. A. M. (1950). Generalized Hamiltonian dynamics. *Canadian Journal of Mathematics,* **2**, 129.

Dixon, L., Friedan, D., Martinec, E., & Shenker, S. (1987). The conformal field theory of orbifolds. *Nuclear Physics,* **B282**, 13.

Dixon, L., Harvey, J. A., Vafa, C., & Witten, E. (1985). Strings on orbifolds. *Nuclear Physics,* **B261**, 678.

Duff, M. J., Nilsson, B. E. W., & Pope, C. N. (1986). Kaluza–Klein supergravity. *Physics Reports,* **130,** 1.

Faddeev, L. D., & Popov, V. N. (1967). Feynman diagrams for the Yang–Mills field. *Physics Letters,* **B25,** 29.

Fradkin, E. S., & Palchik, M. Ya. (1978). Recent developments in conformal invariant quantum field theory. *Physics Reports,* **44,** 249.

Freeman, M. D., & Olive, D. I. (1986). BRS cohomology in string theory and the no-ghost theorem. *Physics Letters,* **B175,** 151.

Frenkel, I. B., Garland, H., & Zuckerman, G. J. (1986). Semi-infinite cohomology and string theory. *Proceedings of the National Academy of Sciences,* **83,** 8442.

Friedan, D. (1982). Introduction to Polyakov's string theory. In *Recent Advances in Field Theory and Statistical Mechanics, Les Houches 1982,* eds. J.-B. Zuber & R. Stora, pp. 839–67. Amsterdam: North-Holland.

Friedan, D. (1985). Nonlinear models in $2 + \epsilon$ dimensions. *Annals of Physics,* **163,** 318.

Friedan, D., Martinec, E., & Shenker, S. (1986). Conformal invariance, supersymmetry, and string theory. *Nuclear Physics,* **B271,** 93.

Gatto, R. (1973). *Scale and Conformal Symmetry in Hadron Physics.* New York: Wiley.

Ginsparg, P. (1988). Curiosities at $c = 1$. *Nuclear Physics,* **B295,** 153.

Ginsparg, P. (1990). Applied conformal field theory. In *Fields, Strings, and Critical Phenomena, Les Houches 1988,* eds. E. Brezin & J. Zinn-Justin, pp. 1–168. Amsterdam: North-Holland.

Ginsparg, P., & Moore, G. (1993). Lectures on 2D gravity and 2D string theory. In *Recent Directions in Particle Theory, TASI 1992,* eds. J. Harvey & J. Polchinski, pp. 277–469. Singapore: World Scientific. E-print hep-th/9304011.

Giveon, A., Porrati, M., & Rabinovici, E. (1994). Target space duality in string theory. *Physics Reports,* **244,** 77. Eprint hep-th/9401139.

Goddard, P., Goldstone, J., Rebbi, C. B., & Thorn, C. (1973). Quantum dynamics of a massless relativistic string. *Nuclear Physics,* **B56,** 109.

Goddard, P., & Thorn, C. B. (1972). Compatibility of the dual Pomeron with unitarity and the absence of ghosts in the dual resonance model. *Physics Letters,* **B40,** 235.

Green, M. B., & Schwarz, J. H. (1985). Infinity cancellations in $SO(32)$ superstring theory. *Physics Letters,* **B151,** 21.

Green, M. B., Schwarz, J. H., & Witten, E. (1987). *Superstring Theory,* in two volumes. Cambridge: Cambridge University Press.

Gross, D. J. (1988). High energy symmetries of string theory. *Physical Review Letters,* **60,** 1229.

Gross. D. J., & Mende, P. F. (1988). String theory beyond the Planck scale. *Nuclear Physics,* **B303**, 407.

Gross, D. J., & Periwal, V. (1988). String perturbation theory diverges. *Physical Review Letters,* **60**, 2105.

Hagedorn, R. (1965). Statistical thermodynamics of strong interactions at high energies. *Nuovo Cimento Supplement,* **3**, 147.

Hamidi, S., & Vafa, C. (1987). Interactions on orbifolds. *Nuclear Physics,* **B279**, 465.

Henneaux, M. (1985). Hamiltonian form of the path integral for theories with a gauge freedom. *Physics Reports,* **126**, 1.

Kato, M., & Ogawa, K. (1983). Covariant quantization of string based on BRS invariance. *Nuclear Physics,* **B212**, 443.

Kawai, H., Lewellen, D. C., & Tye, S.-H. H. (1986). A relation between tree amplitudes of closed and open strings. *Nuclear Physics,* **B269**, 1.

Kazakov, V.A., & Migdal A. A. (1988). Recent progress in the theory of noncritical strings. *Nuclear Physics,* **B311**, 171.

Kiritsis, E. (1997). *Introduction to Superstring Theory.* Leuven: Leuven University Press, in press. E-print hep-th/9709062.

Klebanov, I. R. (1992). String theory in two dimensions. In *String Theory and Quantum Gravity, Trieste 1991,* eds. J. Harvey, *et. al.,* pp. 30–101. Singapore: World Scientific.

Kugo, T., & Ojima, I. (1979). Local covariant operator formulation of non Abelian gauge theories and quark confinement problem. *Supplement to Progress in Theoretical Physics,* **66**, 1.

LeClair, A., Peskin, M. E., & Preitschopf, C. R. (1989a). String field theory on the conformal plane. 1. Kinematical principles. *Nuclear Physics,* **B317**, 411.

LeClair, A., Peskin, M. E., & Preitschopf, C. R. (1989b). String field theory on the conformal plane. 2. Generalized gluing. *Nuclear Physics,* **B317**, 464.

Lewellen, D. C. (1992). Sewing constraints for conformal field theories on surfaces with boundaries. *Nuclear Physics,* **B372**, 654.

Lovelace, C. (1971). Pomeron form factors and dual Regge cuts. *Physics Letters,* **B34**, 500.

Lüst, D., & Theisen, S. (1989). *Lectures on String Theory.* Berlin: Springer-Verlag.

Mandelstam, S. (1973). Interacting-string picture of dual-resonance models. *Nuclear Physics,* **B64**, 205.

Mandelstam, S. (1974). Dual-resonance models. *Physics Reports,* **13**, 259.

Mansfield, P. (1987). Nilpotent BRST invariance of the interacting Polyakov string. *Nuclear Physics,* **B283**, 551.

Marcus, N., & Sagnotti, A. (1982). Tree-level constraints on gauge groups for type I superstrings. *Physics Letters*, **B119**, 97.

Martinec, E. (1987). Conformal field theory on a (super-)Riemann surface. *Nuclear Physics*, **B281**, 157.

Martinec, E. (1992). An introduction to 2d gravity and solvable string models. In *String Theory and Quantum Gravity, Trieste 1991*, eds. J. Harvey, *et. al.*, pp. 1–29. Singapore: World Scientific.

McClain, B., & Roth, B. D. B. (1987). Modular invariance for interacting bosonic strings at finite temperature. *Communications in Mathematical Physics*, **111**, 539.

Misner, C. W., Thorne, K. S., & Wheeler, J. A. (1973). *Gravitation*. San Francisco: Freeman.

Moore, G., & Nelson, P. (1986). Measure for moduli. *Nuclear Physics*, **B266**, 58.

Mumford, D. (1983). *Tata lectures on theta*, in three volumes. Boston: Birkhauser.

Myers, R. C. (1987). New dimensions for old strings. *Physics Letters*, **B199**, 371.

Narain, K. S. (1986). New heterotic string theories in uncompactified dimensions $D < 10$. *Physics Letters*, **B169**, 41.

Narain, K. S., Sarmadi, M. H., & Witten, E. (1987). A note on toroidal compactification of heterotic string theory. *Nuclear Physics*, **B279**, 369.

Neveu, A., & Scherk, J. (1972). Connection between Yang–Mills fields and dual models. *Nuclear Physics*, **B36**, 155.

Ooguri, H., & Yin, Z. (1997). Lectures on perturbative string theories. In *Fields, Strings, and Duality, TASI 1996*, eds. C. Efthimiou & B. Greene, pp. 5–82. Singapore: World Scientific. E-print hep-th/9612254.

Peskin, M. (1987). Introduction to string and superstring theory. In *From the Planck Scale to the Weak Scale, TASI 1986*, ed. H. Haber, pp. 277–408. Singapore: World Scientific.

Polchinski, J. (1986). Evaluation of the one loop string path integral. *Communications in Mathematical Physics*, **104**, 37.

Polchinski, J. (1988). Factorization of bosonic string amplitudes. *Nuclear Physics*, **B307**, 61.

Polchinski, J. (1996). What is string theory? In *Fluctuating Geometries in Statistical Mechanics and Field Theory, Les Houches 1994*, eds. F. David, P. Ginsparg, & J. Zinn-Justin, pp. 287–422. Amsterdam: North-Holland.

Polchinski, J. (1997). TASI lectures on D-branes. In *Fields, Strings, and Duality, TASI 1996*, eds. C. Efthimiou & B. Greene, pp. 293–356. Singapore: World Scientific. E-print hep-th/9611050.

Polchinski, J., & Cai, Y. (1988). Consistency of open superstring theories. *Nuclear Physics,* **B296**, 91.

Polyakov, A. M. (1981). Quantum geometry of bosonic strings. *Physics Letters,* **B103**, 207.

Polyakov, A. M. (1987). *Gauge Fields and Strings.* Chur, Switzerland: Harwood.

Quigg, C. (1983). *Gauge Theories of the Strong, Weak, and Electromagnetic Interactions.* Reading: Benjamin-Cummings.

Rebbi, C. (1974). Dual models and relativistic quantum strings. *Physics Reports,* **12**, 1.

Roček, M., & Verlinde, E. (1992). Duality, quotients and currents. *Nuclear Physics,* **B373**, 630. E-print hep-th/9110053.

Scherk, J. (1975). An introduction to the theory of dual models and strings. *Reviews of Modern Physics,* **47**, 123.

Scherk, J., & Schwarz, J. H. (1974). Dual models for non-hadrons. *Nuclear Physics,* **B81**, 118.

Schwarz, J. H. (1973). Dual resonance theory. *Physics Reports,* **8**, 269.

Schwarz, J. H. (1982). Superstring theory. *Physics Reports,* **89**, 223.

Schwarz, J. H. (1985). *Superstrings. The First 15 Years of Superstring Theory,* in two volumes. Singapore: World Scientific.

Shapiro, J. A. (1972). Loop graph in the dual-tube model. *Physical Review,* **D5**, 1945.

Shenker, S. (1991). The strength of nonperturbative effects in string theory. In *Random Surfaces and Quantum Gravity, Cargese 1990,* eds. O. Alvarez, E. Marinari, & P. Windey, pp. 191–200. New York: Plenum.

Siegel, W. (1988). *Introduction to String Field Theory.* Singapore: World Scientific.

Sonoda, H. (1988a). Sewing conformal field theories. 1. *Nuclear Physics,* **B311**, 401.

Sonoda, H. (1988b). Sewing conformal field theories. 2. *Nuclear Physics,* **B311**, 417.

Sonoda, H. (1989). Hermiticity and CPT in string theory. *Nuclear Physics,* **B326**, 135.

Susskind, L. (1993). String theory and the principle of black hole complementarity. *Physical Review Letters,* **71**, 2367. E-print hep-th/9307168.

't Hooft, G., & Veltman, M. (1973). Diagrammar. CERN report, reprinted in *Under the Spell of the Gauge Principle,* collected work of G. 't Hooft, pp. 28–173. Singapore: World Scientific, 1994.

Thorn, C. B. (1987). A detailed study of the physical state conditions in covariantly quantized string theories. *Nuclear Physics,* **B286**, 61.

Thorn, C. B. (1989). String field theory. *Physics Reports,* **175**, 1.

Tseytlin, A. A. (1989). Sigma model approach to string theory. *International Journal of Modern Physics,* **A4**, 1257.

Vafa, C. (1987). Conformal field theory and punctured surfaces. *Physics Letters,* **B199**, 195.

Veneziano, G. (1974). An introduction to dual models of strong interactions and their physical motivations. *Physics Reports,* **9**, 199.

Weinberg, S. (1964). Photons and gravitons in S matrix theory. *Physical Review,* **135**, B1049.

Weinberg, S. (1979). Ultraviolet divergences in quantum theories of gravitation. In *General Relativity. An Einstein Centenary Survey,* eds. S. W. Hawking & W. Israel, pp. 790–831. Cambridge: Cambridge University Press.

Weinberg, S. (1989). The cosmological constant problem. *Reviews of Modern Physics,* **61**, 1.

Witten, E. (1986). Noncommutative geometry and string field theory. *Nuclear Physics,* **B268**, 253.

Yoneya, T. (1974). Connection of dual models to electrodynamics and gravidynamics. *Progress of Theoretical Physics,* **51**, 1907.

Zwiebach, B. (1993). Closed string field theory: quantum action and the Batalin-Vilkovisky master equation. *Nuclear Physics,* **B390**, 33. E-print hep-th/9206084.

Glossary

(0, 2) compactification a generic heterotic string vacuum having $d = 4$, $N = 1$ supersymmetry. The world-sheet CFT has $N = 2$ right-moving superconformal invariance.

(2, 0) theory in current usage, this refers to a family of nontrivial fixed point theories with tensionless strings and $d = 6$ $(2, 0)$ supersymmetry. These arise on coincident M5-branes and IIA NS5-branes, and on the IIB theory at an A–D–E singularity.

(2, 2) compactification one of a special subset of $d = 4$, $N = 1$ heterotic string vacua, which includes the Calabi–Yau compactifications. The world-sheet CFT has both right-moving and left-moving $N = 2$ superconformal invariance. In the type II string theories, these CFTs give vacua with $d = 4$, $N = 2$ supersymmetry.

A–D–E singularity a singularity of a four-(real)-dimensional complex manifold, resulting from the collapse of one or more two-spheres to zero volume. The terminology A–D–E refers to the Dynkin diagrams of the simply-laced Lie algebras, which describe the intersection numbers of the collapsed spheres.

Abelian differential a globally defined holomorphic (1,0)-form on a Riemann surface.

abstruse identity one of a set of quartic theta function identities due to Jacobi, it implies the degeneracy of bosons and fermions in GSO-projected string theories as required by supersymmetry.

affine Lie algebra see *current algebra*.

anomaly the violation of a classical symmetry by quantum effects. A *gravitational anomaly* is an anomaly in coordinate invariance. A *global anomaly* is an anomaly in a large symmetry transformation (one not continuously connected to the identity).

anomaly polynomial a formal $(d + 2)$-form in d-dimensions, which encodes the gauge and gravitational anomalies.

asymptotically locally Euclidean (ALE) space a space which at long distance approaches flat Euclidean space identified under a discrete group. This is the geometry in the neighborhood of an orbifold fixed point (or blown-up fixed point).

Atiyah–Drinfeld–Hitchin–Manin (ADHM) construction a method for the construction of all Yang–Mills field configurations having self-dual field strength.

auxiliary field a nonpropagating field, one whose field equation is algebraic rather than differential. In many supersymmetric theories, the transformations can be simplified by introducing such fields.

axion a Goldstone boson associated with spontaneously broken PQ symmetry. The *model-independent axion* appears in every perturbative string theory, and is closely related to the graviton and dilaton.

bc **CFT** a free CFT of anticommuting fields with an action of first order in derivatives. There is a family of such CFTs, parameterized by the weight $h_b = 1 - h_c$. For $h_b = 2$ this CFT describes the Faddeev–Popov ghosts associated with conformal invariance.

$\beta\gamma$ **CFT** a free CFT of commuting fields with an action of first order in derivatives. There is a family of such CFTs, parameterized by the weight $h_\beta = 1 - h_\gamma$. For $h_\beta = \frac{3}{2}$ this CFT describes the Faddeev–Popov ghosts associated with superconformal invariance.

Batalin–Vilkovisky formalism an extension of the BRST formalism, for quantizing more general theories with constraints. This has been useful in string field theory.

Becchi–Rouet–Stora–Tyutin (BRST) invariance a nilpotent symmetry of Faddeev–Popov gauge-fixed theories, which encodes the information contained in the original gauge symmetry.

Beltrami differential the derivative with respect to the moduli of the complex structure of a Riemann surface.

Berezin integration a linear operation taking functions of Grassmann variables to complex numbers, with many of the key properties of ordinary integration.

beta function 1. in quantum field theory, the derivative of the effective strength of an interaction with respect to length scale; 2. a special function involving a ratio of gamma functions, which appears in the Veneziano amplitude.

Betti numbers the number of nontrivial p-forms in de Rham cohomology, denoted B_p.

black hole entropy a quantity S proportional to the area of the horizon of a black hole, $S = 2\pi A/\kappa^2$. This has the properties of a thermodynamic entropy: it is nondecreasing in classical general relativity, and the sum of the black hole entropy and the ordinary entropy is nondecreasing even with the inclusion of Hawking radiation. To find a statistical mechanical derivation of this entropy has been a major goal, partly realized in recent work.

black hole evaporation the emission of thermal (Hawking) radiation by a black hole, due to pair production near the horizon.

black hole information paradox a conflict between quantum mechanics and general relativity. Information falling into a black hole is lost and does not reappear when the black hole evaporates; this is inconsistent with ordinary quantum mechanical evolution. It apparently requires either a significant modification of quantum mechanics, or a significant breakdown of the usual understanding of locality.

black p-brane a p-dimensional extended object with an event horizon: a space that is translationally invariant in p directions and has a black hole geometry in the remaining directions.

blow up to deform a singular manifold into a smooth manifold.

Bogomolnyi–Prasad–Sommerfield (BPS) state a state that is invariant under a nontrivial subalgebra of the full supersymmetry algebra. Such states always carry conserved charges, and the supersymmetry algebra determines the mass of the state exactly in terms of its charges. BPS states lie in smaller supersymmetry representations than non-BPS states, so-called *short representations*. When there are short representations of different sizes, one also distinguishes *ultrashort representations,* which are the smallest possible (generally their dimension is the square root of the non-BPS dimension).

Borel summation a method of defining the sum of a divergent series. This has been used as a means of studying nonperturbative effects in field and string theories, but it should be understood that most nonperturbative effects are not usefully studied in terms of the perturbation series.

Born–Infeld action a generalization of the usual gauge field action which is nonpolynomial in the gauge field strength. This was originally proposed as a possible short-distance modification of electromagnetism. It arises as the low energy effective action of the gauge fields on D-branes.

bosonization the exact equivalence of a theory of fermionic fields and a theory of bosonic fields, possible in two dimensions. The boson is a fermion–antifermion pair; the fermion is a coherent state of bosons.

c-map a method for constructing the hypermultiplet moduli space of a type II string theory compactified on a Calabi–Yau three-fold from the vector multiplet moduli space of the other type II theory on the same three-fold.

c-theorem the existence, in unitary CFTs in two dimensions, of a positive quantity c that is monotonically nonincreasing with increasing length scale, and which at fixed points is stationary and equal to the central charge. This is a strong constraint on the global form of the renormalization group flow; no simple analog seems to exist in higher dimensions. Also known as the *Zamolodchikov c-theorem.*

CPT symmetry the combined operation of parity-reversal, time-reversal, and charge conjugation, which is a symmetry of all Lorentz-invariant local quantum field theories.

Calabi–Yau manifold a Kähler manifold with vanishing first Chern class. A *Calabi–Yau n-fold* has $2n$ real $= n$ complex coordinates. Yau's theorem guarantees the existence of a Ricci-flat metric of $SU(n)$ holonomy.

canceled propagator argument a general principle implying, under broad conditions, the vanishing of surface terms on the moduli space of Riemann surfaces and therefore the decoupling of unphysical states in string amplitudes. Such amplitudes are defined by analytic continuation from a regime where the integrand falls rapidly at the boundary and the surface term is identically zero; its continuation is therefore also identically zero.

Casimir energy a shift in the ground state energy of a quantum field theory due to boundary conditions on the fields.

center-of-mass mode the zeroth spatial Fourier component of a quantum field.

central charge an operator (which might be a constant) that appears on the right-hand side of a Lie algebra and commutes with all operators in the algebra. Prominent examples include the constant term in the Virasoro algebra and the charges appearing on the right-hand sides of many supersymmetry algebras.

Chan–Paton degrees of freedom degrees of freedom localized at the endpoints of open strings. These are now interpreted as designating the D-brane on which the string ends.

Chan–Paton factor the vertex operator factor for the state of the Chan–Paton degrees of freedom.

Chern–Simons term a term in the action which involves *p*-form potentials as well as field strengths. Such a term is gauge-invariant as a consequence of the Bianchi identity and/or the modification of the *p*-form gauge transformation. These terms usually have a close connection to topology and to anomalies.

chiral 1. acting in a parity asymmetric fashion; see *chiral multiplet, chiral symmetry, chiral theory, chirality, extended chiral algebra*; 2. invariant under part of the supersymmetry algebra; see *chiral field, chiral multiplet, chiral primary, chiral ring*.

chiral field in supersymmetry, a local operator that is invariant under part of the algebra: the operator analog of a *BPS state*.

chiral multiplet the multiplet of $d = 4$, $N = 1$ supersymmetry with two real scalars. The quarks and leptons are contained in such multiplets. This multiplet is connected to both senses of *chiral*: it contains a fermion with chiral couplings, and the integral of the associated superfield is invariant under half of the supersymmetry algebra.

chiral primary in an $N = 2$ SCFT, a primary field that is also annihilated by one of the rigid supersymmetries $G^{\pm}_{-1/2}$.

chiral ring the closed OPE algebra of chiral fields.

chiral symmetry a symmetry whose action on spinor fields is not parity-symmetric.

chiral theory a gauge theory in which the gauge couplings are not parity-symmetric.

chirality in $d = 2k$ dimensions, the eigenvalue of the operator Γ^d which anticommutes with the Γ^μ. This eigenvalue distinguishes the two Weyl representations, which are related to one another by parity.

Christoffel connection in general relativity, the connection that is constructed from the metric.

critical behavior the behavior of a quantum field theory at an IR fixed point with massless fields, and the approach to this behavior.

closed string a string with the topology of a circle.

cocycle in a vertex operator, an operator-valued phase factor which multiplies the creation–annihilation normal ordered exponential. This in needed in some cases in order to give the operator the correct commuting or anticommuting property.

coefficient functions the position-dependent coefficients of local operators appearing in the expansion of an operator product.

cohomology in any vector space with a nilpotent operator Q (one such that $Q^2 = 0$), the kernel of Q modulo the image of Q. That is, the space of *closed* states (those annihilated by Q) with the *exact* states (those of the form $Q\psi$) defined to be equivalent to zero. *De Rham cohomology* is the cohomology of the exterior derivative d acting on differential forms. On a complex manifold, *Dolbeault cohomology* is the cohomology of ∂ and $\bar\partial$ (the $(1,0)$ and $(0,1)$ parts of d) on (p,q)-forms. *Homology* is the cohomology of the boundary operator. *BRST cohomology* is the cohomology of the BRST operator, and defines the physical space of a gauge-invariant theory.

Coleman–Weinberg formula the expression for the vacuum energy density of a free quantum field, from the renormalized sum of the zero-point energies of its modes.

collapsing cycle a cycle whose volume vanishes in a limit, usually giving rise to a singular manifold.

collective coordinate in quantizing a soliton or other extended object, the degrees of freedom corresponding to its position or configuration.

compact CFT a CFT in which the number of states with energy less than any given value is finite. This is defined by analogy with the spectrum of a differential operator on a compact space.

compactification scale the characteristic mass scale of states whose wavefunctions have a nontrivial dependence on the compact dimensions.

compactify to consider a field theory or string theory in a spacetime, some of whose spatial dimensions are compact.

complex manifold a manifold with an assigned system of complex coordinates, modulo holomorphic reparameterizations of these coordinates.

complex structure an equivalence class of complex coordinates. A given differentiable manifold may have many inequivalent complex structures.

complex structure moduli the moduli that parameterize the inequivalent complex structures on a manifold. In compactification on a Calabi–Yau 3-fold, these are associated with $(2, 1)$-forms.

conformal block in CFT, the contribution of a single conformal family to a sum over states.

conformal bootstrap the partially successful program to construct all CFTs by using only symmetry and consistency conditions.

conformal family the set of states obtained by acting on a highest weight state with Virasoro raising generators in all inequivalent ways; or, the corresponding set of local operators. A *degenerate conformal family* contains null states, which are orthogonal to all states in the family.

conformal field theory (CFT) a conformally invariant quantum field theory.

conformal gauge a choice of coordinates in two dimensions, such that the metric is proportional to the unit metric.

conformal Killing vector a globally defined infinitesimal diff×Weyl transformation that leaves the metric invariant.

conformal transformation a mapping of Euclidean or Minkowski space to itself that leaves the flat metric invariant up to a position-dependent rescaling; equivalently, the subgroup of diff × Weyl that leaves invariant the flat metric. In $d \geq 3$ dimensions this has $\frac{1}{2}(d + 1)(d + 2)$ parameters. In two dimensions it is the set of all holomorphic maps. Finite transformations require the inclusion of points at infinity, as in the case of the Möbius transformations of the sphere.

This usage has become standard in string theory and quantum field theory, but in general relativity *conformal transformation* is defined to be any position-dependent rescaling of the metric, now called a *Weyl transformation* in string theory.

conifold a Calabi–Yau manifold with a singular complex structure, corresponding to the collapse of a three-cycle. The string theory on this space is singular; this is now understood to be due to the quantum effects of a massless 3-brane wrapped on the cycle.

conifold transition a change of topology due to condensation of massless 3-brane fields. Under appropriate conditions, the potential for the massless 3-brane fields on a conifold with multiple collapsed cycles has a flat direction for these fields; this corresponds to a change of topology, blowing up a 2-cycle rather than the collapsed 3-cycle.

constraint a symmetry generator whose matrix elements are required to vanish in physical states, either in the BRST or OCQ sense. These can usually be understood as arising from a gauge symmetry (for so-called *first class constraints,* which are all that we consider), and consist of those gauge symmetry generators that do not vanish by the equations of motion.

coset CFT a CFT constructed as one of the factors of a known CFT, when the energy-momentum tensor of the latter can be written as a sum of commuting pieces. In the classic example the full CFT G is a current algebra, as is one of the factors (H). This can also be thought of as gauging the symmetry H.

cosmological constant the energy density of the vacuum. In a nonsupersymmetric quantum theory (including one with spontaneously broken supersymmetry), there are many effects that give rise to such an energy density. The *cosmological constant problem* is the problem that the cosmological constant in nature is many orders of magnitude smaller than known nonzero effects.

critical dimension the dimension in which a perturbative string theory is consistent in flat Minkowski spacetime; the value is 26 for the bosonic string and 10 for the supersymmetric string theories.

current algebra in quantum field theory, the algebra of the currents associated with a continuous symmetry group g (or of their Fourier modes). As used here, it is the specific algebra that occurs in two-dimensional CFTs, with the energy-momentum tensor defined to be of Sugawara form. The terms *affine Lie algebra* and *affine Kac–Moody algebra* are also used for this algebra, though like *current algebra* they both have broader definitions as well. The term *affine* refers to the c-number (Schwinger) term. An *untwisted* current algebra is the algebra of periodic currents, with integer modes. An algebra can be *twisted* by any automorphism of g.

cycle a topologically nontrivial submanifold (in the sense of homology); a p-cycle is p-dimensional. The *A*- and *B-cycles* are a standard basis for the nontrivial one-cycles on a Riemann surface.

D-brane in the type I, IIA, and IIB string theories, a dynamical object on which strings can end. The term is a contraction of *Dirichlet brane.* The coordinates of the attached strings satisfy Dirichlet boundary conditions in the directions normal to the brane and Neumann conditions in the directions tangent to the brane. A *Dp-brane* is p-dimensional, with p taking any even value in the IIA theory, any odd value in the IIB theory, and the values 1, 5, and 9 in the type I theory; a D9-brane fills space and so corresponds to an ordinary Neumann boundary condition. The Dp-brane is a source for the $(p + 1)$-form R–R gauge field. The mass or tension of a D-brane is intermediate between that of an ordinary quantum or a fundamental string and that of a *soliton.* The low energy fluctuations of D-branes are described by supersymmetric gauge theory, which is non-Abelian for coincident branes.

D-instanton an object localized in (Euclidean) time as well as space, defined by Dirichlet conditions on all coordinates of attached strings. This is similar to a field-theoretic instanton, corresponding to a tunneling process that changes the value of an R–R field strength. More generally, the $(p + 1)$-dimensional world-volume of a Dp-brane, when localized in time and wrapped on a $(p + 1)$-cycle in space, has similar effects.

D-string a D1-brane, in the type I and IIB string theories.

***D*-term** 1. in gauge theories with four or eight supersymmetries, the auxiliary field in the gauge multiplet; 2 the potential term proportional to the square of this auxiliary field, which depends only on the gauge couplings and, in the $U(1)$ case, the value of the Fayet–Iliopoulos parameter.

Del Guidice–Di Vecchia–Fubini (DDF) operators operators satisfying an oscillator algebra, which create a complete set of physical states in OCQ.

descendant a state obtained by acting on a highest weight state with Virasoro raising generators.

diagonal modular invariant a modular-invariant CFT formed by imposing common boundary conditions on the left- and right-moving fields.

diff invariance general coordinate (reparameterization) invariance, usually applied to the world-sheet coordinates.

dilaton the massless scalar with gravitational-strength couplings, found in all perturbative string theories. An exactly massless dilaton would violate limits on nongravitational interactions, but a mass for the dilaton is not forbidden by any symmetry and so dynamical effects will generate one in vacua with broken supersymmetry (the same holds for other moduli). The string coupling constant is determined by the value of the dilaton field.

dimensional reduction in the simplest cases, toroidal compactification retaining only the states of zero compact momentum. More generally (and less physically) the construction of a lower-dimensional field theory by requiring all fields to be invariant under a set of symmetries; this may have no interpretation in terms of compactification.

Dirac quantization condition for an electric and a magnetic charge, the condition that the product be quantized, $\mu_e \mu'_m = 2\pi n$. For two dyons, which have both charges, the condition is $\mu_e \mu'_m - \mu_m \mu'_e = 2\pi n$. These conditions generalize to objects of dimension p and $d - p - 4$ in d dimensions, where one is the source of a $(p + 2)$-form field strength and the other of the Poincaré dual $(d - p + 2)$-form field strength.

Dirac spinor the unique irreducible representation of the algebra of Dirac matrices, which is also known as a Clifford algebra. This is also a representation of the Lorentz group; in even dimensions it is reducible to two Weyl representations of the Lorentz group. The Dirac spinor is complex; in certain dimensions a further Majorana (reality) condition is compatible with Lorentz invariance.

Dirichlet boundary condition the condition that the value of a field be fixed at a boundary. This is the relevant usage in string theory, but in other contexts only the tangent derivative need be fixed.

discrete torsion in forming a *twisted CFT*, a change in the phases of the path integral sectors and therefore in the projection on the Hilbert space.

doubling trick the representation of holomorphic and antiholomorphic fields on a manifold with boundary by holomorphic fields alone, on the doubled copy of the manifold obtained by reflecting through the boundary.

dual resonance model a phenomenological model of the strong interaction, which developed into string theory. The *dual* refers here to *world-sheet duality*.

duality the equivalence of seemingly distinct physical systems. Such an equivalence often arises when a single quantum theory has distinct classical limits. One classic example is *particle–wave duality*, wherein a quantum field theory has one limit described by classical field theory and another described by classical particle mechanics. Another is the high-temperature–low-temperature duality of the Ising model. Here, low temperature is the statistical mechanical analog of the classical limit, the Boltzmann sum being dominated by the configurations of lowest energy. See *Montonen–Olive duality*, *S-duality*, *string–string duality*, *T-duality*, *U-duality*, *world-sheet duality*.

effective field theory the description of a physical system below a given energy scale (or equivalently, above a given length scale).

Einstein metric the metric whose leading low energy action is the *Hilbert action*

$$\frac{1}{2\kappa^2} \int d^d x \, (-G)^{1/2} R \; ;$$

this is independent of other fields. Here κ is the gravitational coupling, related to the Planck length by $\kappa = (8\pi)^{1/2} L_P$. This metric is related to other metrics such as the *sigma-model metric* by a field-dependent Weyl transformation. The existence of distinct metrics would appear to violate the equivalence principle, but when the dilaton and other moduli are massive the distinction disappears.

electroweak scale the mass scale of electroweak symmetry breaking, roughly 10^2 GeV.

enhanced gauge symmetry a gauge symmetry appearing at special points in moduli space, which is not evident in the original formulation of a theory. The classic examples are the gauge symmetries that arise at special radii of toroidal compactification, whose gauge bosons are winding states. Many other mechanisms are now known: D-branes and black p-branes wrapped on collapsing cycles, F-strings or D-strings stretched between various branes in the limit that the latter become coincident, and the gauge symmetry appearing on a zero size $SO(32)$ instanton.

Euclidean having a metric of strictly positive signature. The original connotation that the metric be flat is somewhat disregarded; thus one refers to *Euclidean quantum gravity*, a conjectured analytic continuation of the Minkowskian theory. For the metric itself, the term *Riemannian* for a curved metric of Euclidean signature is more precise.

Euclidean adjoint in a Euclidean quantum theory, the Hermitean adjoint combined with time-reversal. The latter operation undoes the time-reversing effect of the adjoint, so that the combined operation is local.

Euler number the topological invariant

$$\chi = \sum_{p=0}^{d}(-1)^p B_p \ ,$$

where B_p is the pth Betti number. It is equal to $2(1 - g)$ for a Riemann surface of genus g. More properly, the *Euler characteristic*.

expectation value a path integral with specified insertions. This is the term that we have chosen to use, but *correlation function* and *correlator* are also in common usage.

extended chiral algebra the full set of holomorphic operators in a CFT.

extended supersymmetry a supersymmetry algebra in which the supercharges comprise more than one copy of the smallest spinor representation for the given spacetime dimension.

F-term 1. the auxiliary field in the chiral multiplet of $d = 4$, $N = 1$ supersymmetry; 2. the potential term proportional to the square of this field.

F theory 1. a systematic description of IIB superstring states with nontrivial dilaton and R–R scalar backgrounds, which relates these fields to the modulus τ of an auxiliary two-torus; 2. a conjectured twelve-dimensional quantum theory underlying the IIB string. The name is inspired by M-theory, with F for father.

Faddeev–Popov determinant the Jacobian determinant arising from the reduction of a gauge-invariant functional integral to an integral over a gauge slice.

Faddeev–Popov ghosts the wrong-statistics quantum fields used to give a functional integral representation of the Faddeev–Popov determinant.

Fayet–Iliopoulos term in $U(1)$ gauge theories with four or eight supersymmetries, a term in the action which is linear in the auxiliary D-term field.

Feigin–Fuchs representation a representation of minimal model expectation values in terms of free fields.

fibration a space which is locally the product of a fiber F and a base B. The geometry of the fiber varies as one moves over the base, and may become singular. Typical fibers are tori and the $K3$ manifold.

first Chern class on a complex manifold, the Dolbeault cohomology class of the Ricci form $R_{i\bar{j}}dz^i d\bar{z}^{\bar{j}}$.

first-quantized description the representation of a quantized particle theory as a sum over particle paths, or of a string theory as a sum over world-sheets. *Second-quantized* refers to the representation in terms of a functional integral over ordinary or string fields. The term second-quantized implies the reinterpretation of the first-quantized *wavefunction* as a *field operator*. This terminology is in common usage, but it has been argued that it is unsatisfactory, in that it implies a deep principle where none may exist.

Since the sum over world-sheets is itself a quantum field theory, one can equally well call it second-quantized, in which case string field theory is *third-quantized*. Third quantization of an ordinary field theory would describe operators that create and destroy universes, a concept which may or may not be useful.

Fischler–Susskind mechanism the cancellation of divergences and anomalies in the world-sheet quantum field theory against divergences and anomalies from integration over small topological features at higher orders of string perturbation theory. This is needed for the consistency of string perturbation theory in a quantum-corrected background.

fixed point 1. (*in geometry*) a point left invariant by a given symmetry transformation. This becomes a boundary point or a singularity if the space is identified under the transformation; 2. (*in quantum field theory*) a quantum theory whose physics is independent of length scale (scale-invariant). Usually such a theory is conformally invariant as well. A *UV fixed point* is the theory governing the short-distance physics of a quantum field theory; an *IR fixed point* is the theory governing the long-distance physics of a quantum field theory. A *trivial IR fixed point* has no massless fields. A *nontrivial IR fixed point* has massless fields with nonvanishing interactions. A theory whose IR limit is a massless free field theory is therefore described by neither of these terms; it is a *noninteracting IR fixed point*.

flat direction in scalar field space, a line of degenerate local minima. The field corresponding to this direction is a *modulus*.

flop a change of topology which can occur in weakly coupled string theory, where a two-cycle collapses and then a different two-cycle blows up.

fractional charge an unconfined particle whose electric charge is not a multiple of that of the electron. These exist in most $d = 4$ string theories, although in many cases all are superheavy.

fractional string theory a proposed generalization of string theory having constraints whose spin is not a multiple of $\frac{1}{2}$. No complete construction exists.

Fuchsian group a discrete subgroup Γ of the $SL(2, \mathbf{R})$ Möbius transformations of the complex upper-half-plane H, with additional conditions such that H/Γ is a manifold and in particular a Riemann surface.

functional integral in our usage, synonymous with *path integral*.

fundamental region in relation to a coset space M/Γ where Γ is a discrete group, a region F such that every point in M is identified with exactly one point in the interior of F or with one or more points on the boundary of F.

fundamental string (F-string) the original string whose quantization defines a weakly coupled string theory, as distinguished from D-strings and solitonic strings.

fusion rule the specification of which conformal families appear in the operator product of any two primary fields in a given CFT.

gauge-fixing the reduction of a redundant (gauge-invariant) description of a quantum theory to a description with a single representative from each equivalence class.

gaugino a spin-$\frac{1}{2}$ fermion in the same supersymmetry multiplet as a gauge boson.

gaugino condensation a strong coupling effect where a product of gaugino fields acquires a vacuum expectation value. This generally breaks a chiral symmetry but does not directly break supersymmetry; however, in combination with other fields it often induces supersymmetry breaking.

generation a family of quarks and leptons, described by spinor fields in a chiral but anomaly-free set of $SU(3) \times SU(2) \times U(1)$ representations. In $SU(5)$ grand unification these become a **5** + $\overline{\textbf{10}}$, in $SO(10)$ they are contained in a **16**, and in E_6 they are contained in a **27**. An *antigeneration* is the conjugate representation; the distinction between generation and antigeneration is a matter of convention.

genus the number g of handles on a closed oriented Riemann surface: $g = 0$ is a sphere, $g = 1$ is a torus, and so on.

Gepner model a string model based on $N = 2$ *minimal model* CFTs.

ghosts see *Faddeev–Popov ghosts*.

Gliozzi–Scherk–Olive (GSO) projection a construction of modular-invariant string theories by summing over R and NS boundary conditions on the fermion fields and projecting onto states of definite world-sheet fermion number. In supersymmetric string theories there are independent GSO projections on the left-movers and right-movers. The diagonal projection, which acts simultaneously on both sides, produces a nonsupersymmetric theory.

goldstino the massless spin-$\frac{1}{2}$ Goldstone fermion associated with spontaneously broken supersymmetry. In supergravity it combines with the gravitino to form a massive fermion.

Goldstone boson the massless scalar corresponding to fluctuations of the direction of spontaneous symmetry breaking.

grand unification the unification of the $SU(3) \times SU(2) \times U(1)$ gauge symmetries in a simple group.

grand unification scale the mass scale of spontaneous breaking of the grand unified group. Proton stability and the unification of the couplings require that it be within two or three orders of magnitude of the gravitational scale.

Grassmann variable the elements θ_i of an algebra with the relation $\theta_i \theta_j = -\theta_j \theta_i$. These are used to give a path integral representation of fermionic fields, and to define *superspace*. They are also called *anticommuting c-numbers*.

gravitational scale the mass scale at which the dimensionless gravitational coupling becomes of order 1, $m_{\text{grav}} = \kappa^{-1} = 2.4 \times 10^{18}$ GeV; this is $(8\pi)^{-1/2}$ times the Planck mass.

gravitino a spin-$\frac{3}{2}$ fermion in the same supersymmetry multiplet as the graviton.

Green–Schwarz mechanism the cancellation of an anomaly by the modified transformation law of a *p*-form potential in a *Chern–Simons term*.

Green–Schwarz superstring a manifestly supersymmetric formulation of the supersymmetric string theories, with a spacetime-fermionic gauge invariance known as *κ symmetry*. There is no simple covariant gauge fixing.

***H*-monopole** a monopole carrying the magnetic charge of the antisymmetric tensor gauge field $B_{\mu n}$.

heterotic 5-brane the 5-brane carrying the magnetic charge of the massless heterotic string 2-form potential. It is obtained as the limit of a zero size instanton in the heterotic string gauge fields. The instanton configuration is localized in four spatial dimensions, and is therefore a 5-brane in nine spatial dimensions.

heterotic string a string with different constraint algebras acting on the left- and right-moving fields. The case of phenomenological interest has a $(0,1)$ superconformal constraint algebra, with spacetime supersymmetry acting only on the right-movers and with gauge group $E_8 \times E_8$ or $SO(32)$.

Hagedorn temperature the temperature at which the thermal partition function of free strings diverges, due to the exponential growth of the density of states of highly excited strings.

hidden sector the fields that couple to the Standard Model only through gravitational-strength interactions. In *hidden sector models,* these include the fields responsible for supersymmetry breaking.

highest weight state in CFT, a state annihilated by all Virasoro lowering operators, or more generally by all lowering operators in a given algebra.

Hodge number the number of nontrivial (p, q)-forms in Dolbeault cohomology, denoted $h^{p,q}$.

holographic principle the conjecture that the states of quantum gravity in d dimensions have a natural description in terms of a $(d-1)$-dimensional theory. This radical departure from local field theory was motivated by the black hole information problem, and has played a role in attempts to formulate M-theory.

holomorphic analytic, as used in the theory of complex variables. The Minkowskian continuation is *left-moving*. An *antiholomorphic* field is analytic in the conjugate variable, and its continuation is *right-moving*.

holomorphic quadratic differential a globally defined holomorphic $(2, 0)$-form on a Riemann surface.

holomorphic vector field a globally defined holomorphic $(-1, 0)$-form on a Riemann surface.

holonomy consider the parallel transport of a vector around a closed loop on a d-dimensional manifold: it returns to an $O(n)$ rotation of its original value. The set of all rotations that are obtained in this way for a given manifold is a subgroup of $O(n)$; this is the *holonomy group*.

homology see *cohomology*.

hyperelliptic surface a Riemann surface with a $\mathbf{Z}_2$ symmetry. This can be represented as a two-sheeted cover of the sphere with branch cuts.

hyper-Kähler manifold a $4k$-dimensional manifold of holonomy $Sp(k) \subset SO(4k)$. This is the geometry of the moduli space of hypermultiplets in $d = 6$, $N = 1$ or $d = 4$, $N = 2$ supersymmetry, in the limit in which gravity decouples.

hypermultiplet in $d = 6$, $N = 1$ or $d = 4$, $N = 2$ supersymmetry, the multiplet whose bosonic content is four real massless scalars.

identify to define two points (or other objects) to be equivalent, thus producing a coset space.

infrared (IR) divergence a divergence arising from long distances in spacetime, usually signifying that one has calculated the wrong thing.

inheritance principle in twisted (orbifold) theories, the principle that the tree-level amplitudes of untwisted states are the same as in the untwisted theory.

insertion the integrand of a path integral, excluding the weight $\exp(iS)$ or $\exp(-S)$.

instanton in a Euclidean path integral, a nonconstant configuration that is a local but not a global minimum of the action. Such configurations are usually localized in spacetime, are usually topologically nontrivial, and are of interest when they give rise to effects such as tunneling that are not obtained from small fluctuations around a constant configuration. *Spacetime instantons* are instantons in the effective field theory in spacetime. *World-sheet instantons* are instantons in the world-sheet quantum field theory, and correspond to world-sheets wrapping around nontrivial two-cycles of spacetime.

intersection number the number of points at which a set of surfaces intersect, weighted by the orientation of the intersection.

irrelevant interaction an interaction whose dimensionless strength decreases with increasing length scale. In perturbation theory, this is equivalent to a *nonrenormalizable* interaction.

K3 manifold the unique nontrivial Calabi–Yau manifold of four (real) dimensions. To be precise, it is topologically unique, but possesses complex structure and Kähler moduli. Its holonomy is $SU(2)$, so that half of the supersymmetries of a theory are broken upon compactification on K3.

Kac determinant the determinant of the matrix of inner products of states at a given L_0 level of a Verma module.

Kac–Moody algebra see *current algebra*.

Kähler form the $(1,1)$-form $G_{i\bar{j}}dz^i d\bar{z}^{\bar{j}}$, formed from a Kähler metric on a complex manifold.

Kähler manifold a complex manifold of $U(n)$ holonomy in n complex dimensions.

Kähler moduli the moduli parameterizing the Kähler form.

Kähler potential the potential $K(z,\bar{z})$, in terms of which the metric of a Kähler manifold is determined, $G_{i\bar{j}} = \partial_i \partial_{\bar{j}} K$. This is not globally defined, being determined only up to a *Kähler transformation* $K(z,\bar{z}) \rightarrow K(z,\bar{z}) + f(z) + f(z)^*$.

Kaluza–Klein gauge field in a compactified theory, a gauge field originating from the metric of the higher-dimensional theory. The gauge group is the isometry group of the compact space.

Kaluza–Klein monopole a monopole carrying the magnetic charge of a $U(1)$ Kaluza–Klein gauge symmetry. The monopole configuration is the smooth *Taub–NUT* spacetime. It is localized in three spatial dimensions, and is therefore a 6-brane in nine spatial dimensions.

Kaluza–Klein states states with nonzero momentum in a compact spatial direction.

Knizhnik–Zamolodchikov (KZ) equation the differential equation determining the expectation values of the primary fields of a current algebra.

Landau–Ginzburg model a scalar field theory which has a nontrivial IR fixed point when the potential is appropriately tuned. In particular, this gives a Lagrangian representation of the minimal model CFTs.

large coordinate transformation a coordinate transformation that is not continuously connected to the identity.

lattice the set Γ of integer linear combinations of n linearly independent basis vectors in n dimensions. Given a Euclidean or Lorentzian metric, an *even lattice* is one whose points have even length-squared. The dual lattice Γ^* is the set of points v such that $v \cdot w \in \mathbf{Z}$ for all $w \in \Gamma$. The *root lattice* is the set of integer linear combinations of the roots of a Lie algebra.

level 1. the quantized c-number term in a current algebra, also known as the *Schwinger term*; 2. the total oscillator excitation number in free field theory; 3. in a conformal family, the difference between the L_0 eigenvalue of a given state and that of the highest weight state.

level-matching the modular invariance condition that $L_0 - \tilde{L}_0 \in \mathbf{Z}$.

Lie algebra an algebra with an antisymmetric product that satisfies the Jacobi identity. A *simple* Lie algebra has no subalgebra that commutes with its complement. A *simply-laced* Lie algebra has all roots of equal length. A *graded Lie algebra* has odd and even elements, with a symmetric product between odd elements.

light-cone gauge in string theory, the choice of world-sheet time coordinate to coincide with a particular spacetime null coordinate. In theories with local symmetries, a gauge choice such that the connection in a given null direction vanishes.

linear dilaton theory a scalar CFT in which the energy-momentum tensor includes a term proportional to the second derivative of the scalar. This arises in string theory when the dilaton is a linear function of position.

linear sigma model a scalar field theory whose kinetic term is field independent, but whose long-distance physics is governed by a nonlinear sigma model.

Liouville field theory the CFT of a scalar field with an exponential interaction. This arises in various situations, including the *noncritical string*. It corresponds to bosonic string theory in a linear dilaton plus exponential tachyon background.

little string theory one of several interacting string theories without gravity, notably found on NS5-branes in the limit of zero string coupling.

loop expansion in quantum field theory, the Feynman graph expansion, which is equivalent to the expansion in powers of $\hbar$. The *string loop expansion* is the sum over Riemann surfaces, with dimensionless string coupling g. The *world-sheet loop expansion* is the nonlinear sigma model perturbation expansion, in powers of α'/R_c^2 with R_c the compactification radius.

Lorentzian having a mixed signature $(-,\ldots,-,+,\ldots,+)$.

lowering operator operators that reduce the energy of a given state. In CFT, operators that reduce the Virasoro generator L_0 (or $\tilde{L}_0$) by n units carry a grading (subscript) n.

M-theory 1. (*narrow*) the limit of strongly coupled IIA theory with eleven-dimensional Poincaré invariance; 2. (*broad; most common usage*) the entire quantum theory whose limits include the various weakly coupled string theories as well as M-theory in the narrow sense. The name is deliberately ambiguous, reflecting the unknown nature of the theory; M has variously been suggested to stand for membrane, matrix, mother, and mystery.

M2-brane the 2-brane of M-theory, which couples to the potential A_3 of eleven-dimensional supergravity.

M5-brane the 5-brane of M-theory, which carries the magnetic charge of the potential A_3 of eleven-dimensional supergravity.

macroscopic string a string whose length is much greater than the characteristic string length scale. In particular, it is sometimes useful to consider an infinite string stretching across spacetime.

Majorana condition a Lorentz-invariant reality condition on a spinor field. This can be imposed only if the spacetime dimension is 1, 2, 3, 4, or 8 (mod 8).

marginal interaction an interaction whose dimensionless strength is independent of the length scale. In general this might hold only to first order in the coupling of the interaction; a *truly marginal* interaction is one that remains marginal even with finite coupling.

matrix models quantum mechanical systems with matrix degrees of freedom, with critical points governed by noncritical string theories.

matrix theory a quantum mechanical system with matrix degrees of freedom and 32 supercharges, obtained by dimensional reduction of $d = 10$ supersymmetric $U(n)$ Yang–Mills theory. In the large-n limit this is conjectured to define M-theory (in the broad sense).

minimal models several families of solvable CFTs, in which every conformal family is degenerate. There are infinite series of unitary minimal models having $N = 0$, $N = 1$, and $N = 2$ superconformal symmetries, which converge from below on the central charges 1, $\frac{3}{2}$, and 3 respectively.

Minkowskian having a signature $(-, +, +, \ldots, +)$.

mirror symmetry an equivalence between string theories compactified on distinct manifolds. The equivalence reverses the sign of the $U(1)$ charge of one $N = 2$ superconformal algebra, and therefore changes the sign of the Euler number of the manifold.

Möbius group the globally defined $SL(2, \mathbf{C})$ conformal symmetry of the sphere; or, the globally defined $SL(2, \mathbf{R})$ conformal symmetry of the disk.

mode operators the spatial Fourier components of a quantum field.

model see *vacuum*.

modular group the group of large coordinate transformations (often applied to a Riemann surface but also applicable to spacetime).

modular invariance the invariance of the string path integral under large coordinate transformations.

moduli 1. the parameters labeling the geometry of a manifold. Notable examples are the parameters for the complex structure of the string world-sheet, and the parameters for the geometry of compactification; 2. the parameters labeling a space of degenerate (and, usually, physically inequivalent) vacua in quantum field theory. This is closely related to the compactification example: in expanding around the classical limit, each compact solution of the field equations gives a vacuum of the quantum theory, to leading order; 3. the massless fields corresponding to position dependence of these parameters. Contrast *Goldstone boson*.

moduli space the space of geometries or vacua, whose coordinates are the moduli.

monodromy for a quantity which is locally single-valued, the multi-valuedness around nontrivial closed paths.

Montonen–Olive duality the weak–strong duality of $d = 4$, $N = 4$ Yang–Mills theory.

Nambu–Goto action a string action, which is proportional to the invariant area of the world-sheet in spacetime.

Narain compactification the abstract description of toroidal compactification in terms of the lattice of left- and right-moving momenta.

naturalness problem the problem of explaining why a constant of nature takes a value much smaller than estimated nonzero contributions. Examples are the Higgs scalar mass, the cosmological constant, and the QCD θ-angle.

Neumann boundary condition the condition that the normal derivative of a field vanish at a boundary; the value of the field is free to fluctuate.

Neveu–Schwarz algebra the world-sheet algebra of the Fourier modes of the supercurrent and energy-momentum tensor, in a sector where the supercurrent is antiperiodic and its moding therefore half-integer-valued.

Neveu–Schwarz (NS) boundary condition the condition that a fermionic field on the world-sheet be antiperiodic, in the closed string or in the double of the open string (see *doubling trick*). Its Fourier moding is then half-integer-valued.

Neveu–Schwarz 5-brane in the type I and type II superstring theories, the 5-brane that carries the magnetic charge of the NS–NS 2-form potential.

Neveu–Schwarz–Neveu–Schwarz (NS–NS) states in type II superstring theories, the bosonic closed string states whose left- and right-moving parts are bosonic. These include the graviton and dilaton, and in the type II case a 2-form potential.

no-ghost theorem 1. the theorem that the OCQ or BRST Hilbert space has a positive inner product; 2. the further theorem that the string amplitudes are well defined and unitary in this space.

no-scale model a field theory that has, to some approximation, a line of degenerate vacua with broken supersymmetry.

Noether's theorem the theorem that an invariance of the Lagrangian implies a conserved quantity.

noncommutative geometry a generalization of ordinary geometry, focusing on the algebra of functions on a space. The noncommutative collective coordinates of D-branes suggest the need for such a generalization.

noncritical string theory 1. a Weyl-noninvariant string theory — one with a measure of world-sheet distance that is independent of the embedding in spacetime. These include strings with an independent world-sheet metric field, and strings with a short-distance cutoff; 2. more recently, the term has been applied to any string theory that does not have a weakly coupled limit with a Weyl-invariant world-sheet theory. In this form it includes various theories with stringlike excitations in which the coupling is fixed to be of order 1, such as the *(2,0) theory* and the *little string theories*. Such a theory does not have a well-defined world-sheet, because processes that change the world-sheet topology cannot be turned off.

nonlinear sigma model a scalar field theory in which the kinetic term has a field-dependent coefficient. This has a natural interpretation in terms of a curved field space, and corresponds to string theory in curved spacetime, or more generally one with position-dependent background fields.

nonrenormalization theorem a theorem restricting the form of quantum corrections to a given amplitude, or to the effective action. It may require that these corrections vanish, that they arise only at specific orders of perturbation theory, or that they arise only nonperturbatively.

normal ordering a prescription for defining products of free fields by specific subtractions of divergent terms. *Conformal normal ordering,* denoted : :, produces operators with simple conformal properties. *Creation–annihilation normal ordering,* denoted ° °, where lowering operators are put to the right of raising operators, produces operators with simple matrix elements. *Boundary normal ordering,* denoted **:** :, is conformal normal ordering with an additional image charge subtraction to produce operators that are finite as they approach a boundary.

null state a physical state that is orthogonal to all physical states including itself. Or, a descendant in a conformal family which is orthogonal to all states in the family.

old covariant quantization (OCQ) a method of quantizing string theory, similar to the Gupta–Bleuler quantization of electrodynamics. It is equivalent to the light-cone and BRST quantizations.

one-loop the leading quantum correction, coming from surfaces of Euler number zero in string perturbation theory.

open string a string that is topologically a line segment.

operator equation in quantum theory, an equality between operators that holds in arbitrary matrix elements; equivalently, an equality that holds when inserted into a functional integral with arbitrary boundary conditions.

operator product expansion (OPE) the expansion of a product of operators as a sum of local operators. This provides an asymptotic expansion, as the separation of the operators vanishes, for an arbitrary expectation value containing the product. In CFT the expansion is convergent.

orbifold 1. (*noun*) a coset space M/H, where H is a group of discrete symmetries of a manifold M. The coset is singular at the fixed points of H; 2. (*noun*) the CFT or string theory produced by the gauging of a discrete world-sheet symmetry group H. If the elements of H are spacetime symmetries, the result is a theory of strings propagating on the coset space M/H. A *non-Abelian orbifold* is one whose point group is non-Abelian. An *asymmetric orbifold* is one where H does not have a spacetime interpretation and which in general acts differently on the right-movers and left-movers of the string; 3. (*verb*) to produce such a CFT or string theory by gauging H; this is synonymous with the third definition of *twist*.

oriented string theory a string theory in which the world-sheet has a definite orientation; world-sheet parity-reversal is not treated as a gauge symmetry.

orientifold a string theory produced by the gauging of a world-sheet symmetry group H, where H includes elements that combine the world-sheet parity-reversal Ω with other symmetries.

orientifold plane a plane (of any dimension p) consisting of fixed points of the orientifold group H (specifically, of an element of H that includes Ω).

p-brane a p-dimensional spatially extended object. Examples are *black p-branes*, *Dp-branes*, *M2-* and *M5-branes*, *NS5-branes*, *heterotic 5-branes*, and (in $d >$ 4) *Kaluza–Klein monopoles*.

p-form a fully antisymmetric p-index tensor, usually written in an index-free notation.

p-form gauge field a generalization of Abelian gauge theory, with a p-form potential A, a $(p-1)$-form gauge parameter λ, and a $(p+1)$-form field strength F. For $p = 0$ this is an ordinary massless scalar; for $p = 1$ it is an Abelian gauge field. A $(p+1)$-form potential couples naturally to a p-brane, through the integral of the form over the world-volume. A *self-dual p-form theory* is one where $*F = F$; this requires the spacetime dimension to be $d = 2p + 2$, and further d must be 2 mod 4 (in the Minkowskian case) in order that $** = 1$.

p-p' string an open string with one endpoint on a Dp-brane and the other on a Dp'-brane.

(p, q)-form on a complex manifold, a tensor that is completely antisymmetric in p holomorphic indices and q antiholomorphic indices.

(p, q) string a bound state of p F-strings and q D-strings in the IIB theory.

parafermion CFT a family of coset CFTs with $\mathbf{Z}_n$ symmetry, generalizing the $\mathbf{Z}_2$-invariant free fermion theory. These describe the generic critical behavior of a system with $\mathbf{Z}_n$ symmetry.

parity transformation an operation that reflects one spatial dimension, or any odd number. One distinguishes *spacetime parity, P* (or β) and *world-sheet parity* Ω.

partition function a sum over the spectrum of a quantum system, weighted by $e^{-H/T}$ where H is the Hamiltonian and T the temperature. Often additional charges are included in the exponent. This is the basic object in equilibrium statistical mechanics. In string theory it is given by a path integral on the torus or the cylinder, and so arises in one-loop amplitudes.

path integral a representation of the transition amplitudes of a quantum system as a coherent sum over all possible histories. In quantum mechanics the history is a particle path; in quantum field theory it is a path in field space; in first-quantized string theory it is the embedding of the string world-sheet in spacetime.

Pauli–Villars regulator a means of regulating quantum field theories by introducing a very massive wrong-statistics field.

Peccei–Quinn (PQ) symmetry an approximate symmetry, violated only by anomalies.

period matrix a $g \times g$ matrix characterizing the complex structure of a genus-g Riemann surface.

perturbation theory the expansion of the amplitudes of a quantum system in powers of the coupling.

physical state in a quantum system with constraints, a state annihilated by the constraints. In OCQ this is a state annihilated by the Virasoro lowering generators and having a specified L_0 eigenvalue.. In BRST quantization it is a state annihilated by the BRST operator. In both these cases, the true physical spectrum is the space of physical states with an additional equivalence relation, physical states differing by a null state being identified.

picture in the RNS superstring, one of several isomorphic representations of the vertex operators. The q-picture consists of vertex operators of $\beta\gamma$ ghost charge q. The natural pictures are $q = -1$ and $-\frac{1}{2}$, with higher pictures including partial integrations over supermoduli space. The *picture changing operator* increases q by one.

Planck length the natural length scale of quantum gravity, $L_\mathrm{P} = M_\mathrm{P}^{-1} = 1.6 \times 10^{-33}$ cm, constructed from $\hbar$, c, and G_N.

Planck mass the natural mass scale of quantum gravity, $M_\mathrm{P} = 1.22 \times 10^{19}$ GeV, constructed from $\hbar$, c, and G_N.

plumbing fixture a procedure for constructing higher genus Riemann surfaces from lower ones by sewing in a handle. The construction includes a parameter q, such that when q goes to 0 the handle degenerates or *pinches*. This gives a canonical representation of the boundary of the moduli space of Riemann surfaces.

Poincaré dual a map from p-forms to $(d - p)$-forms, given by contraction with the completely antisymmetric tensor.

Poincaré invariance the invariance group of the flat metric, consisting of translations and Lorentz transformations.

point group the orbifold group H, with translations ignored (applicable only for orbifolds having a spacetime interpretation).

Polyakov path integral a representation of first-quantized string theory as a path integral with an independent world-sheet metric. A local Weyl symmetry guarantees that the classical degrees of freedom are the same as those of the Nambu–Goto theory.

primary field in CFT, a local operator annihilated by all of the lowering generators of a given algebra, such as the Virasoro algebra. The corresponding state is a *highest weight state*.

projective space a compact n-dimensional space constructed from a linear $(n + 1)$-dimensional space by identifying points under the overall rescaling $(x_1, \ldots, x_{n+1}) \cong (\lambda x_1, \ldots, \lambda x_{n+1})$. For x_i and λ real this produces RP^n, and for x_i and λ complex it produces CP^n (which has n complex $= 2n$ real dimensions).

pseudospin in a current algebra, an $SU(2)$ subalgebra not contained in the Lie algebra of the center-of-mass modes.

puncture a marked point on a Riemann surface, the position of a vertex operator.

QCD string a reformulation of non-Abelian gauge theory as a string theory, conjectured to exist at least in the limit of a large number of colors.

quaternionic manifold a $4k$-dimensional manifold with the holonomy group $Sp(k) \times SU(2) \subset SO(4k)$, with specific $SU(2)$ curvature. This is the geometry of the moduli space of hypermultiplets in $d = 6$, $N = 1$ or $d = 4$, $N = 2$ supergravity.

R symmetry a symmetry that acts nontrivially on the supercurrent(s).

raising operator operators that reduce the energy of a given state. In CFT, operators that increase the Virasoro generator L_0 (or $\tilde{L}_0$) by n units carry a grading (subscript) $-n$.

Ramond algebra the world-sheet algebra of the Fourier modes of the supercurrent and energy-momentum tensor, in a sector where the supercurrent is periodic and its moding therefore integer-valued.

Ramond (R) boundary condition the condition that a fermionic field on the world-sheet be periodic, in the closed string or in the double of the open string (see *doubling trick*). Its Fourier moding is then integer-valued.

Ramond–Neveu–Schwarz (RNS) superstring the formulation of type I and II superstrings that has superconformal invariance but not manifest spacetime supersymmetry. The latter emerges after imposing the GSO projection on the string Hilbert space.

Ramond–Ramond (R–R) states in type I and type II superstring theories, the bosonic closed string states whose left- and right-moving parts are fermionic. These include p-form potentials C_p, with p taking all odd values in the IIA string and all even values in the IIB string.

rank 1. the maximal number of commuting generators of a Lie algebra; 2. the number of indices on a tensor.

rational CFT a CFT with a finite number of primary fields under an extended chiral algebra, a generalization of the *minimal models*. Such CFTs are highly constrained.

refermionization after bosonization, the construction of new spin-$\frac{1}{2}$ fields from linear combinations of the bosonic fields. These fermions are nonlocal, nonlinear functions of the original fermions.

Regge behavior in scattering at large center-of-mass energy-squared s and fixed momentum transfer-squared $-t$, the scaling of the amplitude as $s^{\alpha(t)}$. The values of t where $\alpha(t) = j$ is a nonnegative integer correspond to exchange of a particle of spin j and mass-squared $-t$.

Regge slope denoted α', the square of the characteristic length scale of perturbative string theory. The tension of the fundamental string is $1/2\pi\alpha'$.

relevant interaction an interaction whose dimensionless strength increases with distance. In perturbation theory, this is equivalent to a *superrenormalizable* interaction.

renormalization group equation the differential equation governing the change of physics with length scale.

renormalization theory the calculus of path integrals.

Riemann–Roch theorem the theorem that the number of metric moduli minus the number of conformal Killing vectors on a Riemann surface is -3χ, with χ being the Euler number of the surface; and, generalizations of this result.

Riemann surface a two-(real)-dimensional complex manifold, equivalent to a Weyl equivalence class of Riemannian manifolds.

root a vector of the eigenvalues of the maximal set of commuting generators of a Lie algebra, associated with a state in the adjoint representation.

S-duality a duality under which the coupling constant of a quantum theory changes nontrivially, including the case of weak–strong duality. Important examples are the $SL(2,\mathbf{Z})$ self-dualities of IIB string theory and of $d = 4$, $N = 4$ supersymmetric Yang–Mills theory. More loosely, it is used for weak–strong dualities between different theories, such as IIA–M-theory (on a circle) duality, $SO(32)$ heterotic–type I duality, and E_8 heterotic–M-theory (on an interval) duality. In compactified theories, the term S-duality is limited to those dualities that leave the radii invariant, up to an overall coupling-dependent rescaling; contrast *T-duality* and *U-duality*.

S-matrix the overlap amplitude between states in the infinite past and states in the infinite future; the scattering amplitude. In coordinate-invariant quantum theories this is generally the simplest invariant. The term usually implies a basis of free particle states; this is problematic in theories with massless particles due to IR divergences, and meaningless in theories at nontrivial IR fixed points.

scale transformation a rigid rescaling of spacetime, or of the world-sheet.

Scherk–Schwarz mechanism the breaking of supersymmetry by dimensional reduction that includes a spacetime rotation.

Schottky group a discrete subgroup Γ of the $SL(2,\mathbf{C})$ Möbius transformations of the sphere S_2, with additional conditions such that S_2/Γ is a manifold and in particular a Riemann surface.

Schwarzian the combination of derivatives appearing in the finite conformal transformation of the energy-momentum tensor.

Schwinger–Dyson equation the operator equations of a quantum theory, expressed as equations for the expectation values.

short multiplet see *BPS state*.

sigma model metric the metric appearing in the string world-sheet action. Also known as the *string metric,* this differs from the *Einstein metric* by a dilaton-dependent Weyl transformation.

simple current an operator J such that, for any primary field $\mathcal{O}$, the operator product $J\mathcal{O}$ contains only a single conformal family.

soliton a state whose classical limit is a smooth, localized, and (usually) topologically nontrivial classical field configuration; this includes particle states, which are localized in all directions, as well as extended objects. By contrast, a state of ordinary *quanta* is represented near the classical limit by small fluctuations around a constant configuration. In a theory with multiple classical limits (*dualities*), solitons and quanta may exchange roles.

special Kähler geometry the geometry of the moduli space of vector multiplets in $d = 4$, $N = 2$ supergravity. It is most simply defined (section B.7) in terms of *special coordinates,* which are fixed up to a symplectic transformation. *Rigid special geometry* is obtained in the limit where gravity decouples and the supersymmetry becomes global.

spectral flow the adiabatic change in the spectrum produced by a continuous change in the boundary conditions.

spin the behavior of a field or a state under rotations. In a CFT this is given in terms of the conformal weights by $h - \tilde{h}$.

spin field the vertex operator for a Ramond ground state, which produces a branch cut in the spinor fields.

spin structure one of a set of inequivalent ways of defining a spinor field globally on a manifold. Roughly speaking, it corresponds to a choice of signs in the square roots of the transition functions.

spurious state in OCQ, a state produced by Virasoro raising operators.

state–operator isomorphism in CFT, a one-to-one isomorphism between states of the theory quantized on a circle and local operators. Also, a one-to-one isomorphism between states of the theory quantized on an interval and local operators on a boundary. In d dimensions the circle becomes a $(d-1)$-sphere and the interval a $(d-1)$-hemisphere.

string coupling the dimensionless parameter g governing the weights of different Riemann surfaces in string perturbation theory, the contribution from surfaces of Euler number χ being weighted by $g^{-\chi}$. The string coupling is related to the dilaton by $g = e^{\Phi}$. This definition corresponds to the amplitude to emit a closed string; the amplitude to emit an open string is proportional to $g^{1/2}$.

string field theory the representation of string theory as theory of fields, the fields being maps from a circle (or interval) into spacetime. This corresponds to an infinite number of ordinary quantum fields. This formalism can reproduce string perturbation theory, but it is unclear whether it can be defined beyond perturbation theory.

string metric see *sigma model metric*.

string scale the mass scale $\alpha'^{-1/2}$ characterizing the tower of string excitations.

string–string duality a term sometimes used to denote a weak–strong duality between different string theories, in particular between the heterotic string compactified on T^4 and the IIA string compactified on K3.

string tension the mass per unit length of a string at rest, related to the Regge slope by $1/2\pi\alpha'$.

Sugawara construction in current algebra, the construction of the energy-momentum tensor as a product of two currents. Originally proposed as a phenomenological model in four dimensions, this was later found to be an exact result in two dimensions.

superconformal algebra an extension of the conformal (Virasoro) algebra to include anticommuting spinor generators. The $(N, \tilde{N})$ superconformal algebra has N left-moving and $\tilde{N}$ right-moving supercurrents.

superconformal current algebra an extension of the conformal transformations to include both spin-$\frac{3}{2}$ and spin-1 currents.

superconformal field theory (SCFT) a quantum field theory that is invariant under superconformal transformations.

supercurrent a conserved spinor current. This includes the world-sheet current T_F associated with superconformal transformations, and the spacetime current associated with spacetime supersymmetry

superfield a field on superspace, with specific transformation properties under a change of coordinates.

supergravity the union of general relativity and supersymmetry, implying also the promotion of supersymmetry to a local symmetry.

supermanifold (or *superspace*) a formal extension of the concept of manifold to include both commuting and anticommuting (Grassmann) coordinates.

supermoduli the anticommuting parameters characterizing a super-Riemann surface.

superpartner scale the mass scale of the superpartners of the Standard Model particles. This is expected to be between 10^2 and 10^3 GeV if supersymmetry solves the naturalness problem of the Higgs scalar mass.

superpotential in $d = 4$, $N = 1$ supersymmetry, the holomorphic function of the superfields that determines the nongauge interactions.

super-Riemann surface a supermanifold defined in terms of superconformal transition functions between patches.

supersymmetry a symmetry whose charge transforms as a spinor, which relates the masses and couplings of fermions and bosons.

supersymmetry breaking scale the mass scale of the expectation value that breaks supersymmetry. The superpartner scale is the supersymmetry breaking scale times the strength of the coupling of the Standard Model fields to the supersymmetry breaking fields.

T-**duality** a duality in string theory, usually in a toroidally compactified theory, that leaves the coupling constant invariant up to a radius-dependent rescaling and therefore holds at each order of string perturbation theory. Most notable is $R \to \alpha'/R$ duality, which relates string theories compactified on large and small tori by interchanging winding and Kaluza–Klein states. More generally it includes shifts of antisymmetric tensor backgrounds and large coordinate transformations in spacetime. Contrast *S-duality* and *U-duality*.

't Hooft–Polyakov monopole a classical solution with magnetic charge, which exists whenever a simple group is spontaneously broken to a group with a $U(1)$ factor.

tachyon a particle (almost always a scalar) with a negative mass-squared, signifying an instability of the vacuum.

tadpole an amplitude for creation of a single particle from the vacuum, induced by quantum effects.

target space the space in which a function takes its values. This is usually applied to the nonlinear sigma model on the string world-sheet, where the target space is itself spacetime.

Teichmüller parameters the moduli for the complex structure of a Riemann surface (strictly speaking, points in *Teichmüller space* are not identified under the modular group).

tensionless string theory an interacting theory with tensionless strings. These can arise as p-branes with $p-1$ directions wrapped on a collapsing cycle, as various 2-branes with one direction stretched between higher-dimensional branes when the latter become coincident, and on zero-size $E_8 \times E_8$ instantons. In general the coupling is fixed to be of order one, so there is no perturbation expansion. *Tensionless* implies that the string tension in units of the gravitational scale goes to zero; it is not applied to the fundamental string, which becomes noninteracting in that limit.

tensor multiplet 1. the multiplet of $d = 6$, $(1,0)$ supersymmetry whose bosonic content is one self-dual tensor and one scalar. This reduces to a vector multiplet of $d = 4$, $N = 2$ supersymmetry; 2. the multiplet of $d = 6$, $(2,0)$ supersymmetry whose bosonic content is one self-dual tensor and five scalars. This reduces to a vector multiplet of $d = 4$, $N = 4$ supersymmetry.

tensor operator in CFT, a local operator whose conformal or superconformal transformation involves only the first derivative of the transformation; synonymous with a *primary field* of the conformal or superconformal symmetry. Such an operator is mapped to a *highest weight state* by the state–operator isomorphism.

tetrad a basis of d orthonormal vector fields in d dimensions. The term tetrad originates in $d = 4$ (as does the equivalent term *vierbein*) but there is no other convenient term for general d.

the theory formerly known as strings see *M-theory*, second definition.

theory see *vacuum*.

theta functions the family of holomorphic functions having simple periodicity properties on a torus.

Thirring model the solvable quantum field theory of a single Dirac fermion with a quartic interaction in $1 + 1$ dimensions. This is equivalent under bosonization to a free scalar field at a general real radius.

threshold correction a correction to the low energy effective action, and in particular to the gauge coupling, due to virtual massive particles.

topological string theory a modification of string theory without local dynamics; all observables are topological.

toric geometry a generalization of the idea of projective space. Roughly speaking, this corresponds to the most general linear sigma model.

toroidal compactification the periodic identification of one or more flat dimensions.

torsion a term applied to various 3-form field strengths, so called because they appear in covariant derivatives in combination with the Christoffel connection.

tree-level the Feynman graphs which become disconnected if one propagator is cut, or the analogous string amplitudes, the sphere and disk. These correspond to classical terms in the effective action.

twist 1. (*verb*) to define a field in a periodic space to be aperiodic by a symmetry transformation h; 2. (*noun*) the aperiodicity h; 3. (*verb*) given a CFT or string theory, to construct a new theory using a symmetry group H. One adds closed strings twisted by any of the elements $h \in H$, and requires all states to be invariant under the transformations in H. This is equivalent to treating H as a world-sheet gauge symmetry. The term *orbifold* is also used as a synonym; 4. (*noun, archaic*) world-sheet parity.

twisted state a closed string with twisted periodicity.

type ... supergravity the low energy supergravity theory of the corresponding string theory. The Roman numeral signifies the number of $d = 10$ supersymmetries, and IIA and IIB distinguish whether the two supersymmetries have opposite or identical chiralities respectively.

type I superstring the theory of open and closed unoriented superstrings, which is consistent only for the gauge group $SO(32)$. The right-movers and left-movers, being related by the open string boundary condition, transform under the same spacetime supersymmetry.

type IIA superstring a theory of closed oriented superstrings. The right-movers and left-movers transform under separate spacetime supersymmetries, which have opposite chiralities.

type IIB superstring a theory of closed oriented superstrings. The right-movers and left-movers transform under separate spacetime supersymmetries, which have the same chirality.

type 0 string a pair of nonsupersymmetric string theories, which have tachyons and no spacetime fermions, with the same world-sheet action as the type II theories but with different projections on the Hilbert space.

U-**duality** any of the dualities of a string theory, usually of a toroidally compactified type II theory. This includes the *S-dualities* and *T-dualities*, but in contrast to these includes also transformations that mix the radii and couplings.

ultrashort multiplet see *BPS state*.

unit gauge a choice of coordinate and Weyl gauges in two dimensions, such that the metric is the identity.

unitary as applied to a quantum system, the property of having a conserved inner product in a positive-norm Hilbert space.

unoriented string theory a string theory in which world-sheet parity-reversal Ω is a discrete gauge symmetry. The perturbation theory includes unoriented world-sheets, and the spectrum is restricted to states with $\Omega = +1$.

UV divergence a divergence arising from short distances in spacetime, usually signifying a limit to the validity of a theory.

vacuum a stable Poincaré-invariant state. The novel feature of systems with unbroken supersymmetries is the frequent appearance of degenerate but physically inequivalent vacua.
While it is now clear that the different string theories are actually different *vacua* in a single theory, it is still common to use the term *theory* for each. Also, different CFTs within a single string theory are sometimes referred to as different *theories* rather than *vacua*. The term *model* is used to refer to string vacua whose low energy physics resembles the Standard Model.

vector multiplet in $d = 4$, $N = 1$ or $d = 6$, $N = 1$ supersymmetry, the multiplet whose bosonic content is a massless vector field. The $d = 6$ multiplet reduces to a vector field plus two real scalars in $d = 4$, $N = 2$ supersymmetry.

Veneziano amplitude the bosonic string tree-level amplitude for four open string tachyons.

Verma module the set of states obtained by acting on a highest weight state with Virasoro raising generators in all inequivalent ways, with the requirement that all such states be linearly independent (the generators act freely). This gives a representation of the Virasoro algebra that depends on the central charge c and on the weight h of the highest weight state. This is very similar to a *conformal family,* except that in the case of a degenerate representation some states in the latter may vanish; the Verma module would in this case give a reducible representation.

vertex operator a local operator on the string world-sheet, corresponding to a string in the initial or final state.

Virasoro algebra in a CFT, the infinite-dimensional Lie algebra of the Fourier modes of the energy-momentum tensor.

Virasoro–Shapiro amplitude the bosonic string tree-level amplitude for four closed string tachyons.

W **algebras** a family of extended chiral algebras with currents of spin greater than 2.

W **string** a proposed generalization of string theory with constraints of spin greater than 2. No complete construction exists.

Ward identity a relation between the divergence of the expectation value of a conserved current and the same expectation value without the current.

weight 1. in CFT, the L_0 or $\tilde{L}_0$ eigenvalue, which determines the behavior of an operator under scale transformations and rotations (= *conformal weight*); 2. in Lie algebra, a vector of the eigenvalues of the maximal set of commuting generators of a Lie algebra, in a state of a given representation.

Wess–Zumino consistency condition the condition that the second derivative of a functional integral with respect to the background fields be symmetric. This strongly constrains the form of possible anomalies in a theory.

Wess–Zumino–Novikov–Witten (WZNW) model a conformally invariant non-linear sigma model on a group manifold with an antisymmetric tensor background.

Weyl anomaly an anomaly in the Weyl transformation, this determines the critical dimension in the Polyakov formalism. This is sometimes called the *trace anomaly* or the *conformal anomaly.*

Weyl condition in even dimensions, the condition that a fermion have definite chirality. This defines a spinor representation of the Lorentz group, which contains half of the components of the Dirac representation of the gamma matrix algebra.

Weyl transformation a position-dependent rescaling of the metric.

Wilson line 1. a gauge field with vanishing field strength but with nontrivial parallel transport (holonomy) around nontrivial paths in spacetime. 2. the gauge-invariant operator that measures such a field: the trace of the path-ordered product of the line integral of the vector potential.

Wilsonian action the action in a low energy effective field theory, which incorporates the effects of higher energy virtual states.

winding state a closed string whose configuration is a nontrivial path in a non-simply-connected spacetime.

world-sheet 1. the two-dimensional surface in spacetime swept out by the motion of a string. 2. the abstract two-dimensional parameter space used to describe the motion of a string.

world-sheet duality the equivalence between Hamiltonian descriptions obtained by cutting open a string world-sheet along inequivalent circles. Important examples are the associativity condition, from the four-point sphere amplitude, and the equivalence of the open string loop and closed string tree descriptions of the cylinder.

wrapped refers to a p-brane, q of whose dimensions are wound on a nontrivial compact submanifold of spacetime, leaving a $(p - q)$-dimensional extended object.

Yang–Mills field non-Abelian gauge field.

Zamolodchikov metric the expectation value of a pair of local operators on the sphere, giving a natural inner product for the corresponding string states. This is the metric that appears in the kinetic term in the spacetime action.

zero modes 1. of a differential operator: eigenfunctions with zero eigenvalue. In a Gaussian functional integral, these would give an infinite factor in the bosonic case and a zero factor in the fermionic case. In general these have a physical origin and the functional integral has an appropriate integrand to give a finite result; 2. center-of-mass modes.

zero-point energy the energy due to vacuum fluctuations of quantum fields.

zero-slope limit the limit $\alpha' \to 0$. Only massless string states, described by low energy field theory, remain.

Index

Page numbers in italics refer to exercises